**Risk Assessment**
http://dx.doi.org/10.5772/intechopen.68673
Edited by Valentina Svalova

**Contributors**

Hana Pacaiova, Riitta Molarius, Milton Montejano-Castillo, Mildred Moreno-Villanueva, Catalin Cioaca, Vasile Prisacariu, Mircea Boscoianu, Stig O. Johnsen, Varun Gopinath, Kerstin Johansen, Johan Ölvander, Israel Gomy, Mitsutoshi Suzuki, Gevorg Tepanosyan, Lilit Sahakyan, David Pipoyan, Armen Saghatelyan, Michaela Ibrion, Kristian Helmholt, Matthijs Vonder, Bram Van Der Waaij, Elena Lazovik, Niels Neumann, Zakiya Hoyett, Safet Kozarevic, Emira Kozarevic, Pasqualina Porretta, Fabrizio Santoboni, Andrea Antonucci, Xiaoxia Ding, Linxia Wu, Peiwu Li, Robert Wålinder, Krystyna Pietrzyk, Ireneusz Czmoch, Kayoko Yamamoto, Zhaoxin Yang, Oscar Andrés Cuanalo Campos

**Published by InTech**
Janeza Trdine 9, 51000 Rijeka, Croatia



**Publishing Process Manager** Slobodan Momcilovic
**Technical Editor** SPi Global
**Cover** InTech Design team

First published February, 2018
Printed in Croatia
Legal deposit, Croatia: National and University Library in Zagreb

Additional hard copies can be obtained from orders@intechopen.com

Risk Assessment, Edited by Valentina Svalova
p. cm.
Print ISBN 978-953-51-3798-6
Online ISBN 978-953-51-3799-3

# RISK ASSESSMENT

Edited by **Valentina Svalova**

For EU product safety concerns:
IN TECH d.o.o., Prolaz Marije Krucifikse Kozulić 3, 51000 Rijeka, Croatia,
info@intechopen.com or visit our website at intechopen.com.

# Meet the editor

Dr. Valentina Svalova is affiliated with the Sergeev Institute of Environmental Geoscience, a part of the Russian Academy of Sciences. She graduated from the Moscow State University with a PhD in Physical-Mathematical Sciences from the Mechanical-Mathematical Faculty. She has published more than 400 scientific publications including a monograph "Landslide Risk: Assessment, Management and Reduction" (2017), and the results of her research have been presented at more than 100 international scientific conferences and congresses in more than 50 countries. Her membership in international and scientific organizations include the following: associate member of the International Informatization Academy; member of the board of directors of the International Geothermal Association; president of the Geothermal Energy Society (GES) of Russia; member of International Association for Mathematical Geosciences (IAMG); a scientific secretary of Geothermal Council of Russia, Russian Academy of Sciences; and the board of representatives of International Consortium on Landslides (ICL). She received the International Best Paper Award "PRESSZVANIE," nomination for "Clean Energy," 2015.

# Contents

# Preface

Risk is a relatively new and not fully explored concept especially for different kinds of events and activities. There are many definitions of risk. And often, a scientific study or a scientific approach to the problem begins with a presentation of the author's position and the choice of the risk definition of natural, technogenic, medical, financial, and other events. This individualistic approach is difficult to avoid. Spores are carried out so far. For example, whether or not there is a risk without material damage to people.

If one of the main systematic approaches to research on hazards and disasters is their classification, even now the concept of risk management can be considered as a new step of scientific development and a new basement for systematic investigations of danger, hazards, and disasters.

The development of the risk concept demands the promotion of the methods for risk assessment and calculation. It makes the risk theory a scientific discipline with good mathematical background. It is necessary to elaborate common approaches to risk calculation for different types of natural hazards and activities. The methods of seismic risk assessment as the most promoted ones must be spread to landslides, karst, suffusion, flooding, pollution, and other types of natural hazards and risks and also to complex and multirisk.

Arising from everyday life, gambling, finance, business, and building the risk concept became the subject for scientific research and the basement for systematic investigations of natural and man-made hazards and disasters, medical and health problems, financial crisis, and environmental catastrophes.

In common sense, risk is the potential of gaining or losing something of value. Values, such as physical health, social status, emotional well-being, and financial wealth, can be gained or lost when taking risk resulting from a given action or inaction and foreseen or unforeseen. Risk can also be defined as the intentional interaction with uncertainty. Uncertainty is a potential, unpredictable, and uncontrollable outcome; risk is a consequence of action taken in spite of uncertainty.

In risk-analysis science, risk is considered as a measure of the probability and severity of an adverse effect to health, property, or the environment. Risk is often defined as the probability of the hazard event multiplied by the possible consequences. One of the most common approaches defines that risk is the expectation of the damage, or expected value of damage.

Risk analysis is the use of available information to calculate the risk to individuals, population, property, or the environment from hazards. Risk analyses generally contain the following steps: hazard identification and vulnerability evaluation.

Vulnerability is the degree of loss of a given element or a set of elements exposed to the occurrence of a natural or man-made hazard. It is expressed on a scale of 0 (no loss) to 1 (total loss).

Risk assessment is considered as the process of making a recommendation on whether existing risks are acceptable and present risk control measures are adequate, and if they are not, whether alternative risk control measures are justified or will be implemented. Risk assessment incorporates the risk analysis and risk evaluation phases.

Sometimes, risk assessment is considered as risk calculation on the base of selected parameters and establishment of ranking risk criteria.

Acceptable risk refers to the level of human and property loss that can be tolerated by an individual, household, group, organization, community, region, state, or nation. For instance, the risk of flooding can be accepted once every 500 years, but it is not acceptable in every 10 years. In other words, it means those risks whose benefits are so great that individuals or groups in society are willing to take or be subjected to the risk. It is usually calculated in actions taken to minimize the disaster risk. The probability of occurrence of an acceptable risk is small. The concept of acceptable risk evolved partly from the understanding that absolute safety is generally an unachievable goal and that even very low exposures to certain toxic substances may confer some level of risk.

Risk management is considered as the complete process of risk assessment and risk reduction. Risk reduction implies some methods and measures, such as legislative, organizing, economic, engineering, information, and others. Sometimes, in a narrow sense, risk management is considered as measures for risk reduction. And, in this sense, the problem of risk management is seen as a series of events leading to risk reduction and avoiding. It includes modeling, monitoring, forecasting, prognosis, engineering works, insurance, and others.

Summarizing the systematic approaches to research on hazards and disasters based on the risk concept, it is possible to present the next steps and scheme to establish criteria for ranking risk posed by different types of natural or man-made hazards and disasters:

1. Hazard identification;
2. Vulnerability evaluation;
3. Risk analysis;
4. Concept of acceptable risk;
5. Risk assessment;
6. Risk mapping;
7. Measures for risk reduction:
   a) legislative;
   b) organizational and administrative;
   c) economic, including insurance;
   d) engineering and technical;
   e) modeling;
   f) monitoring;
   g) information.

It is the responsibility of the local governments to establish rules meant to reduce the effects of environmental hazards and disasters. Land-use regulations and policies are required in areas that are prone to natural hazards. The absence of such regulations and destructive human activities is among the main factors that favor environmental risk.

For example, whenever a landslide occurs, no matter if it is caused by slope saturation with water, seismic activity, or a volcanic eruption, the damages are disastrous. Thousands of households may be swept away or buried in mud and tens to hundreds of people could lose their lives.

This apocalyptic image should make local governments pay more attention to the prevention of such natural phenomena. It is important for a local government to know which areas are prone to landslides and take appropriate measures in order to reduce vulnerability to such hazards.

Vulnerability to landslides depends on location, frequency of landslide events, type of human activity in the area, and other factors.

The effects on people and buildings can be lessened if hazardous areas are avoided or if activities in such areas are restricted or deployed under certain conditions. Local governments are responsible for land-use policies and other regulations meant to reduce the risks for landslides to take place.

Exposure to hazards may be reduced if individuals educate themselves on the past history of these phenomena. Departments of local governments that are responsible with planning and engineering may help a lot with their advice.

People can also benefit from the professional services of engineering geologists, civil engineers, or geotechnical engineers, all qualified to evaluate the potential of a hazardous site.

Due to the huge losses that landslides imply, their prevention is of maximum importance for all the people living in the area of hazard. Preventing a landslide from causing material damage and human losses should be a main goal of local authorities.

Similar situation occurs in any kind of human activity and life.

The book reflects the state-of-the-art problem and addresses the risk assessment to establish the criteria for ranking risk posed by different types of natural or man-made hazards and disasters, medical and health problems, financial crisis, and environmental catastrophes; to quantify the impact that hazardous event or process has on person, population, environment, and structures; and to enhance the strategies for risk reduction and avoiding.

**Dr. Valentina Svalova**
Sergeev Institute of Environmental Geoscience
Russian Academy of Sciences
Moscow, Russia

Section 1

# Natural Risk Assessment

Chapter 1

# Landslides: Methodology to Select Stabilizing Construction Works

Oscar Andrés Cuanalo Campos

Additional information is available at the end of the chapter

http://dx.doi.org/10.5772/intechopen.70797

**Abstract**

In landslide areas, after assessing the risk level, the obligatory questions from government authorities, communities, civil protection managers, and researchers are: What can we do? What should we do? What must we do? There are different strategies to reduce the vulnerability and risk: (a) increasing the knowledge of the population, (b) establishing an early warning system, and (c) selecting and constructing structures. The aim of this chapter is to present the methodology to select stabilizing construction works to avoid a landslide, through the "valuation factors," which are parameters to assess the intrinsic and trigger instability factors (morphology, geology, hydrogeology, vegetation, rainfall, earthquake, erosion, human activity, etc.). The *valuation factors* are presented in graphs, equations, and tables; based upon them, the different construction works are selected, including (a) geometric adjusting for reducing destabilizing forces; (b) reinforcement elements, anchors, and pile barriers to increase the resistive forces; (c) drainage for eliminating surface runoff water or lowering the hydrostatic pressure; (d) retaining walls to support the horizontal pressure; and (e) surface protection to prevent rock falls and reduce erosion and infiltration. The methodology has been used successfully in several mountainous regions: Puebla, Hidalgo, Chiapas, Baja California in México, and Ocaña in Colombia.

**Keywords:** landslides, construction works, valuation factors

## 1. Introduction

The landslides often cause disasters and damage to people and their properties at the mountainous areas around the world; these disasters cause casualties and economic losses, such as housing, infrastructure, public services, roads, bridges, hospitals, etc., and the interruption of the normal activities of the region, such as agriculture, livestock, commerce, tourism, financial transactions, etc. A fundamental problem to solve is to make the investments for reconstructing and rehabilitating the destroyed places, which must be obtained from other social investment programs, donations from other countries, and/or sources of external financing that lead to indebtedness and impoverishment of communities, regions, or countries [1].

The first step to establish adequate strategies for prevention, reduction, management, and risk mitigation is to assess landslide hazard, vulnerability, and risk, the latter in terms of casualties and economic losses. The rational solution to landslide problem is to relocate exposed and vulnerable people to secure sites, but acquiring land in mountainous regions is very difficult; besides the majority of population is rooted to its origin place, and it cannot be relocated so easily.

### 1.1. The landslide problem

Landslide is a failure through a surface in which shear resistance has been exceeded; it is featured by the movement of slope materials that slide downhill.

Landslides can occur due to natural and human factors (intrinsic and trigger factors); although many landslides are triggered by natural phenomena (heavy rains, earthquakes, volcanism, freezing and thawing, erosion and scouring, etc.), it can be estimated that many of those that have caused deaths and injuries can be attributed to human activity impact.

The landslide material may include from a simple rock that falls to a great slide of several hundreds or thousands of cubic meters of material dragged in an avalanche or in a debris flow. They also range in extent some affect only a very small area while others entire regions. The distance that the material travels during movement can also differ significantly, with displacements ranging from a few cm to many km in length, depending on the volume of material, its water content, and the slope inclination. The velocity of a landslide may range from a slow, almost undetectable, gradual movement that remains active for a long period of time (displacements of a few cm per year), to sudden rapid collapse.

### 1.2. The landslide disasters

Landslides are increasingly affecting our planet, like earthquakes, volcanic eruptions, hurricanes, floods, avalanches, etc., amplifying their intensity by different human activities that modify the delicate balance in nature.

In some mountainous regions at Central and South America, there are many communities belonging to ethnic groups that inhabit areas classified as high and very high poverty, whose features, among others, include localities of less than 2500 inhabitants, illiterate population in a large percentage, very low income, precarious social infrastructure, and houses built with fragile materials such as plastic, cardboard, and/or wood. In other places, there also are displaced populations due to social conflict, forcing them to move and take refuge in very vulnerable areas. In all these places, usually natural phenomena cause real disasters to impact on highly vulnerable communities [2].

## 2. Landslide instability factors

Landslide instability factors can be divided into two large groups—*intrinsic and triggers*—the first ones depend on the internal properties of slopes material and have a close relation with the type of failure and the susceptibility of the slope to a specific movement. The second ones, known also as external factors, are directly influenced by the climatic conditions, by extreme events such as earthquakes and volcanism, and the impact by human activities [3, 4].

## 3. Risk analysis

In order to assess the risk, a detailed analysis of the landslide hazard and the vulnerability of exposed people is required. **Figure 1** shows a sequential scheme that summarizes the different steps to be taken into account; a brief description is given below [5].

### 3.1. Hazard

Historical records of landslides in the study area, including their geographical location, magnitude, intensity, degree of affectation or damage, and their frequency, must be investigated. From these data, a catalog or inventory of landslides that includes the type of movement and the intrinsic and trigger factors of the instability is elaborated [6].

### 3.2. Vulnerability

The authors of this chapter propose to evaluate the population vulnerability from the exposure level (EL) and the expected damage degree (EDD): the first value according to the height of the slope and the safety factor obtained from the geotechnical stability analysis and the second based on the type of structures constructed (degree of fragility) and velocity of landslide [7, 8].

### 3.3. Risk

The risk assessment should result in the number of people affected and the cost of damages caused by the occurrence of the phenomenon under study [9].

## 4. Slope stability analysis

The landslides and slope instability are among the most common failure of earth masses or rocks. The weight of the land mass and their water content is the main force that produces the failure, while the shear strength of the terrain, diminished by the water pressure, is the main strength. Analysis of the slope stability is a problem of plastic equilibrium; when the mass is about to fail, the forces that produce the movement have become equal to the resistance that opposes the mass to be moved. A slight increase in forces is sufficient to produce a continuous deformation that can end in the general failure.

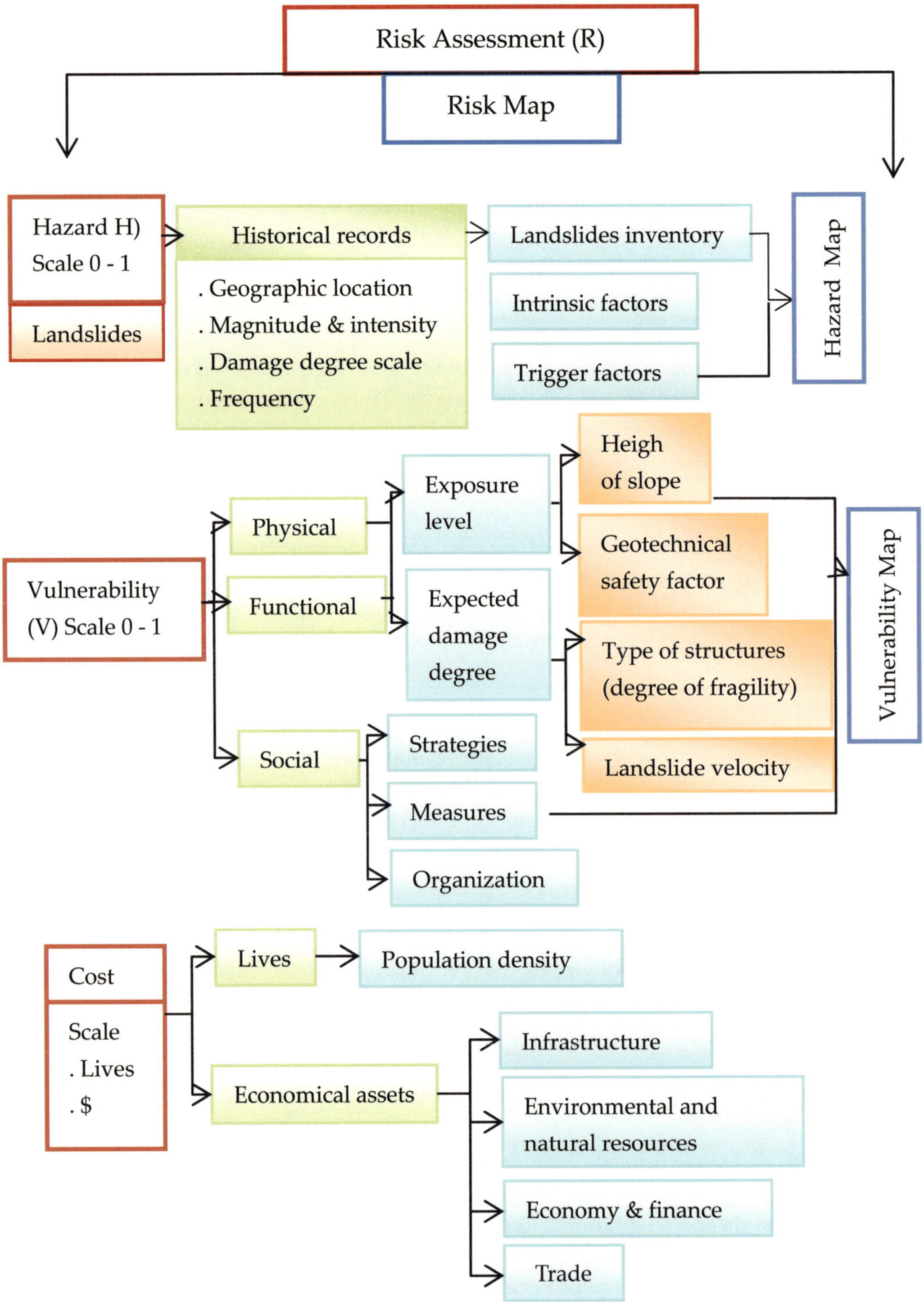


**Figure 1.** Sequential scheme to assess landslide hazard, vulnerability, and risk [5].

| Risk score | Concentration measured (ppm) | | | |
|---|---|---|---|---|
| | <900 | 900–2099 | 2100–16,000 | >16,000 |
| Insignificant | | | | |
| Low | | | | |
| Medium | | | | |
| High | | | | |

**Table 10.** Interpretation of monitoring data.

**Figure 3.** Testing weather sensors: (a) photo and (b) capture from mission planner.

| Humidity (%) | Temperature (°C) | Flammable gases (ppm) |
|---|---|---|
| 41 | 19 | 366 |
| 41 | 19 | 331 |
| 41 | 19 | 322 |
| 41 | 19 | 298 |
| 41 | 19 | 296 |
| 41 | 19 | 291 |
| 41 | 19 | 299 |
| 41 | 19 | 305 |
| 41 | 19 | 315 |
| 41 | 19 | 983 |
| 41 | 19 | 797 |
| 41 | 19 | 640 |

**Table 11.** Test results.

## 4. Decision support platform

The information provided in the risk analysis and event monitoring (if the option is activated) are integrated into a decision support platform using the Delphi 7 program. The decisional architecture was designed to meet both user requirements (modular, flexible, scalable) as well as the need for overlapping with the stages of the risk management process.

The platform is designed to: easily identify the infrastructure asset of the climate vector in real-time and to identify the medium-term evolution; to update the risk indicators (**Figure 4**); to visualize the level of risk; to be able to save a risk report (data about the climatic vector, the infrastructure asset, the level of risk, the severity of the impact, and the adaptation strategies).

User input values are essential to delimit the risk classes according to the scales from **Tables 4–6**. For situations of inapplicability, the value 1 is entered. To facilitate the accurate determination of the vulnerability level, scenario analysis is performed.

For the correct interpretation of the results provided by the support decision-making platform, it is very important to understand the following: the growth or decrease trend of the climatic vector in the next 10 or 25 years, respectively; the projected climate change for the next 10 and 25 years; the level of uncertainty reported for each climatic vector; communicating best practices identified at regional level from other critical infrastructures. For the event monitoring section, data are available to the user after executing the mission/monitoring in .txt or.avi format. Once downloaded, they are automatically retrieved and displayed in the Monitoring Report (**Figure 5**).

This section provides the possibility to save a risk report by creating a text version (.txt). For the correct interpretation of the results provided by the support decision-making platform, it is very important to understand the following: the growth or decrease trend of the climatic

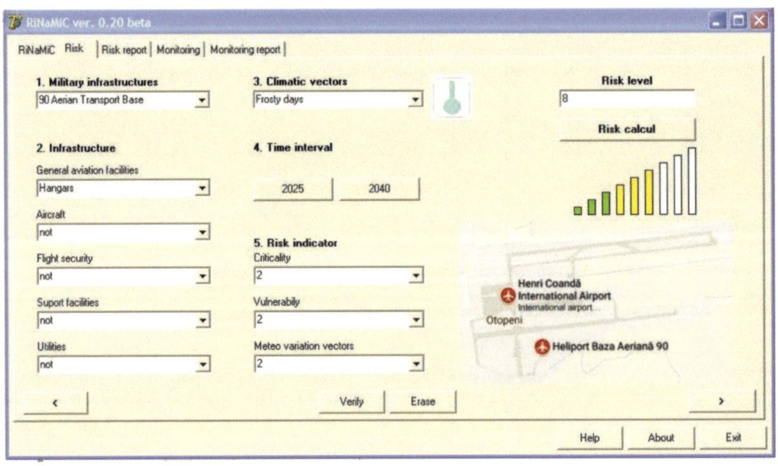


**Figure 4.** Definition of risk indices.

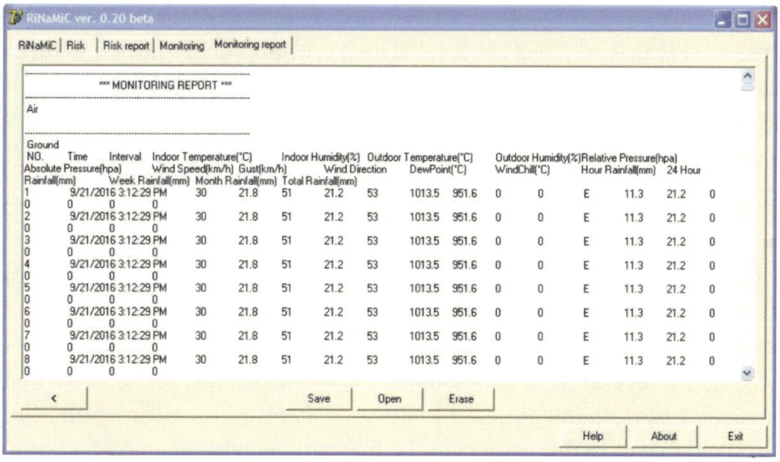


**Figure 5.** Display air monitoring report.

vector in the next 10 or 25 years; the projected climate change for the next 10 and 25 years; the level of uncertainty reported for each climatic vector; communicating best practices identified at regional level to and from other military objectives.

## 5. Conclusions

The impact created by such a scientific approach beyond the benefits of active innovation-based learning can also be assessed in potentially applicative terms: increasing the resilience of critical military infrastructures to natural events with catastrophic effects in the context of climate change. Future research directions include the removal of the current limitations of this innovative solution: it is necessary to improve the databases on infrastructure assets, the potential impact of the climatic vectors and the adaptation strategies; the application does not evaluate the cumulative effects of multiple vectors; data recorded by sensors integrated in aerial vectors are not available in real time.

The impact of this scientific approach is found in four main directions: increasing the capacity to understand the dynamics of natural risks from climate change and its impact on infrastructure assets/specific operations in the Romanian Air Force infrastructures; assessing the effectiveness of current plans in the context of future needs; optimizing the allocation of resources needed to initiate the climate change adaptation planning process; and increasing the resilience of critical military infrastructures in the Air Force to natural events with catastrophic effects in the context of climate change.

The solution can also significantly improve the planning process at the level of each military objective, which is currently taking place within an adaptive-reactive management. Existing planning documents do not provide opportunities to incorporate adaptation strategies to climate change in the short, medium, and long term. These documents are drawn up in the form of action orders whose triggering factor is the occurrence of an event or alerts issued by the NMA.

From the point of view of adaptive-preventive management, it is necessary to develop Climate Change Adaptation Plans (PASCs) at the level of each critical military infrastructure. The military objectives that have been endowed or are in the process of endowment with modern military technique represent an emergency. This process involves the initiation of some infrastructure projects.

This risk management approach in critical military infrastructures also has a number of limitations that constitute future research directions: for transformation into a functional model, it is necessary to improve the built databases on infrastructure assets, the potential impact of climatic vectors and adaptation strategies; information is not exhaustive, for special situations (e.g., hangars design) further studies and analysis are required; the model does not assess the cumulative effects of several vectors on the same infrastructure or specific operation; data recorded by weather sensors integrated into the air vector (temperature, humidity, flammable gas) should be available in real time also to the institutions with responsibilities in emergency management.

## Acknowledgements

This research was supported by the Romanian National Authority for Scientific Research CNCSUEFISCDI in MAPIAM project "Modular Aircraft Platform for Intelligent Atmospheric Monitoring," Program 2—Increasing the Competitiveness of the Romanian Economy through Research, Development and Innovation, Domain 3—Energy, Environment and Climate Change. We are thankful to our colleague Sebastian Pop who provided expertise that greatly assisted the research in the field of aerial vector operation.

## Author details

Catalin Cioaca*, Vasile Prisacariu and Mircea Boscoianu

*Address all correspondence to: catalin.cioaca@afahc.ro

"Henri Coanda" Air Force Academy, Brasov, Romania

## References

[1] Romanian Government. Decision no 718 for the Approval of the National Strategy on Critical Infrastructure Protection [Internet]. 2011 [Updated: 2011]. Available from: http://ccpic.mai.gov.ro/docs/HGR718_2011.pdf [Accessed: April 23, 2016]

[2] Romanian Government. Decision no 1198 Regarding he Designation of National Critical Infrastructures [Internet]. 2012 [Updated: 2012]. Available from: http://ccpic.mai.gov.ro/docs/HGR%201198_2012.pdf [Accessed: May 4, 2016]

[3] Romanian Environment and Forests Ministry. National Strategy of Romania on Climate Change 2013-2020 [Internet]. 2012 [Updated: 2012]. Available from: http://www.mmediu.ro/beta/wp-content/uploads/2012/10/2012-10-05-Strategia_NR-SC.pdf [Accessed: May 16, 2016]

[4] General Inspectorate for Emergency Situations. National Disaster Risk Assessment Methodology (RO-RISK) [Internet]. 2016 [Updated: 2016]. Available from: https://www.igsu.ro/documente/RO-RISK/Metodologia%20de%20evaluare%20unitara%20a%20riscurilor%20-%20versiune%20finala.pdf [Accessed: November 12, 2016]

[5] European Commission. Risk Assessment and Mapping Guidelines for Disaster Management [Internet]. 2010 [Updated: 2010]. Available from: https://ec.europa.eu/echo/files/about/COMM_PDF_SEC_2010_1626_F_staff_working_document_en.pdf [Accessed: October 12, 2016]

[6] Transportation Research Board. Climate Change Adaptation Planning: Risk Assessment for Airports [Internet]. 2015 [Updated: 2015]. Available from: https://www.nap.edu/download/23461# [Accessed: February 19, 2017]

[7] Busuioc A. Scenarios for Changing the Climate Regime in Romania for 2001-2030 [Internet]. 2006 [Updated: 2006]. Available from: http://mmediu.ro/new/wp-content/uploads/2014/02/2012-04-23_schimbari_climatice_schimbareregimclimatic2001_2030.pdf [Accessed: February 22, 2017]

[8] Cazac V, Boian I, Prepeliță A. The main types of natural hazards and their impact on the environment and society. Natural and Anthropogenic Hazards. 2005;**5**(23):18-25

[9] National Infrastructure Protection Centre. Risk Management: An Essential Guide to Protecting Critical Assets [Internet]. 2002 [Updated: 2002]. Available from: http://www.iwar.org.uk/comsec/resources/risk/risk-mgmt.pdf [Accessed: March 4, 2017]

[10] Ministry of Environment and Climate Change. Romania's National Strategy on Climate Change 2013-2020 [Internet]. 2013 [Updated: 2013]. Available from: http://www.mmediu.ro/beta/wp-content/uploads/2013/10/2013-10-01_SNSC.pdf [Accessed: April 7, 2017]

[11] Loghry JD, Veach CB. Enterprise risk assessments: Holistic approach provides companywide perspective. Professional Safety Magazine. 2009;**15**(2):31-35

[12] Danila S. Risk Management in Air Security. Bucharest: Military Publishing House; 2016. p. 164

[13] Abt E, Rodricks J, Levy J, Zeise L, Burke T. Science and decisions: Advancing risk assessment. Risk Analysis. 2010;**30**(3):1028-1036. DOI: 10.1111/j.1539-6924.2010.01426.x

[14] Stoneburne G, Goguen A, Feringa A. Risk Management Guide for Information Technology Systems [Internet]. 2002 [Updated: 2007]. Available from: http://csrc.nist.gov/publications/nistpubs/800-30/sp800-30.pdf [Accessed: March 12, 2017]

[15] Cioacă C, Ciufudean C, Boşcoianu M. Applications of cost-benefit analysis in designing security strategies for aviation systems. In: 15th International Conference on Automatic Control, Modelling & Simulation (ACMOS13); WSEAS-NAUN; Brasov. 2013:283-287

Chapter 3

# Earthquake Culture: A Significant Element in Earthquake Disaster Risk Assessment and Earthquake Disaster Risk Management

Michaela Ibrion

Additional information is available at the end of the chapter

http://dx.doi.org/10.5772/intechopen.70434

**Abstract**

This book chapter brings to attention the dramatic impact of large earthquake disasters on local communities and society and highlights the necessity of building and enhancing the earthquake culture. Iran was considered as a research case study and fifteen large earthquake disasters in Iran were investigated and analyzed over more than a century-time period. It was found that the earthquake culture in Iran was and is still conditioned by many factors or parameters which are not integrated and do not work harmoniously towards building and sustaining an earthquake culture in Iran. A historical possibility of an earthquake culture in Iran was mainly severed by culture, especially beliefs, strong geopolitics in Iran and in Middle East, a complex and dynamic political landscape in Iran, foreign invasions and wars. However, there is a great potential in Iran for the earthquake culture to be built and developed. The earthquake culture is recommended to be integrated within earthquake disaster risk assessment and earthquake disasters risk management studies which are performed and carried out in Iran and other countries at seismic risk. The contribution of this book chapter is towards the earthquake disasters studies and policies for the countries at earthquake risk.

**Keywords:** earthquake, earthquake culture, earthquake disaster risk assessment, earthquake disaster risk management, Iran, earthquake disaster risk reduction

## 1. Introduction

Earthquakes are the manifestation of a living Earth [1] and occur at "unpredicted times and in unpredicted places" ([2], p. 352). Their rapid and dramatic effects on nature, culture and on unprepared communities and societies can cause the "archetypal sudden-impact disaster" with high death tolls, injuries and massive destructions ([3], p. 42). Earthquake disasters claim the highest number of lives among other type of disasters in the world. Nowadays, earthquake disaster risk reduction is a key term within disaster research arena. Earthquake disaster preparedness is a highly nonlinear phenomenon and can bring huge positive results on earthquake disaster risk reduction. Even small steps can significantly reduce the death tolls and injuries. Moreover, the focus needs to be shifted from response and recovery to prevention and mitigation, building resilience, reduction of risk to acceptable or tolerable levels, implementation of lessons from past disasters, past experiences and failures [4–6].

This research study brings to attention the dramatic impact of large earthquake disasters on society and local communities in Iran and particularly emphasizes the necessity of investigation and the importance of building an earthquake culture in Iran. Moreover, highlights the necessity to integrate the earthquake culture within the framework of earthquake disaster risk assessment and earthquake disaster risk management.

The following sections provide theoretical insights about risk, risk assessment, risk management, and earthquake culture. Further on, an investigation of 15 case studies of earthquakes and earthquake disasters in Iran over more than a century time-period (1909–2014) is presented. For investigation of the earthquake culture in Iran, a geo-historical and socio-cultural study was performed with an emphasis on an interdisciplinary approach. Narratives of earthquake disasters survivors, archival documents, earthquake field reports, cartographic and various academic studies were used for this study.

## 2. Risk

Risk can be basically defined as the likelihood multiplied by consequences or Hazard x Potential worth of loss [7]. Furthermore, with reference to the fundamental definitions of risk, Faber [8] and Lacasse et al. [5] advised that risk is seen as an equation such as $R = P \cdot C$, where R is Risk, P is Probability that an event will occur multiplied by expected Consequences or severity of adverse effects to life, health, properties and environment.

We live in a "world risk society" ([9], p. 1) where it is out of reality and practice to eliminate the risk. The disaster risk can never be eliminated, but it can be reduced to levels which are acceptable or tolerable by society [5, 10]. Moreover, Lacasse and Nadim [7] and Lacasse [10] warned that societies proved to be very less tolerant of an event where a huge number of lives is lost all of a sudden, compared with the same number of lives which is lost over the time in a number of separate events. Earthquake disasters are example of such events which test the tolerance and acceptance limits from communities and society [11, 12]. Risk can be expressed as individual risk and societal risk to people and assets. Societal risk can be presented through F-N diagrams or F-N curves expressing the risk level that a society is apparently willing to accept. Acceptable risk is the level of risk which a society aims to achieve it, and the tolerable risk refers to the risk level that can be reached after compromises in order to gain some benefits. The societal acceptable risk levels vary among national codes and standards around the world. The F-N curves relate to the annual or other temporal occurrence probability (F) of an event capable for causing (N), or more fatalities. The term N can be replaced by other quantitative measures of consequences, for instance, costs. As a note, the F-N curves offer statistical observations and not the acceptable or tolerable thresholds. The F-N curves can be also used to describe the safety levels of particular facilities. Moreover, the F-N curves offer a good illustration for comparison of the calculated probabilities with observed frequencies of failure for facilities [5, 7, 10].

Very often, the risk analysis is seen as a magic wand that will transform the unpredictable in predictable [13]. "The risk of risk analysis" and the ontological matter that an uncertainty even analyzed and calculated is still an uncertainty was also emphasized ([14], p. 810). Woo [15] highlighted that uncertainty is an inexorable feature of human world and brought to attention the requirement to consider both the epistemic and aleatory uncertainty. The relation between risk, space and time needs to be also considered as a risk analysis "only reflects the situation at a certain point in time, and for a given location with a distinct spatial dimension of scale" ([16], p. 194). Interdisciplinary approaches have been seen as essential strategies for risk analysis and disaster risk reduction (DRR).

## 3. Risk assessment

Lacasse [10] highlighted that risk assessment and reliability analysis are important tools for informed decision-making. According to Faber [8], risk assessment is seen as a comparison between estimated risk and accepted risk which initially was stated in risk acceptance criteria. Lacasse and Nadim [7] and Lacasse [10] warned that one of the most difficult tasks in risk assessment and risk management is the selection of risk acceptance criteria. Societies which experience very frequently geohazards may have a different risk acceptance level comparative with those societies which experience them rarely. In case that risk is not acceptable within the specified risk acceptance criteria, there are various ways for risk treatment: risk mitigation, risk reduction, risk transfer and risk acceptance [8].

Risk analysis and risk evaluation are basically seen as component parts of risk assessment. Risk analysis comprises hazard and risk identification, risk estimation and risk calculation. For robustness, increased accuracy and value of risk assessments it is required among others, to address the interconnection between physical and human systems, to consider both spatial and temporal scales, to address, analyze and communicate the uncertainties. Various formal and informal tools are utilized within the risk assessment [7, 10, 17]. Multi-risk assessments which studied the interaction and amplification of risk, cascading hazards, dynamic vulnerabilities to multi-risk have stirred a great interest, but still is a field yet to be developed and requires intensive collaboration and expertise from various disciplines [5, 7, 10].

With reference to the assessment of disaster risk in various countries around the world, there is the danger of oversimplifying and misunderstanding problems, priorities and concerns of people at risk. Moreover, there is a permanent struggle between the quantitative or the so-called technical risk approaches and the socio-cultural risk [18]. A cooperation among these approaches would be beneficial to disaster risk assessments.

## 4. Risk management

Risk management integrates recognition and assessment of risk with development and implementation of adequate strategies of risk mitigation and risk reduction. Risk management represents a systematic application of policies, procedures and practices to the tasks of communication, consultation, establishing the context, identification, analysis, evaluation, monitoring and implementation of risk mitigation measures. Due to epistemic and aleatory uncertainty, risk management is decision-making under the condition of uncertainty. Likewise risk assessment, risk management requires also a multi-disciplinary approach [5, 7, 10].

Disaster risk management (DRM) confronted around the world various difficulties linked to political will, governance, available budget, implementation of legislation. Moreover, Okada [19] advised that DRM needs to have pre-disaster orientation instead of being focus on post-disaster phase, to take in consideration multiple hazards, to be closely linked to urban planning and management and to be inclusive and not limited only to governmental organizations and institutions, but to engage citizens, Non-Governmental Organizations (NGO)s, private companies, local communities, individuals. DRM needs to start from the local level or community level. Furthermore, for the present century, the integrated disaster risk management (IDRIM) is a necessary and required perspective in dealing with disaster risk.

## 5. State of the art on the earthquake culture

Earthquake culture is particularly linked to earthquake hazard, earthquake risk, and earthquake disasters and refers to the capacities of communities and society of knowing to live with earthquake risk. Seismic culture, local seismic culture, seismic prevention cultures are examples of other interchangeable terms with the earthquake culture. The earthquake culture concept has roots in theoretical insights and research studies carried out during the time in different geographical places around the world.

Mileti and Darlington [20] analyzed the existence of an earthquake culture in the San Francisco Bay Area, years after the Loma Prieta 1989 earthquake. Earthquakes were not frequent in that specific area, but earthquake risk was known and local culture was abundant of accounts of the last and previous other earthquake disasters. Earthquake preparedness and readiness was seen as a part of local culture. Helly [21] emphasized that origins and development of local seismic cultures are influenced by frequency of earthquakes, their intensities, death toll and injuries, and extent of damages. Ferrigni [22] highlighted that a visible cultural adaptation to earthquakes is the seismic architecture or vernacular seismic architecture which develops over the time. However, Ferrigni [22] advised that earthquakes do not always generate a local seismic culture, particularly connected to earthquake resistant buildings. Nevertheless, according to Pierotti [23], the seismic culture can develop during the time, and the case of Japan is given as an example. In Japan, the memory of earthquake disasters has been kept alive through written records, oral accounts, legends, stories, through the lessons from earthquake disasters. Homan ([24], pp. 1–2) employed the hypothesis launched by the European

University Centre for Cultural Heritage (CUEBC) that there is a correlation among frequency of earthquakes and local building practices. Consequently, two types of seismic cultures might develop: a "seismic prevention culture" when earthquakes are frequent and a "seismic culture of repairs" when earthquakes are low in frequency.

Homan and Eastwood ([25], p. 629) analyzed the seismic culture in Turkey, after the Kocaeli (Izmit) 1999 earthquake disaster, and advised that "Seismic cultures could be described as being the knowledge (both pragmatic and theoretical) that has built up in a community exposed to seismic risks through time." Degg and Homan [26] emphasized that the earthquake vulnerability in the Middle East was seen possible to be reduced through the seismic cultures. Karababa and Guthrie ([27], p. 32) analyzed the seismic culture, and particularly "the seismic construction culture" for the Lefkada Island, situated on the west part coast of Greece. They warned that "… the seismic construction culture expressed tangibly by buildings and tacitly in the local know-how is only a small subset of the seismic culture." Karababa and Guthrie ([27], p. 32) advised that a seismic culture can include "all the activities, attitudes, behaviors, and perceptions of the local population regarding earthquakes." Halvorson and Hamilton ([28], p. 322) analyzed the seismic culture for Mountainous Central Asia, in Pakistan, Afghanistan, Tajikistan and Kyrgyzstan and defined the term of "seismic culture as a broad concept that encompasses a range of cultural adaptations to seismic risk and hazard."

For Iran, other terms were used in connection with culture and earthquake disaster, for instance, seismic safety culture or culture of earthquake safety. Moreover, in Iran, the emphasis has been on disaster education, the culture of safety, public awareness and preparedness, especially education and trainings. Parsizadeh et al. [29] highlighted the importance of earthquake awareness and preparedness for all school levels and their contribution for building a culture of earthquake safety. Formal and informal school earthquake education, safety of school buildings and development of the national earthquake safety drill have been seen as priorities.

Berberian [1] with concern of many earthquake disasters in Iran, highlighted the necessity of creating a culture of prevention in Iran. Furthermore, Berberian [1] examined for a time period of many centuries, various cultural aspects linked to earthquakes and earthquake disasters in Iran. Berberian [1, 30] and Berberian and Yeats [31] emphasized that demographic changes, rapid urbanization, the raise of mega-cities, increased seismic urban risk, poor construction of buildings, the corrupted building industry, inaction, ignorance and non-accountability with regards to enforcement of building codes and land-use severely have impacted the earthquake disaster preparedness and have highly increased the seismic risk in Iran.

Ibrion et al. [4, 11, 12, 32, 33] and Parsizadeh et al. [34] investigated various aspects of the earthquake culture in Iran with focus on cultural aspects of resilience and earthquake disaster risk reduction. Ibrion et al. [11] explored and analyzed several aspects of the meanings and perceptions of time with reference to earthquakes and earthquake disasters in Iran and how the earthquake culture is built over the time. Ibrion et al. [4, 12, 32] and Parsizadeh et al. [34] explored how the earthquake culture was impacted by different places of Iran and more precisely, by the landscapes and cultural landscapes of arid, semi-arid and mountainous areas of Iran. Aspects of the intricate relationships between the cultural landscapes of arid and semi-arid areas, cultural beliefs, earthquake disasters and the communities' earthquake risk perceptions and resilience were further investigated by Ibrion et al. [4, 12, 32] and Parsizadeh et al. [34]. Ibrion et al. [12] particularly analyzed how the beliefs, cultural traditions and rituals impacted the handling of the dead people, the earthquake disaster risk management and the resilience of survivors after large earthquake disasters in Iran. The impact of lessons and socio-cultural learning from large earthquake disasters in Iran on the earthquake culture was examined by Ibrion et al. [4, 33].

## 6. Earthquake culture in Iran

15 earthquake disasters which affected different places of Iran were investigated as research case studies: *Silakhor 1909, Salmas 1930, Torud 1953, Buyin Zahra 1962, Dasht-e Bayaz 1968, Ferdows 1968, Karzin-Qir 1972, Tabas 1978, Golbaf 1981, Sirch 1981, Rudbar 1990, Zirkuh (Qa'enat) 1997, Bam 2003, Ahar 2012* and *Shonbeh-Bushehr 2013,* see **Figure 1**.

All of these 15 earthquakes and earthquake disasters in Iran over more than a century time-period (1909–2014) can be considered as important wake-up calls toward building an earthquake culture in Iran. Their massive destruction, injuries and particularly, death tolls require long-term sustainable strategies for earthquake disaster risk reduction Iran, see **Figure 2**.

In Iran, there is no linear correlation between magnitude and number of dead people, as even a medium magnitude earthquake caused one of the highest number of death (e.g., Bam 2003), see **Figure 2**. Furthermore, it was observed that all the 15 earthquakes over more than a century time-period have the magnitude $M_w$ less than 7.5. Moreover, over the time, the number of dead people on a given magnitude can largely vary, and many examples can be identified: Buyin Zahra 1962 and Sirch 1981, both with $M_w$ 7.0, but with different death tolls, Golbaf 1981 and Bam 2003, both with $M_w$ 6.6, Salmas 1930 and Dasht-e Bayaz 1968, both with $M_w$ 7.1, and Silakhor 1909 and Tabas 1978, both with $M_w$ 7.4, see **Figure 2**. The demography of Iran suffered many changes in the last and present century and major urban areas and many towns and villages are situated in the proximity of fault lines in Iran [1, 30]. Tehran with a population of almost 15 million people, a well documentation of historical earthquakes, and close proximity to at least eight adjacent and other inner city active faults lines is at risk from a moderate-magnitude to a large-magnitude earthquake, more precisely in the range of approximately 6.5 till 7.4 $M_w$. Tehran is at risk from "an earthquake time bomb" and a large earthquake disaster [31]. This high seismic disaster risk in Tehran, the 15 earthquake disasters in Iran over more than a century time-period (1909–2014), and many other earthquake disasters over the centuries can be basically considered as required conditions for the existence and development of an earthquake culture in Iran.

The earthquake culture is highly motivated by frequency of earthquakes, their intensities, death tolls, injuries and extent of damages and destruction. However, in Iran, over more than a century time-period (1909–2014) the earthquake time became equivalent with the earthquake

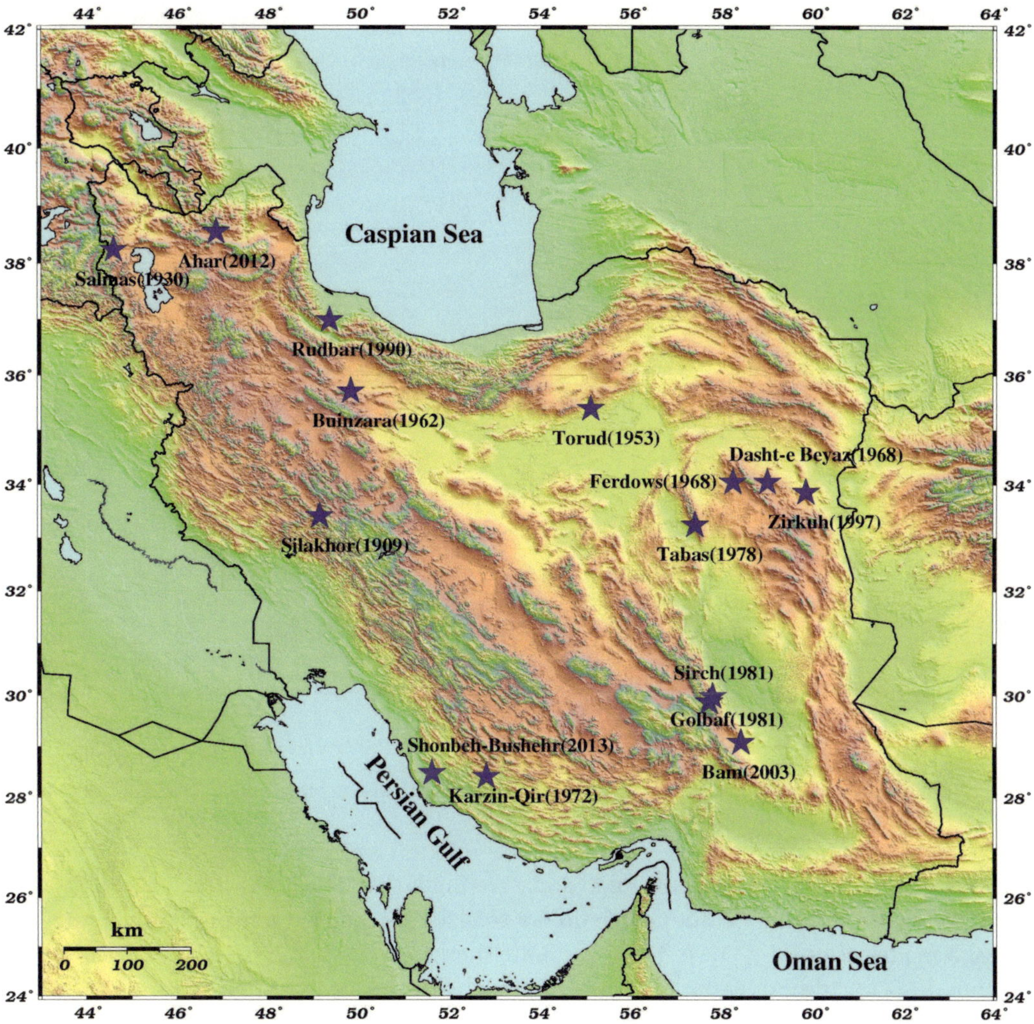


**Figure 1.** Earthquakes and earthquake disasters in Iran over more than a century time-period (1909–2014) and Iran map. Source: Prof. Mohammad Mokhtari and Mr. Arash Islami, International Institute of Earthquake Engineering and Seismology (IIEES).

disaster time. Many lessons from earthquake disasters in Iran have remained ignored and forgotten. When an earthquake disaster occurs, the forgotten and ignored old lessons emerge together with new lessons, and all are categorized under the label of "Lessons-Learned." An amalgam of old and new lessons from earthquake disasters is being repeated again and again, over the time, in different places of Iran. The next earthquake disasters are only a matter of time, if the lessons from earthquake disasters are yet pending to be learned and implemented [4, 33]. Moreover, if no earthquake preparedness is in place and resilience of communities is not improved, the next earthquakes will be again followed by earthquake disasters. Almost 10 years after Bam 2003, an earthquake disaster survivor declared "If an earthquake happens

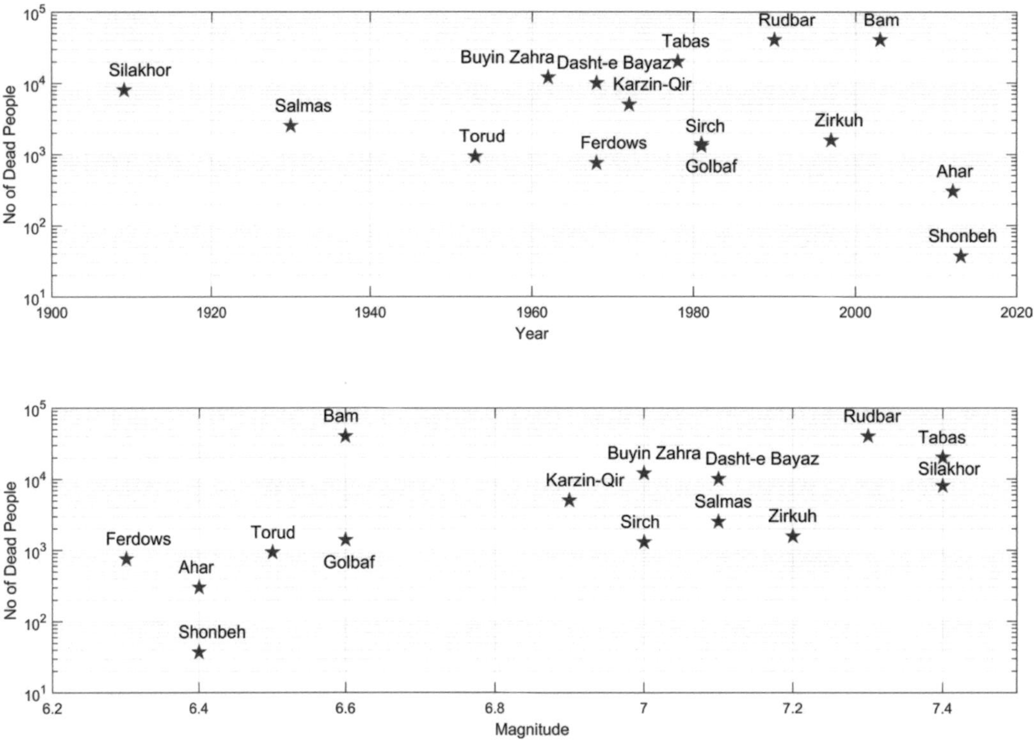


**Figure 2.** Earthquakes and earthquake disasters in Iran over more than a century time-period (1909–2014), number of dead people versus year and magnitude.

again, the same things will be repeated" and "We think that disaster is in the news, disaster is not for us, it is for others. But people do not realize that the next person might be them. It is just a matter of time" ([11], p. 16). Over more than a century time-period (1909–2014), the rhythm of socio-cultural learning from earthquake disasters or even large earthquake disasters is very slow in Iran. Furthermore, Ibrion et al. [4, 33] advised that the socio-cultural learning from earthquake disasters needs to become integrated and to involve various levels of participation and accountability for, see **Figure 3**.

Learning from earthquake disasters in Iran over more than a century time-period is a dynamic and complex process which requires long-term strategies, responsible earthquake disaster risk management and especially, a sustainable framework such as a culture of resilience and earthquake disaster risk reduction or an earthquake culture [4, 33].

Strong politics and geopolitics in Iran and in the Middle East, complex and dynamic power structures in Persia/Iran, and dramatic history of Persia/Iran [30, 35–39] have contributed during the time and still contributes nowadays to the erosion of an earthquake culture in Iran. The earthquake culture has not been seen as a top priority in Persia/Iran over more than a century time-period (1909–2014). The focus of Persia/Iran and the resources of country have been concentrated on other more important and dramatic priorities and events including and not limited to the effervescent events during the last kings Qajars, implementation of

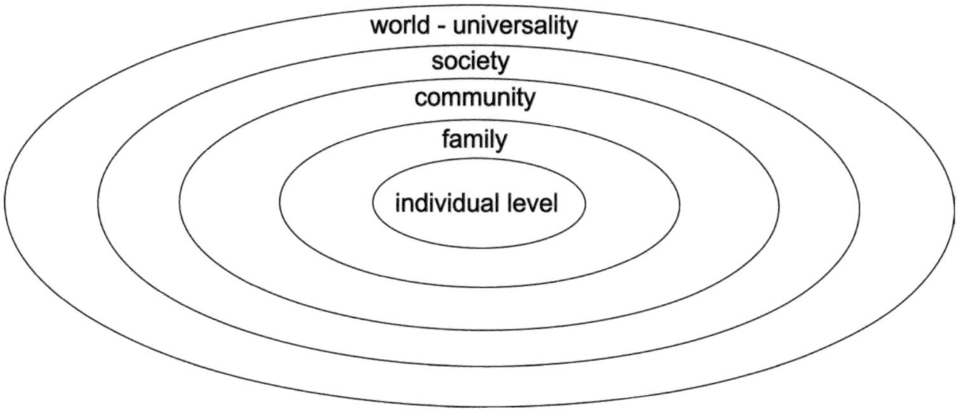


**Figure 3.** Integrated socio-cultural learning from earthquake disasters—levels of participation and accountability.

the Anglo-Russian Convention in 1907, military occupation of country after the second war, the forced abdication and exile of Reza Shah Pahlavi, the 1953 coup d'etat in Iran, political turbulence in the country and the collapse of Mohammad Reza Shah Pahlavi, the Iranian Revolution, the Islamic Republic in 1979, the 8 years Iran-Iraq war (1980–1988), national safety and security of Iran, dynamic political landscape in Iran, economical-political sanctions, and many others, and then, far from a sustainable and long-term social-cultural learning from large earthquake disasters and building an earthquake culture in Iran. In this entire effervescent context, the earthquake disaster preparedness was not truly seen as a high priority in Iran. Many large earthquake disasters in Iran occurred during or around tumultuous political events and turbulent geopolitical arena in the area. The earthquake disasters of Silakhor 1909, Torud 1953, Tabas 1978, Golbaf 1981, Sirch 1981, Rudbar 1990, Bam 2003, Ahar 2012 and Shonbeh-Bushehr 2013 are just few examples. The building and development of the earthquake culture in Iran was overshadowed by strong geopolitics and various interests of the global and regional powers in Iran and in the area of the Middle East or West Asia.

The discovery of oil in Iran in 1908 was the first oil discovery in the Middle East area [40]. After more than 100 years of exploration, production and export, Iran still has "gigantic energy reserves", vast oil deposits and "mammoth reserves of gas" ([41], p. 48, 59). According to Abbaszadeh et al. [42], Iran has the fourth largest oil reserves and the second largest natural gas reserves in the world. However, over more than 100 years' time-period (1909–2014), and despite rich oil and gas resources of Iran and high economic revenues from oil over the years, the number of earthquake disasters in Iran increased together with a high raising death toll.

Indeed, there is more than one century from the discovery of oil in Iran, but the revenues from oil exports began to have an important role in the Iranian economy just after the 1960s [43]. Reasons for this situation were linked to the fact, that for almost a period of 50 years, from 1908 till late 1953, Iran received very low level of royalties from the foreign oil company which operated and controlled all oil operations in Iran; more precisely, from the Anglo-Persian Oil Company (APOC) and later, the Anglo-Iranian Oil Company (AIOC) with a majority of shares owned by the British Government. Nationalization of oil industry in Iran and establishment of the National Iranian Oil Company (NIOC) by Mohammad Mosaddeq was followed by tough British sanctions and coup d'etat in 1953, in Iran, organized by a close collaboration between the American Central Intelligence Agency (CIA), British Military Intelligence section 6 (MI6), governmental institutions of USA and Britain, USA Embassy in Tehran, and with the support from Mohammad Reza Shah Pahlavi, Palace, Majlis members, clerics, notably Abul Qassem Kashani, merchants, and other supporters inside of Iran. After the 1953 coup, the oil concession in Iran was given to a consortium of major international companies which gained full control over production, refining, management and distribution of oil; 40% of shares went to the Anglo-Iranian Oil Company which was renamed British Petroleum (BP) and 40% went to a group of American companies. This consortium was supposed to give 50% of profits to Iran and this was the type of agreement signed by USA and Britain with other countries in Middle East at that time [35]. Creation of OPEC in 1960, and various geopolitics in the Middle East impacted the production and prices of oil in Iran. The oil revenues started to increase significantly for Iran, especially, in the period 1960–1978. A high dependence on oil incomes of the Iranian government, a higher volatility of international prices, a risen inflation, a political effervescence in Iran, and strong geopolitics in Iran and Middle East area, contributed to the downfall of the Pahlavi dynasty and to the 1979 events. After the 1979 Islamic Revolution, the production of oil was reduced by the political choice. Following the invasion of Iran by Iraq and the Iran-Iraq war (1980–1988), the production of oil was considerably reduced and affected by war and its massive destruction. High oil revenue volatility and many waves of economic sanctions applied on Iran, by United States and later on, by the European Union had also their impact on the Iranian economy.

A high dependency on oil revenues, the complexity of the political landscape in Iran and a very reduced transparency has negatively affected the accountability of state, governmental institutions and other many organizations in Iran and negatively impacted the economy of Iran [35, 42, 43].The accountability toward the earthquake disaster risk reduction in Iran was also negatively impacted. Over more than a century time-period (1909–2014), despite some registered progress, the rhythm of socio-cultural learning from earthquake disasters is slow in Iran and an earthquake culture is yet pending to be developed. An integrated earthquake disaster risk management in Iran requires urgent and critical strategic measures and actions. However, if responsible and sustainable planning and actions are applied and carefully monitored, the abundance of oil and gas resources of Iran can have a highly positive impact on the earthquake culture and earthquake disaster risk reduction in Iran.

Earthquake culture in Iran is definitely strongly influenced by the Iranian culture. The culture in Persia/Iran has been strongly articulated through the power of beliefs system which is very deeply rooted and has a strong influence over local communities, governance, rites and rituals, meanings of the time and place, earthquake disaster preparedness and mitigation, daily life, just to name few [1, 11, 12, 30, 32, 34, 37, 44, 45]. Double impact of beliefs on the resilience of survivors, earthquake disaster risk management and earthquake disaster risk reduction in Iran was highlighted by Ibrion et al. [12] and Parsizadeh et al. [34]. Moreover, the earthquake risk perception in Bam, before the Bam 2003 earthquake disaster was negatively impacted by the strong cultural beliefs linked to the cultural landscape of Bam, represented by Arg-e Bam, Qanats and gardens of khorma trees [32, 34].

Earthquakes and earthquake disasters are present in geo-mythology, legends, stories, oral traditions, poems, spiritual texts, inscriptions of Persia/Iran [1, 30] and theoretically, they have the role to indicate the existence of an earthquake culture in Iran. However, in Iran, myths, legends, stories, poems, and many other cultural manifestations are considered to be just part of the Iranian culture and not linked at all to an earthquake culture. Moreover, in practice, they do not have an active role toward earthquake disaster mitigation and particularly, earthquake disaster awareness and education. These cultural parameters/factors such as myths, legends, stories, oral traditions, poems, spiritual texts, inscriptions, etc., are interpreted as a sign of the cultural resilience of Persia/Iran over the centuries and they are just passive and not active toward building an earthquake culture in Iran. This was identified also by Parsizadeh et al. [34] for the case of Bam 2003 earthquake disaster. Moreover, Parsizadeh et al. [34] recommended that myths, legends, poems, stories, oral traditions, and various other oral and written accounts cultural manifestations need to be integrated within earthquake disaster awareness and the efforts of building an earthquake culture in Iran, see **Figure 4**.

In the world, it seems that a successful way of learning from earthquake disasters and mega-disasters was shown by Japan and its status of "earthquake nation" [46, 47]. Moreover, in Japan, there is a mature earthquake culture and of a culture of disaster prevention and the

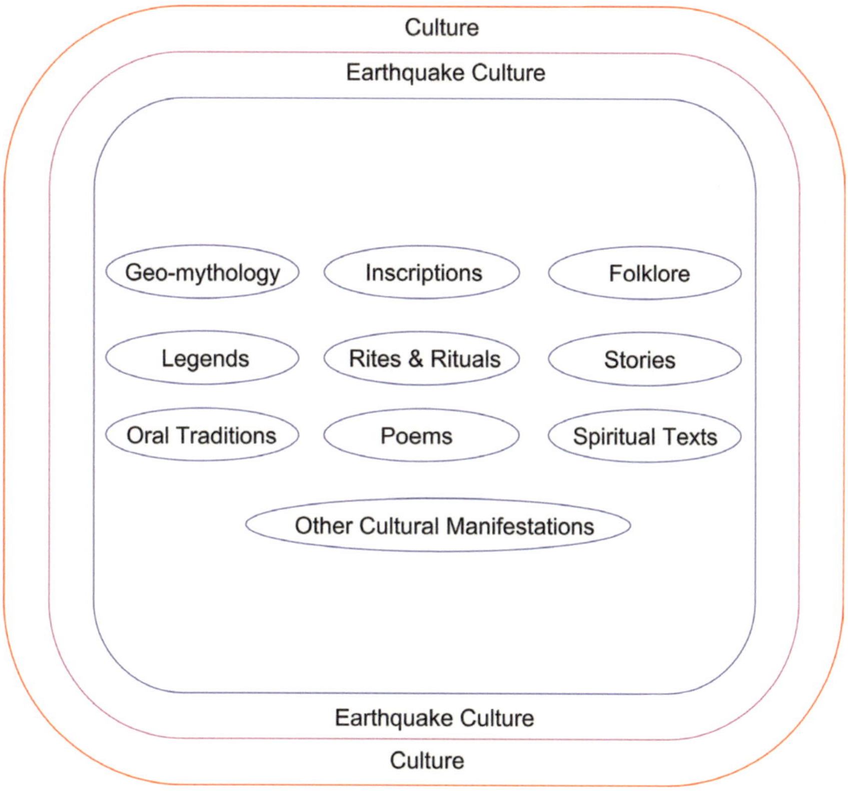


**Figure 4.** The framework of culture, earthquake culture and cultural parameters.

rhythm of socio-cultural learning from large earthquake disasters is very high. However, the March 2011 cascading disasters highlighted for Japan and other countries around the world that "Preventive investments pay, but be prepared for the unexpected" as it is important to understand that "...the risks from natural hazards can never be completely eliminated..." ([47], p. 5, 6). After a necessary adaptation to the Iranian local context, the Japanese culture of continuous learning from past earthquake disasters, continuous improvements of the earthquake disaster preparedness, the status of a strong earthquake culture and the progress of integrated earthquake disaster risk management can serve as learning models for Iran [4, 33].

Over more than a century time-period (1909–2014), the existence and development of the earthquake culture in Iran has been conditioned by a large array of parameters or factors [4, 11, 12, 32–34], see **Figure 5**.

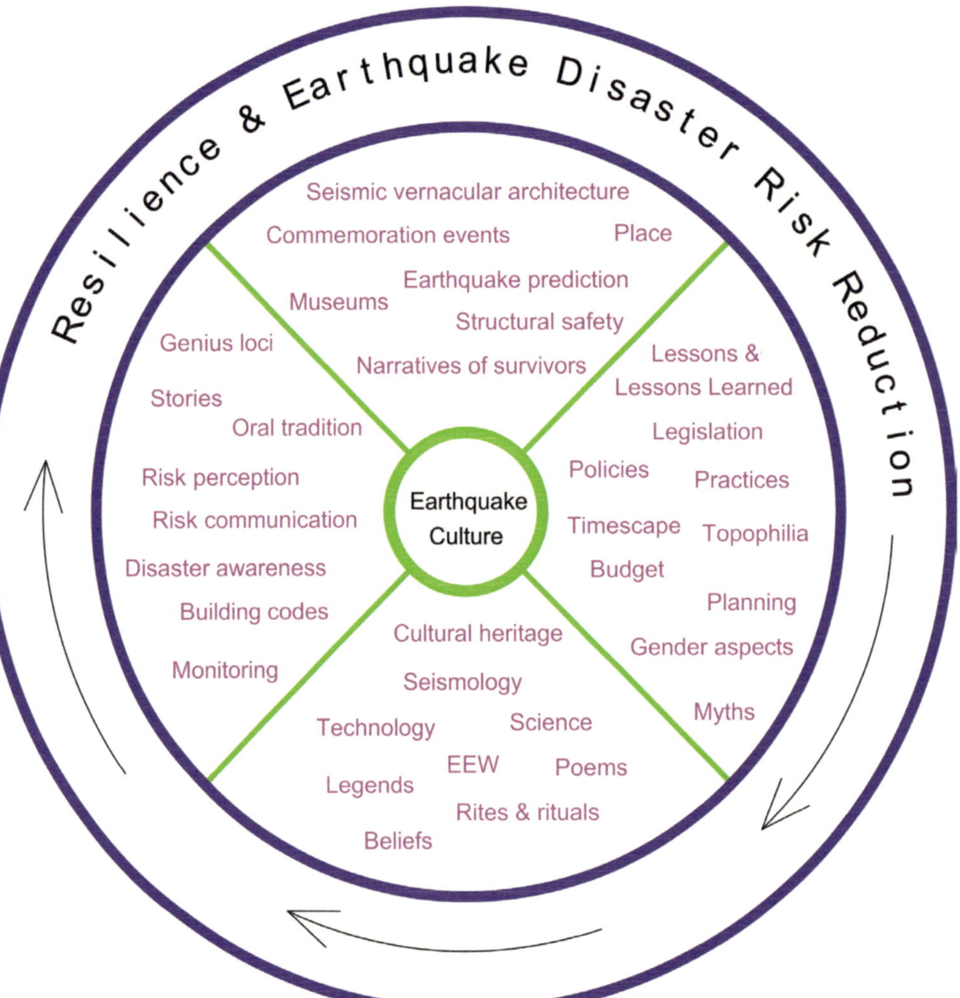


**Figure 5.** The wheel of earthquake culture and its parameters/factors.

All of these parameters or factors exist till a degree in Iran, but they are not integrated and do not work in harmony toward building and sustaining an earthquake culture. Moreover, the development and even existence of such parameters/factors is hindered by inadequate long-term planning, evaluation and monitoring, improper budget, lack of accountability, socio-cultural and political will and implementation of sustainable strategies and actions.

An assessment of the earthquake culture's existence and development required to be incorporated into the earthquake disaster risk assessment. Assessment of the earthquake culture status needs to be considered as one of the important criteria within risk analysis, and risk evaluation phases. Moreover, the earthquake culture needs to be also highly considered as part of the risk treatment strategies and plans and to be carefully monitored over the time, see **Figure 6**.

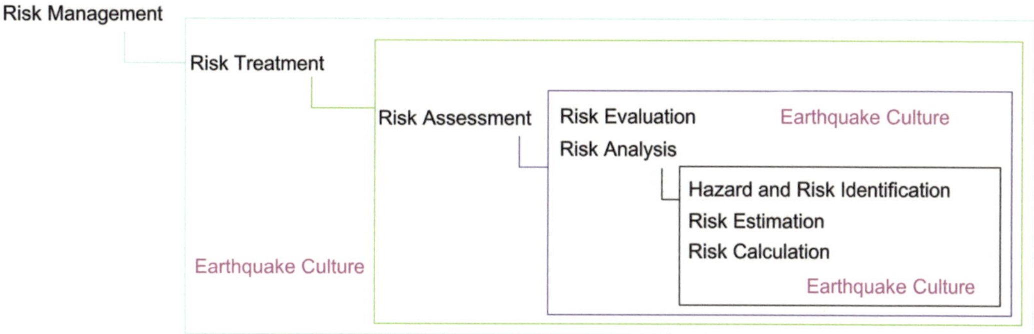


**Figure 6.** Earthquake disaster risk assessment, earthquake disaster risk management and earthquake culture.

The risk estimation and risk calculation considering the earthquake culture's parameters require awareness, transfer of knowledge, communication, a feasible time framework and further research investigations. Moreover, while interdisciplinary approaches are not easy to achieve, they are nonetheless essential and need to be applied within the investigation of the earthquake culture and within the earthquake disaster risk assessment and management.

## 7. Concluding remarks

Earthquake culture in Iran is highly motivated by the frequency of earthquake disasters over the last and present century and particularly, by the high death tolls, injuries, and massive damages and destructions. Based on the investigated case studies over more than a century time-period (1909–2014), the earthquake culture in Iran is yet pending to become a coherent and well-functioning *present reality*, but there is a great potential in Iran for the earthquake culture for being build and developed.

A *historical possibility* of an earthquake culture in Iran was mainly severed by the powerful influence of culture, especially beliefs systems, strong politics and geopolitics in Persia/

Iran and in Middle East area, complex and dynamic power structures in Iran, wars, foreign invasions, massive destructions and even by existence of rich oil and gas resources in Iran. The last twentieth century and even the beginning of present century made no exemption from this dramatic course of the Persian/Iranian history. The earthquake culture was not seen as a national top priority in Persia/Iran over more than a century time-period (1909–2014). The focus and the resources of the country have been concentrated on other more important and dramatic priorities and events very tightly connected with the national safety and security, dynamic and complex political and geopolitical landscape in Iran and in Middle East area, waves of economical-political sanctions and far from building an earthquake culture in Iran.

Earthquake culture requires time to develop, but it is also over the time that the earthquake culture can become forgotten and lost. Consequently, long-term sustainable actions, transdisciplinary approaches, accountability and integrated efforts from local communities, various institutions and organizations, researchers, practitioners, policy makers, governance, society, legal frameworks, appropriate budget, planning and monitoring are recommended for fostering an earthquake culture in Iran and to prioritize it at national, regional and local levels. Place specific strategies for the earthquake disaster risk reduction, enhancing resilience and building an earthquake culture are also recommended. Iran is a cultural nation, but not yet a nation with an earthquake culture. Iran needs to integrate the earthquake nation within its framework of cultural nation. The collective cultural memory represented by geo-mythology, legends, stories, oral traditions, poems, spiritual texts, inscriptions and other oral and written accounts about earthquakes and earthquake disasters needs to play an active role, to be integrated and to actively contribute to the earthquake disaster awareness and mitigation and to the efforts of building an earthquake culture in Iran. The rhythm of implementation of lessons and the socio-cultural learning from large earthquake disasters is very slow in Iran. Living with earthquake disasters in Iran needs to be replaced by learning to live with earthquakes in Iran.

Existence and assessment of the earthquake culture's development stages needs to be incorporated as an important step into the earthquake disaster risk assessment and earthquake disaster risk management. To encourage and support the earthquake culture as a *future probability* is highly recommended, if the reduction of the high earthquake disaster risk is strongly aimed for in Iran. Demography and a high urbanization in Iran together with a high seismic vulnerability of buildings demand the existence and development of an earthquake culture in Iran. The rich oil and gas resources of Iran can have a high positive impact and contribution to the building and development of the earthquake culture in Iran.

## Acknowledgements

Dr. Farrokh Nadim, Prof. Amir R. Nejad, Prof. Mohammad Mokhtari, Dr. Farokh Parsizadeh and Dr. Manuel Berberian are greatly acknowledged and thanked. The financial support from NTNU Publishing Fund, Trondheim, Norway is highly acknowledged and thanked.

## Author details

Michaela Ibrion

Address all correspondence to: mibrion5@gmail.com

Department of Geography, Faculty of Social and Educational Sciences, Norwegian University of Science and Technology (NTNU), Trondheim, Norway

## References

[1] Berberian M. Earthquakes and Coseismic Surface Faulting on the Iranian Plateau, A Historical, Social and Physical Approach. Amsterdam: Elsevier; 2014

[2] Mokhtari M. Earthquake prediction activities and Damavand earthquake precursor test site in Iran. Natural Hazards. 2010;**52**:351-368

[3] Alexander D. Natural Disasters. Dordrecht, The Netherlands: Kluwer Academic Publishers; 1999

[4] Ibrion M, Mokhtari M, Nadim F. Earthquake disaster risk reduction in Iran: Lessons and "Lessons Learned" from three large earthquake disasters; Tabas 1978, Rudbar 1990, and Bam 2003. International Journal of Disaster Risk Science. 2015;**6**(4):415-427

[5] Lacasse S, Nadim F, Hoeg K. Risk assessment and mitigation in geo-practice. In: Rollins K, Zekkos D, editors. Geotechnical Engineering State of the Art and Practice: Keynote Lectures from GeoCongress 2012; March 25-29 2012. Oakland, California: American Society of Civil Engineers; 2012

[6] Woo G. Operational earthquake forecasting and risk management. Seismological Research Letters. 2010;**81**(5):778-782

[7] Lacasse S, Nadim F. Learning to live with Geohazards: From research to practice. In: Juang CH, Phoon KK, Puppala JA, Green AR, Fenton AG, editors. Geo-risk 2011: Risk Assessment and Management, Proceedings of GeoRisk 2011; 26-28 June 2011. Atlanta, Georgia: American Society of Civil Engineers; 2011. pp. 64-116

[8] Faber MH. Risk and Safety in Civil Surveying and Environmental Engineering, Lectures Notes. Switzerland: Swiss Federal Institute of Technology, ETHZ; 2003

[9] Beck U. World at risk: The new task of critical theory. Development and Society. 2008;**37**(1): 1-21

[10] Lacasse S. Hazard, reliability and risk assessment-research and practice for increased safety. In: Keynote Lectures at the 17th Nordic Geotechnical Meeeting; 25-28 May 2016. Reykjavik, Iceland; 2016

[11] Ibrion M, Mokhtari M, Parsizadeh F, Nadim F. Timescape of the earthquake disasters in Iran: The intricacies of earthquake time and earthquake disaster risk reduction. Geografiska Annaler: Series A, Physical Geography. 2015;**97**(1):197-216

[12] Ibrion M, Parsizadeh F, Pakdaman Naeini M, Mokhtari M, Nadim F. Handling of dead people after two large earthquake disasters in Iran: Tabas 1978 and Bam 2003—Survivors' perspectives, beliefs, funerary rituals, resilience and risk. International Journal of Disaster Risk Reduction. 2015;**11**:60-77

[13] Beck U. Risk Society, Towards a New Modernity. Great Britain: Sage Publication; 1992

[14] Hagmann J. Fukushima: Probing the analytical and epistemological limits of risk analysis. Journal of Risk Research. 2012;**15**(7):801-815

[15] Woo G. Calculating Catastrophe. London: Imperial College Press; 2011

[16] Fuchs S, Keiler M. Space and time. Coupling dimensions in natural hazard risk management?. In: Muller-Mahn D, editors. The Spatial Dimension of Risk. How Geography Shapes the Emergence of Riskscapes. USA and Canada: Routledge; 2013. pp. 189-201

[17] Crozier M, Glade T. Hazard assessment for risk analysis and risk management. In: Alcantara-Ayala I, Goudie AS, editors. Geomorphological Hazards and Disasters Prevention. NY, USA: Cambridge University Press; 2010. pp. 221-232

[18] Hewitt K. Regions of Risk: A Geographical Introduction to Disasters. New York, USA: Routledge, Taylor and Francis Group; 2014

[19] Okada N. Urban diagnosis and Integrated Disaster Risk Management. Journal of Natural Disaster Science. 2004;**26**(2):49-54

[20] Mileti DS, Darlington JD. The role of searching in shaping reactions to earthquake risk information. Social Problems. 1997;**44**(1):98-103

[21] Helly B. Local seismic cultures: A European research program for the protection of traditional housing stock. Annali di Geofisica. 1995;**XXXVIII**(5-6):791-794

[22] Ferrigni F. The recovery of the local seismic culture as preventive action. In: Ferrigni F, Helly B, Mauro A, Mendes Victor L, Pierotti P, Rideaud A, Teves Costa P, editors. Ancient Buildings and Earthquakes, The Local Seismic Culture Approach: Principles, Methods, Potentialities. Bari, Italy: Edipuglia; 2005. pp. 215-242.

[23] Pierotti P. History and praxis. In: Ferrigni F, Helly B, Mauro A, Mendes Victor L. Pierotti P, Rideaud A, Teves Costa P, editors. Ancient Buildings and Earthquakes, The Local Seismic Culture Approach: Principles, Methods, Potentialities. Bari, Italy: Edipuglia; 2005. pp. 29-51

[24] Homan J. Seismic Cultures: Myth or Reality? [Internet]. 2004. Available from: http://www.grif.umontreal.ca/pages/papers2004/Paper%20-%20Homan%20J.pdf [Accessed: 22-06-2010]

[25] Homan J, Eastwood JW. The 17 August 1999 Kocaeli (Izmit) earthquakes: Historical records and seismic culture. Earthquake Spectra. 2001;**17**(4):617-634

[26] Degg M, Homan J. Earthquake vulnerability in the Middle East. Geography. 2005;**90**(1): 54-66

[27] Karababa SF, Guthrie MP. Vulnerability reduction through local seismic culture. IEEE Technology and Society Magazine. 2007;**26**(3):30-41

[28] Halvorson SL, Hamilton PJ. Vulnerability and the erosion of seismic culture in mountainous Central Asia. Mountain Research and Development. 2007;**27**(4):322-330

[29] Parsizadeh F, Ghafory-Ashtiany M, Heshmati V, Seif A-E. Iran's school earthquake safety initiative. Regional Development Dialogue, United Nations Centre for Regional Development, Nagoya, Japan. 2007;**28**(2):33-46

[30] Berberian M. Development of geological perceptions and explorations on the Iranian plateau: From Zoroastrian cosmogony to plate tectonics (ca. 1200 BCE to 1980 CE). In: Sorkhabi R, editors. Tectonic Evolution, Collision, and Seismicity of Southweast Asia: In Honor of Manuel Berberian's Forty-Five Years of Research Contributions: Geological Society of America. USA: The Geological Society of America; 2016

[31] Berberian M, Yeats R. Tehran: An earthquake time bomb. In: Sorkhabi R, editors. Tectonic Evolution, Collision, and Seismicity of Southweast Asia: In Honor of Manuel Berberian's Forty-Five Years of Research Contributions: Geological Society of America. USA: Geological Society of America; 2016. Special Paper. p. 525

[32] Ibrion M, Lein H, Mokhtari M, Nadim F. At the crossroad of nature and culture in Iran: The landscapes of risk and resilience of seismic space. International Proceedings of Economics Development and Research. 2014;**71**(8):38-44

[33] Ibrion M, Mokhtari M, Parsizadeh F, Lein H, Nadim F. Towards a culture of resilience and earthquake disaster risk reduction in Iran—"Lessons-Learned" from earthquake disasters. In: Proceedings of the 7th International Conference on Seismology and Earthquake Engineering; 18-21 May 2015; Tehran, Iran. Tehran, Iran: IIEES; 2015

[34] Parsizadeh F, Ibrion M, Mokhtari M, Lein H, Nadim F. Bam 2003 earthquake disaster: On the earthquake risk perception, resilience and earthquake culture—Cultural beliefs and cultural landscape of Qanats, gardens of Khorma trees and Arg-e Bam. International Journal of Disaster Risk Reduction. 2015;**14**(4):457-469

[35] Abrahamian E. A History of Modern Iran. New York: Cambridge University Press; 2008

[36] Axworthy M. Empire of the Mind, A History of Iran. Philadelphia, USA: Basic Books; 2008

[37] Katouzian H. The Persians, Ancient, Medieval and Modern Iran. New Haven and London: Yale University Press; 2009

[38] Mansel P. Sultans in Splendor, Monarchs of the Middle East 1869-1945. London, UK: Parkway Publishing; 1988

[39] Milani A. The Shah. New York, USA: Palgrave MacMillan; 2011

[40] Sorkhabi R. The First Oil Discoveries in the Middle East, Flashback on the First Discoveries of Oil Fields in the Middle East. Geo Expro; 2010

[41] Amuzegar, J. Iran's oil as a blessing and a curse. The Brown Journal of World Affairs. 2008;**15**(1):47-61

[42] Abbaszadeh P, Maleki A, Alipour M, Kanani Maman Y. Iran's oil development scenarios by 2025. Energy Policy. 2013;**56**:612-622

[43] Mohaddes K, Pesaran MH. One Hundred Years of Oil Income and the Iranian Economy: A Curse or a Blessing? CESifo Working Paper no. 4118, Category 12: Empirical and Theorethical Methods. Center for Economic Studies and Ifo Institute; Munich, Germany. 2013

[44] Babayan K. Mystics, Monarchs, and Messiahs, Cultural landscape of early modern Iran. USA: Harvard University Press; 2002

[45] Ghafory-Ashtiany M. View of Islam on earthquakes, human vitality and disaster. Disaster Prevention and Management. 2009;**18**(3):218-232

[46] Clancey G. Earthquake Nation, the Cultural Politics of Japanese Seismicity, 1868-1930. Berkeley and Los Angeles: University of California Press; 2006

[47] Ranghieri F, Ishiwatari M. Learning from Megadisasters, Lessons from the Great East Japan Earthquake. Washington, DC: The World Bank; 2014

Section 2

# Technogenic Risk

Chapter 4

# Mexico City after September 2017: Are We Building the Right City?

Milton Montejano-Castillo and
Mildred Moreno-Villanueva

Additional information is available at the end of the chapter

http://dx.doi.org/10.5772/intechopen.72499

**Abstract**

Due to its destructive effect, a disaster always raises questions about its causes. In the case of the earthquake that occurred in Mexico City on September 19, 2017, one of the most surprising and astonishing situations was buildings that were damaged or collapsed by the earthquake, but which had been recently constructed. These had been built 9 months up to 12 years before, and others were still not inhabited. On the other hand, as in 1985, public spaces have been playing a key role both in the emergency phase and in the reconstruction phase. However, the new public spaces that accompany the most recent housing projects have lost much of their quality. What factors have influenced these urban processes? What are the stakeholders that produce both the new urban forms and the new public spaces? Are there ways to measure the quality of these new public spaces? We depart from the hypothesis that the recomposition of territories of opportunity in Mexico City has been based on the adoption of trends supported by the economy, rather than in the needs of the population, resulting in exclusionary and uninhabitable public spaces in case of disaster.

**Keywords:** Mexico City, earthquake, public space, urban form, disaster risk

## 1. Introduction

On September 19, 2017, a 7.1-magnitude earthquake shook Mexico City, which was known to happen because of the determinants of its territory, but it was not known when or the size of the disaster. Due to its evacuation protocols and the coincidence of the day commemorating 32 years of one of the most devastating earthquakes in this city (September 19, 1985), the citizenship mobilized nimbly to attend the emergency stage.

The open city was immediately occupied: streets, squares, gardens and street lane dividers; first as safe places to safeguard life as a reaction during the earthquake and then became centers for the collection of tools for rescuers, food and medicine collection, healthcare centers, psychological assistance to citizens with information on missing persons, centers for the collection of food for pets, veterinary care for rescue dogs and pets found, digital attention, but they also became a life opportunity as temporary shelters.

On the other side, in the post-disaster phase, one of the most surprising and astonishing situations was buildings that were damaged or collapsed by the earthquake, but which had been recently constructed. These had been built 9 months up to 12 years before, and others were still not inhabited. These housing buildings have been the result of the so-called real estate boom that has been changing on the one hand the verticality of the city and, on the other, the occupation of the territory, which has been monopolized without leaving sufficient reserves of open spaces; so it seems that the risk conditions have been being built.

For this reason, some questions arise as: How are these new urban forms being created in Mexico City? That is, are the policies for the growth of the City not being respected or are these the ones that are allowing this verticality with little public space? Are we creating safe and habitable cities or are we exchanging safety for built space at high costs? And, in this sense, is our public space inclusive or exclusive, so that in case of disaster is useful to citizens? Is 32 years sufficient to dilute the memory of the disaster and the preventive aspects: density and role of public space in the event of disaster?

The objective of this chapter is to analyze the condition of the public space in Mexico City in quantitative and qualitative terms, understanding public space as parks, public squares and walks. Hence, some public spaces are compared, evaluating the instruments, actions and public interventions for the creation and improvement of them.

To undertake this analysis, we take as case study the neighborhoods Granada and Ampliación Granada, both located in the Municipality of Miguel Hidalgo in Mexico City, that since the end of the first decade of this century have had a reconfiguration in their use from industrial land to residential use. A contrasting case is Polanco, an adjoining neighborhood of success since its creation in the early twentieth century, which supports the new real estate image of the mentioned neighborhoods.

In the recent reconfigured areas, there was an opportunity to create habitable public spaces, but especially the opportunity to enhance public space as an eventual resource in the emergency and reconstruction phase of the city—given its seismic nature and propensity to flood—but this was not done.

We depart from the hypothesis that the recomposition of territories of opportunity in Mexico City has been based on the adoption of trends supported by the economy, rather than in the needs of the population, resulting in exclusionary and uninhabitable public spaces in case of disaster.

## 2. Urban form and seismic risk in Mexico City

In 1985, an 8.1-magnitude earthquake shook Mexico City leaving an official balance of more than 3000 fatalities and hundreds of buildings collapsed. As many factors came into play, we limit here to underline the relationship between the natural hazard (seismic waves) and the physical vulnerability reflected in one of the most characteristic morphological features of the city, its verticality and the existence and/or absence of public spaces as a support resource in the emergency and recovery phase.

According to Meli [1], the statistical analysis of damages after 1985 revealed that the collapse of buildings was not at random. Regardless of age, materials and structure of buildings, there were certain types of buildings that particularly collapsed, having in common the number of floors (see **Table 1**): buildings between 7 and 12 floors had more collapses than low-rise buildings. Such a finding made sense when the natural conditions of the soil were revised.

The area of the largest number of collapsed buildings had been the area of the former lake. That is, the lacustrine nature of Mexico City (it was founded on a lake which was later artificially dried out) impregnated the soil with certain characteristics, resulting in an area with three types of soils: (a) the area of the lake (where the lake was formerly located); (b) a transition zone (with part of hard and soft ground) and (c) a zone of hills (with a high resistance capacity) (see **Figure 1**).

The seismic waves that affected Mexico City in 1985 were produced in the coast of the State of Michoacán and traveled 400 km, but upon arriving in Mexico City and coming into contact with the clay soil area, the oscillation period of the waves was amplified. After the earthquake of 1985, the studies carried out on the collapse of the buildings revealed that the causes of this collapse were not so much a function of the age of the construction and the type of structure, but of the height itself due to a natural phenomenon known as "resonance" [1, 2], which causes the seismic movement to be amplified due to the coincidence of the frequency of vibration of the ground with that of the building. By matching the periods of oscillation of ground and buildings, the waves were amplified (reinforced), resulting in inertial forces that ended up causing the collapse of buildings of certain heights.

As a consequence of this phenomenon, the construction regulations in Mexico City were modified, making sure that the buildings were calculated considering the oscillation periods depending on the type of soil. When deciding the number of levels of the buildings (with a more or less constant period of oscillation per floor), this number of floors and its corresponding period of oscillation should not coincide with the period of oscillation of the soil in that area to avoid the phenomenon of resonance.

| Number of storeys | Cases of damage (percentage of that range of buildings) |
|---|---|
| 1–2 | 0.9 |
| 3–5 | 1.3 |
| 6–8 | 8.4 |
| 9–12 | 13.6 |
| 7–12 | 10.5 |

Source: Meli [1], p. 135.

**Table 1.** Percentage of collapsed or severely damaged buildings according to number of storeys in Mexico City after the earthquake of 1985.

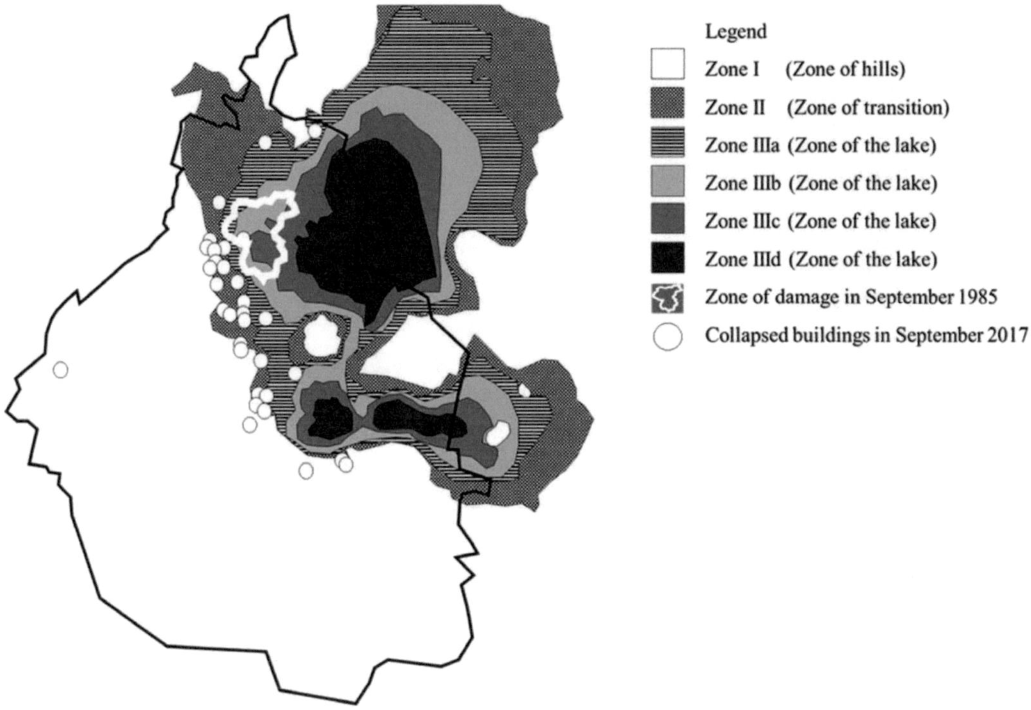


**Figure 1.** Seismic zoning of Mexico City published after the September 2017 earthquake showing the damaged zones in 1985 and 2017. Source: Own elaboration based on the official map of seismic zoning (http://www.atlas.cdmx.gob.mx/zonificacion_sismica.html).

On the other hand, the structure of the building should be sufficiently "flexible" and "ductile" enough to dissipate the seismic energy (thereby making the building less vulnerable). If high-rise buildings were built in a soft ground, engineering design should ensure that energy dissipates—with the help of seismic dampers for example—before the higher floors begin to oscillate, a fact that was achieved in the buildings that could afford this technology.

The earthquake on September 19, 2017 of 7.1 degrees, −which caused 228 fatalities and the collapse of 38 buildings[1], brought more elements and hypothesis to be added about the damage to occur. One of the points of discussion and analysis was that the most impacted area was not the one of the former lake as it was in 1985. On this time, although the earthquake was of lesser magnitude, the epicenter was located closer to Mexico City (120 km away), causing the amplification of the waves not in the area of the lake but in the area of transition, causing the collapse of buildings from four to seven levels, thus revealing "a complex pattern of movement and very variable in the space [3]. To the latter, it should be added the question about the correct application of land use zoning and dubious authorizations for the construction of buildings for residential use, since many of the collapsed and damaged buildings were just beginning their useful life (see **Table 2**).

On the other hand, public space played again a fundamental role both in the emergency phase and in the reconstruction stage. It is no coincidence that the spaces that were used in 1985 and 2017 correspond to projects where public space was, from the beginning, the most

[1]Without taking into account 24 buildings that officially will have to be demolished due to the damage they suffered [4].

http://dx.doi.org/10.5772/intechopen.72499

important component. An example of this balance between housing and public space is the Colonia Hipódromo Condesa, built in 1926 and designed by the architect José Luis Cuevas Pietrasanta (see **Figure 2**). The land was an old racecourse and the architect simply continued with its original shape giving a radial structure and at the center a large park and a green belt. Despite the densification that this area has been subject to, public space remains as an invaluable resource at the time of the emergency. In September 2017, this space was used to organize search and rescue activities, medical service, psychological care, pet care and collection of donations (food, tools, etc.). At the same time, other damaged areas of the city and recent real

| Municipality/address | Number of storeys | Approximated time of being inhabited | Type of damage | Explicative hypothesis of the damage |
|---|---|---|---|---|
| Benito Juarez/General Emiliano Zapata 56 | 7 | 9 months | The back part of the building collapsed | Corruption in the process of land use permission and the number of storeys was higher than the authorized |
| Benito Juárez/Bretaña 90 | 7 | 4 months | One of the towers collapsed and the other was severely damaged | Two seven-storey towers were built above old existent buildings (one with more than 50 years old). The number of storeys was higher than the authorized |
| Benito Juárez/Calzada de Tlalpan 1234 | 10 | 11 months (departments still on sale) | Partial damages and cracks | It was built with low quality construction materials and the number of storeys was higher than the authorized |
| Benito Juárez/Benito Juárez 29 | 6 | 4 years | Partial damages, collapse of roofs and balconies, cracks in the walls | It was built with low quality construction materials and the number of storeys was higher than the authorized |
| Benito Juárez/Calzada de Tlalpan 550 | 17 | 5 years | Severe fractures in the building and cracks in the walls. | Failure to comply with the buildings regulations. The number of storeys was higher than the authorized |
| Benito Juárez/Eje Central 521 | 8 | Departments were not sill inhabited. | Sever fractures in the building | Failure to comply with the buildings regulations. The number of storeys was higher than the authorized |
| Gustavo A. Madero/ Insurgentes 1260 | 12 | 2 years | Visible cracks | It was built with low quality construction materials and the number of storeys was higher than the authorized |
| Cuauhtémoc/San Antonio Abad 66 | 11 | 2 years | Severe fractures in the building and cracks in the walls | It was built with low quality construction materials and the number of storeys was higher than the authorized |

Source: *Revista Obras* no. 538, Oct 2017 and Najar A. "Las razones por las que colapsaron tantos edificios en CDMX (y no todas son el sismo)", *Animal político* (internet). http://www.animalpolitico.com/2017/10/las-razones-las-colapsaron-tantos-edificios-ciudad-mexico-no-todas-terremoto/

**Table 2.** Residential use buildings damaged or collapsed during the earthquake of September 19, 2017 in Mexico City.

estate development projects lacked these spaces, making especially difficult the moment of evacuation (see **Figure 3**). The configuration of such new projects is the combination of several factors and conditions described below.

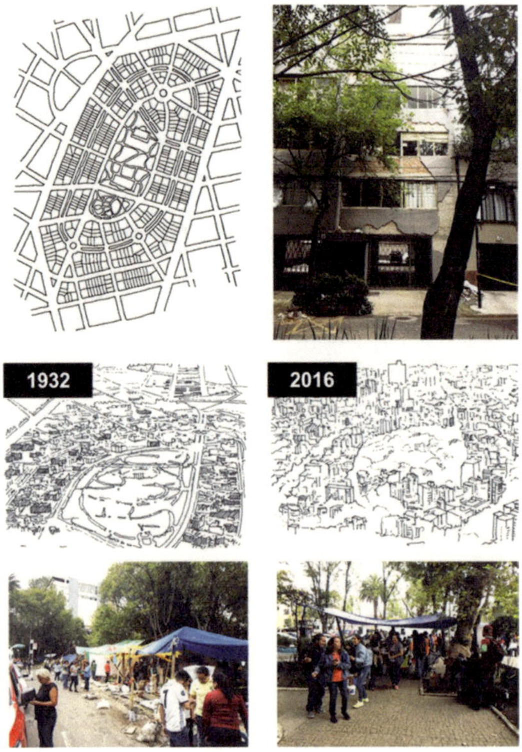


**Figure 2.** España Park plan (top left); process of verticalization 1932–2016 (at the center) and the use of the park during the earthquake of September 2017 (bottom). Image of damages (top right). Photos by the authors. Own drawings based on aerial photographs from Fondo Aerofotográfico Acervo Histórico Fundación ICA.

**Figure 3.** Collapsed buildings in Mexico City in the September 2017 earthquake. Despite the density of the buildings, the absence of public space in the surroundings is evident. Photo: Rosa Lilia Pedraza Vázquez.

## 3. Recent transformations in Mexico City: actors and factors

In the case of Mexico City, the last decades of the twentieth century brought a change in public policies and a depopulation of the central parts, especially due to the process of deindustrialization and the earthquake of 1985. This meant a reinvention of the city for this century, through standards that call for a redensification and the opportunity to occupy spaces that were attractive to the private sector during the first decade. This meant that the city exceeded its limits, gentrifying spaces and consequently producing poorly rehabilitated residual public spaces or the creation of reduced spaces.

### 3.1. New policies

The instrument for urban development policies called "Bando Dos," proposed to redensify the city with the specific objective of ordering the urban growth of Mexico City, preventing the construction of more housing in the outskirts of the city. The instrument was presented on December 7, 2000 by the then head of government (Andrés Manuel López Obrador). It had different objectives for the ordering of Mexico City, such as: to stop disordered growth; to safeguard the preservation of soil of the then Federal District (now Mexico City DOF ), preventing the growth of the urban areas and thus avoiding covering the recharge zones of aquifers. It was determined that the districts that had suffered considerable depopulation were mainly four: Cuauhtémoc, Benito Juárez, Miguel Hidalgo and Venustiano Carranza, all located in the central area of the city. It was also assessed which had been disorderly populated, predominating the south and east. It was determined that there is little infrastructure in the city for a strong real estate development [5].

Among the policies implemented was the promotion of population growth toward the districts of Benito Juárez, Cuauhtémoc, Miguel Hidalgo and Venustiano Carranza to take advantage of the infrastructure and services that are currently underutilized, and the construction of housing for the lower income classes [5]. However, in these central districts, such as Benito Juárez, the project did not work as expected. At first, there was a real estate boom, but if it was not successful it was because of the high cost of housing and the poor infrastructure. In different neighbors, there was a wild transformation of the city landscape by cutting down trees and constructing big buildings: where there had been houses for six to eight people, now there appeared buildings with eight to ten floors for many families. In these new buildings, however, not all apartments were sold.

As a part of the first consequences, in 2010, the government of the Federal District at the time, together with the Ministry of Social Development (SEDESOL), the National Council of State Housing Entities (CONOREVI), The Autonomous University of Mexico (UNAM), the Housing Fund of the Institute of Security and Social Services for State Workers (FOVISSSTE), the National Workers Housing Fund Institute (INFONAVIT) and the Federal Mortgage Society (SHF), published the Guide for residential redensification in the internal city [6], in which they present a methodology to identify redensification scenarios, as well as instruments to favor it so as to join the smart city growth system and position Mexico in the international environment in this respect, for which they are planned to address a series of issues, such as increase in the costs of displacements of the inhabitants of said areas; greater consumption of fuels and greater production of emissions polluting the atmosphere; loss of preservation areas, aquifer recharge zones and agricultural production areas; higher costs of urbanization that represent a significant burden for local governments and social and economic segregation of urban space [6].

The approximate 10-year delay for this guide to be published—to take measures on matters of redensification policies—caused for constructions to be carried out during that time in different zones that lack integration with the social fabric, for it has been seen that elite zones are created, which keeps the population dissatisfied and afraid of being displaced. There was an unlimited number of claims derived from the implementation of the Bando 2, caused by the fear of the modification of the environment, decrease of the quality of life, of safety, of the value of real estate, feelings of dispossession or feelings of injustice, for decisions were made that affected the territory without the main interested parties being informed, taken into account or heard, a loss of confidence of the population in the authorities and experts that promoted the project, above all when there is a tradition of local organization and mobilization, risk perception and a feeling of uncertainty. The technical and scientific studies that validated the project were questioned [7].

### 3.2. Deindustrialization

With the economic opening abroad with the 1988 free trade agreement, there was a shift in the activities of the manufacturing industry that caused a process of deindustrialization. The industries were moved toward the outskirts of the city or even toward other territories [8]. This process is not yet finished. There are still areas of the city with disappearing industries. With this movement and the change toward a tertiary economy, the reconfiguration of the city was affected on one hand due to the opportunity of land within the city, seized by the real estate power, and on the other hand due to the change of policies that did not work as expected. In Mexico City, some of the areas that have passed through the process of deindustrialization at the end of the twentieth century were the municipalities (delegaciones) Benito Juárez, Cuauhtémoc, Miguel Hidalgo, Venustiano Carranza, as well as Azcapotzalco and Gustavo A. Madero [9]. In recent years, the mass production of housing has captured some of these areas, leading them to transformations that are a result of the inclusion-exclusion struggle that is reflected in the absent public space. An example of this is the case of the neighborhoods Granada and Ampliación Granada, in the Miguel Hidalgo delegation, which has been a categorical place throughout History. From an economic point of view, we could say that it has gone through three sectors: agricultural, industrial and tertiary.

In 1920, the lands of the Hacienda de los Morales were divided, playing a significant role in the urbanization of the city of Mexico due to the fact that part of the space was used for the colonia Polanco assigned to upper middle housing, in which the neighborhood project of the first half of the twentieth century was based on public space. This was a key as it grew until it was divided into five sections, sharply contrasting the neighborhoods Granada and Ampliación Granada, which began to be industrially established without public spaces. The following were some of the factories in the place: the General Motors Factory in 1923; the Mexico glass factory in that same year; the Modelo Brewery and the General Popo in 1925; the Tabiques La Universal Factory, whose year of establishment is unknown; the Chrysler Factory in 1939 and thereafter until 1961; the Palmolive Factory; the Halaxtoc textile factory; Laminadora LMMSA; pharmaceutical industries; Factory in Lago Andrómaco Street; Bolt Factory; factories in lake Neuchatel; Furniture and Steel Factory and another Cotton factory [10].

### 3.3. Actors and programs in the production of public space

In Mexico, there are various governmental instances responsible for intervening in or making public space such as The Department of Urban Development and Housing (SEDUVI), the Public Space Authority (AEP) or the different municipalities (former Delegaciones). However, when public space shows specific characteristics and values for which it has been cataloged as equity, the instances for intervening in it change or they are accompanied by certain strict guidelines for their regeneration, such as the INAH (National Institute for Anthropology and History), the INBA (National Institute for Fine Arts), the Historical Center Authority or the UNESCO, according to the case. Each one of the aforementioned instances intervenes in public space from different perspectives and with various actors. The Department of urban Development and Housing, for example, is responsible for designing policies applicable to the city, attempting for them to integrate society when acting and interacting with it, so as to transform the city in an inclusive manner. It creates the Programs of Delegations, Partial Programs and the Urban Development Program for the purpose of ordering the city in all its aspects: mobility, public space, housing, urban infrastructure, basic services, always with the idea of improving and positioning it as a safe city.

On the other hand, there is the Public Space Authority (AEP), which is a decentralized entity of the SEDUVI. It not only designs policies to apply them to urban space but also directly intervenes through the design of the space and the contracting and subcontracting of construction and design companies. Some of its programs and projects are as follows: Ecopark, Bajo puentes (underbridges), Pasos seguros (safe steps), publicidad exterior (advertising), Parques de Bolsillo (pocket parks) and Parques lineales (linear parks), among others. The AEP was created in 2008. It works on the various projects with different companies, for example, CTS Embarq with the Model street, GABANA engineering and GCB Construcciones y Servicios for the refurbishment of the street Torcuato Tasso, Proyecsa e Ingenieros, ANACE Construcciones, Grupo Q and B and Servicios integrados RUBE for the regeneration of the Alameda Central, Grupo Velasco, JM Constructora, Kassar Construcciones, 128 Arquitectura and Diseño Urbano para Espacios Públicos de Bolsillo, to mention a few.

With respect to the organization of the Historic Center of Mexico City, there is another decentralized entity called the Historic Center Authority, created in 2007, which proposes public policies for integration and promotes the refurbishment of public spaces located in this square. However, there are various actors that participate in the intervention and construction of public space. Even when the aforementioned entities are present, the participation of the citizens is already contemplated in almost the majority. Participating in the modifications of the urban environment means a social commitment more than a political one, but the action surpasses that which is social, political and economic.

Concerning the programs for public space in Mexico City, since the first decade of the 21st century, a series of urban projects were implemented by the Department of Urban Development and Housing (SEDUVI) and the Public Space Authority (AEP) to create or intervene in spaces with characteristics of deterioration and abandonment in some cases, including economic activity, which addressed the demands of the inhabitants. On the one hand, among the newly created public space projects were those that had a renewed design, with the minimum characteristics necessary to be used and enjoyed, such as low bridge projects, public pocket parks or bonds of friendship. On the other hand, are the projects of improvement and refurbishment of public spaces, in which there are improvements of spaces with an inclusive design, refurbishment of heritage spaces, pedestrianization and semi-pedestrianization of streets, illuminate your city program and ecoparq and refurbishments of monuments (see **Table 3**).

| **Newly created public space programs** | | **Public space refurbishment programs** | |
|---|---|---|---|
| Public pocket parks | Design of social interaction, identity and economic activity, in remaining streets or spaces between buildings | Refurbishment of monuments | Its purpose is to rescue sculptural monuments, integrate them harmoniously into public space and recover them for interaction |
| Bonds of friendship | Project in the development of cultural and political relationships between the two countries, through the donation of a sculpture placed in a newly created public space | Improvement of spaces with inclusive design | Improve pedestrian accessibility and the vehicular flow of the avenue that was inadequately designed for the intense pedestrian and automobile capacity |
| Underbridges | This seeks to rescue abandoned or under-used public spaces, providing them infrastructure with high technical specifications to address the basic needs of the population, including spaces for commerce. | Illuminate your City Program | This unifies public lighting in primary and secondary roads to prevent the "zebra effect" from being produced, which is a phenomenon that creates variations in the intensity of the lighting of the streets |
| Pedestrianization and semi-pedestrianization | Consolidate the pedestrian section of Public space of the Historic Center, promote sustainable mobility, optimize vehicular and pedestrian travel times, provide universal accessibility and optimize the heritage value of the area | Ecoparq | Recovery of public spaces through the installation of parking meters. This improves the mobility of the city |
| Mobile park | Spaces assembled in trailer parks, equipped with game tables for children, a rest area, green areas, with natural vegetation and chairs called Parkes. These are placed in spaces that are generally used as parking lots | Refurbishment of heritage spaces | This complements the recovery of public spaces of the historic center, and additionally promotes the use of heritage spaces by optimizing their social function and spacing in benefit of the inhabitants |

Source: SEDUVI.

**Table 3.** Public space programs activated in the twenty-first century in Mexico City.

http://dx.doi.org/10.5772/intechopen.72499

### 3.4. Verticalization and public space in the new urban territories

By the start of the nineteenth century, the neighborhoods Granada and Ampliación Granada were changing their morphology, land use and population. The main change was the use of industrial land to residential land, which was attractive for real estate developers, who saw that its potential was supported by the urban image of the bordering sector Polanco. The two neighborhoods were given different informal names following the first interventions: Ampliación Polanco, Polanco Bis, Polanco II or the Nuevo Polanco; however, a series of contrasts have been seen between Polanco and the more recently built neighborhoods (Granada and Ampliación Granada). The most significant difference between said neighborhoods is the type of public space. In spite of the luxurious residential buildings that broke the specification

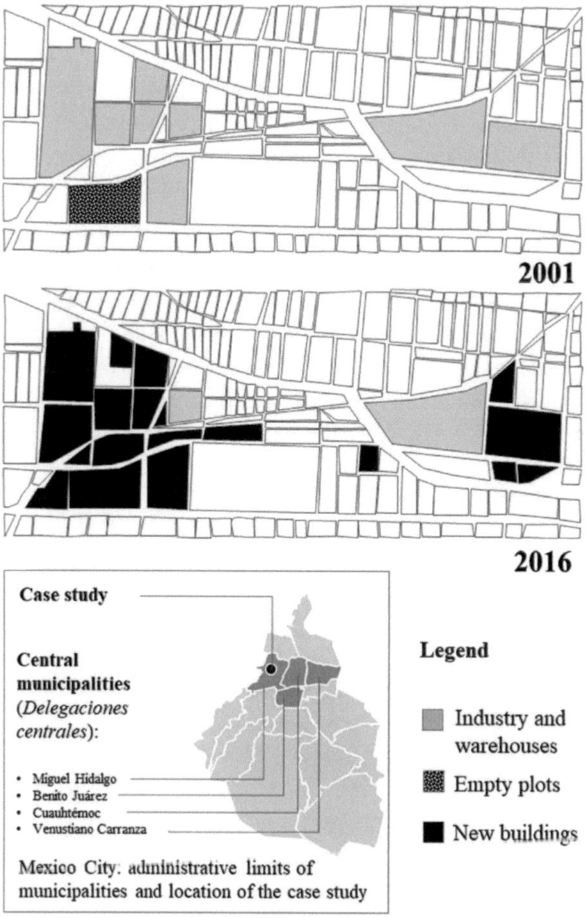


**Figure 4.** Urban transformations occurred in the twenty-first century in the territory of Ampliación Granada (expansion of Granada Neighborhood) and Granada Neighborhood in the period between 2001 and 2016. Source: Own elaboration based on Google Earth images from 2001 to 2007. Information from 2008 to 2016 is based on own field survey illustrated on Google Earth maps.

of the Bando Dos and the norm 26 to create housing construction of social and popular interest on urban land and thereby redensify the zones of Mexico City in which there is a certain lack of population, they lost the opportunity to create housing with high quality public spaces (see **Figure 4**).

Due to the rapid and disordered growth in some areas of Mexico City, in 2013, the implementation of the norm that proposed the redensification was detained due to the abuse of the land use and its changes in the type of housing that should be implemented. However, in that same year, the Action through Cooperation System (SAC) was created, which is an instrument to manage and create policies that include public action, the intervention of the State, as well as the private party, that is, the participation of land owner companies to interact with each other in the interest of improving the city for which the Department of Urban Development and Housing (SEDUVI) is responsible.

One of the main characteristics of the area is that, at its pace of development, it has not only been activated housing for the elites, but commercial and service activity has also been developed, creating large office buildings or shopping centers with foreign brand stores. It has become common in the area for small shopping centers with convenience stores, minisupers, restaurants, cafes and bars to be built in the lower part of housing buildings. The main problem was that there were no public spaces. However, far from providing a solution, due to the new constructions, trees have also had to be cut down, and trees have been changed for ornamental plants that represent consequences for the environment and deterioration in life quality. Thus, the place only has what are now the public spaces of the twenty-first century, such as pocket parks (three on the Cuernavaca Railroad), linear parks (that of the Cuernavaca Railroad) and low bridges (that of San Joaquín Avenue at the intersection with Moliere Street). On the opposite case we find wide parks and walkways in the area of Polanco (see **Figure 5**).

**Figure 5.** Public spaces in the neighborhoods of Granada (left at the top) and Polanco (left at the bottom) and their location. Source: Own elaboration.

## 4. Bases for a context-sensitive assessment of public space in Mexico City

With the assumption that it is essential for urban studies to include different approaches and to pay attention to the processes that transform the city, three views are taken into account for the understanding and analysis of the public space: 1. The habitability of public space, i.e. the human condition of public space. 2. The vision of inclusion regarding physical and social aspects of public space. 3. A vision in the globalized sense of the trends reflected in the space. At the end of the section, we present the main variables that could be the basis of a model to analyze the quality of public space in this city. This model is applied then to the above-mentioned case with the intention to compare the qualities of different public spaces of neighbor areas but produced in different historical periods.

### 4.1. The habitability of public space

When we talk about desirable public spaces, it could be seen as something subjective. Each human being thinks differently and according to their cultural characteristics, and to that extent, needs could vary. But even in the same country, the geographical or economic situation of each family would imply different demands. Something is very certain, however, and that is that we all have the need to co-inhabit. Each species on this planet has its natural habitat, fishes in the water, monkeys in the jungles and forests and lions in the savannahs. Habitat is the space where species are born, grow, reproduce and die, that is, the space on earth where they meet all their needs. Even when human beings are governed by this general rule, there are two fundamental elements that make them different from other species: the first and most important is that their habitat is not natural, but artificial; and secondly, apart from the physiological needs they need to satisfy, they are also creative beings [11].

Habitability is defined here as the capacity of a place to meet human needs [12], and although several authors consider that habitability refers only to the material and structural conditions of built spaces [13–15], without taking into account the social aspect in the outside [16], habitability for man would be as much within the architectural element as outside of it. Habitability goes beyond the door of the house to the street, toward the public space, where the social function, the community, comes into play, because it is there where "the expression and social identification of the others is built," based on the expression and symbolic construction of the space [17], we leave our house behind to find a huge machinery concentrating the totality of our culture, but which also encapsulates international movements and trends we must incorporate during our journey.

The habitable public space is one that maintains a balance between the material and immaterial elements that intervene in the places of free access for all human beings, regardless of gender, religion, race or social class in order to satisfy the collective needs. Elements of habitability in the public space can be measured and diminished, as appropriate, taking into account the global and local transformations, and the determinants of type of settlement, but … How do we know if the public space is to a greater or lesser extent livable? For this, we consider three theories:

According to the theory of human need by Len Doyal and Ian Gough, cited by Reyes [18], needs are constructed socially and derived from the cultural environment. The authors take into account indices to measure the welfare between nations based on the needs of: appropriate health care, security, economic safety, clean water, adequate food, shelter as a mean of protection from the elements, relationships of recognition, safe working environments and relationships of recognition and belonging. The needs proposed by this theory are general and can be considered basic in different territories and different social groups. It should be taken into account, however, that the cultural and natural environment, the new technologies and even the policies for urban space make human requirements more complex and even different. This is the case of multicultural cities and the public space should regard it as a principle to meet the needs mentioned above.

Based on Max Neef's theory about human needs, Reyes [18] analyzes the habitability of public space and combines criteria from existential and axiological categories, where existential categories focus on needs of being referred to personal or collective attributes, having, which contains the mechanisms and laws required, doing, as personal or collective actions, and being, in those spaces of action and construction of needs, satisfactors and economic goods; while the axiological categories cover the requirements of subsistence, protection, affection, understanding, participation, creation, identity and freedom. This refers us, in terms of the existential category, to social action that allows us to build axiological relations that give meaning to space.

Schiller's theory, cited by Valladares et al. [19], is that of the qualities of the habitable public space where, from variables with a specific meaning and value, he measures the habitable public space, and the qualities space should cover for habitability are as follows: permeability to allow open connections in the urban fabric by measuring them according to the size of typical urban blocks and the elements that can limit them such as railroad tracks or other types of barriers; vitality as a characteristic of the spaces to be places of social interaction measured through the activity there; variety to encourage the complementary uses of the city, variation of typologies and uses; readability to facilitate social and spatial relations from the variable use and density of those who use the city; and robustness which allows an adequate combination and variety of uses at any time of the day with the ability to adapt the space.

According to the theories above, analysis of the habitable public space must be made taking into account physical elements and the design of the space and also considering the social elements of basis subsistence and even the more complex ones such as identity and legal duty. Therefore, we can examine the public space in two dimensions, where the different needs of humans can be encapsulated for the analysis of habitability in the public space, the first, the physical or material dimension, and the second, the intangible dimension, which goes from the social to the spiritual.

In the physical or material dimension, it is possible to concentrate the tangible and quantitative elements that are presented in the urban space, such as public water services, drainage and light, street furniture, transport infrastructure with subway systems, rapid transit busses, light rails, suburban trains, busses, collective transport, bicycle-taxis, bicycles, recreation areas, roads, streets, avenues, circuits, highways, communications infrastructure, public telephones, internet, police officers, security modules and road safety. It is important to mention that the city also has infrastructure for housing, education and health, among others. Similarly, in the immaterial dimension, which goes from the social to the spiritual, it would be the one where we find intangible elements such as the urban social identity, symbolic interactionism, perception of security, culture and social exchange.

### 4.2. The vision of inclusion

The 'inclusive' public space is the place where activities and discussions are open to all. It is the place where authorities have the responsibility to guarantee the existence of a public space where people express their opinions, assert their claims and use it for their purposes [20]. However, if there is this concern about inclusion, it means that there are elements that make cities exclusionary so that inclusion-exclusion are studied in a dual way. To this end, two aspects of study are taken into account: 1. social inclusion by exclusion and 2. physical or design inclusion.

*Social inclusion by exclusion.* Public space historically has been valued as a factor of social inclusion and as an inescapable instrument for urban planning. However, the loss of protagonism due to the weakening of previous forms of sociability (resulting in social inequalities and fragmentation) and the emergence of alternative forms of relationship (of communications and encounters introduced by technology, the feeling of insecurity) have sharpened the barrier between recreational and leisure spaces that are used by different social groups. Not forgetting that people of higher income go to private places to recreate, using the street just to circulate, not caring about the state and the quality of public space, which often remains in the background and helps to generate what Bauman calls "ghettos of exclusion," cited by Acuña et al. [21].

Ramírez Kuri and Ziccardi identify discriminatory practices in the labor market, such as access to goods and services; the weakening of social cohesion; luxury consumption activities that can be dissolved by making effective economic, social, cultural and sustainable rights which encourage the integration of the society with the city; informal activities and social conflicts [22].

And on the other hand, we have *physical inclusion or inclusion by design*. In the search to determine the components that public space has for inclusion, we return to the studies that have been carried out to identify the components of exclusion that Ramirez Kuri and Ziccardi analyze, such as the location of the place to determine the quality of services and their infrastructure; the informal and established commerce that pervades the urban space and which fosters crime and the deterioration of the public space and its accessible design [22]. However, these elements are taken in reverse, that is, on the positive side of that which the public space must have to be considered inclusive, such as enough urban infrastructure.

In the design of inclusive public spaces, it is essential to take into account the physical components that foster social integration. From the perspective of Sergio Zermeño, the following are identified as components of exclusion: inaccessible primary and secondary roads; public spaces of richer classes appropriated by needy sectors; crossroads, roads, squares, parks, sidewalks, etc. which operate as frontiers, excess of surveillance and corridors watched by guards, police officers and cameras, and he also identifies social components such as high risk of violence and virtual walls [23].

### 4.3. The question of globalization trends reflected in urban space

Public-owned spaces must be able to adapt and survive to global transformations, which by their very contrasting nature absorb these changes in different ways, depending on their environments and the impacts public places are constantly having. Globalization, one of the strongest influences on a city in every sense, whether to its society, space or culture, reinvents them as great scenarios with strong economic and political rather than cultural and social alterations which irreversibly impact on the city's inhabitants. In this sense, the overall composition of the public space is witnessed in two aspects: The public space as an alienable resource, in the sense of appropriation and privatization; and the public space affected by its constructions, in the sense of transformations.

*Public space as an alienable resource through appropriation and privatization of the space in a non-legal way*. This causes scarcity of public spaces, mainly because of the wide commercialization of everything, a reflection of the globalization, bad economy, excessive appropriation and high delinquency, as this is fundamentally brought about by street vendors or informal establishments that create pervaded scenarios. The transformation of Latin American cities and their spaces are a consequence of social, cultural and technological phenomena. These changes create a new form of social organization, a new cultural model, which can be called postmodernity, globalization or neoliberal culture. This regards the space as a resource, a product, with social, sensual and symbolic policies, which appropriate, use and transform the spaces of cities [24].

It is evident that the production of public space in current cities has changed, the measures for its construction and even its activities are different, but, what is the cause? Although the causes can be many, there is still an ongoing search for the logic that gives us elements to understand the urban transformations that have been tried to be defined with names that are sometimes even difficult to pronounce, composed or decomposed words or more than one to name what is happening: redensification, urbanization, consolidation, gentrification, multiculturalism, and people participation, among others.

In Latin America, the study of processes such as gentrification is recent. Although it is true that the bases defining this concept are not new, the term itself is relatively young, invented by the British sociologist Ruth Glass [25], who observed the differences in social structure from the establishment of higher cost housing in specific areas of Central London, thus examining the invasion of middle and upper classes on working class neighborhoods, displacing and changing the social fabric.

Later, the sociologists Bruce London and John Palen in 1984 tried to explain gentrification by means of five theories that involve different aspects of the life in the city: the ecological-demographic theory, which refers to population and generational statistical aspects (baby boomers); the sociocultural theory, seen from the values, feelings, attitudes, ideas and beliefs of society;

the political economic theory, which is based on two approaches: the traditional and the Marxist ones; the community network theory: the community lost and the community gained and finally the theory of social movements and the influence of counter movements [26].

On the other hand, in 1987, Neil Smith's view proposed two theories to explain gentrification by observing the phenomenon from the economic and social point of view with the "production-side theory" and the "consumption-side theory." These theories address the problem of the automobile, urban expansion, changes in lifestyles, depopulation of the city center, transport and pedestrian spaces, where human relations are diminished, but above all, he focuses his research on the results of increased employment in business districts. The interest of this geographer in these elements is an answer to the very elements that have caused the greatest problems in recent decades and have been part not only of gentrification but also of the processes of redensification, rehabilitation and the numberless patches made cities [27].

For gentrification to exist, it must be in a specific geographical space and it is considered to be happening when there is a process of investment and reinvestment of capital, when there are a series of transformations in the urban landscape due to the settlement of higher income social groups in these specific geographies and when there is a direct or indirect displacement of the existing social groups [28]. In the current debate, Michael Janoschka addresses gentrification with six points: 1. Neo-liberal policies of Gentrification, all types of public policies that establish an alliance with the capital that is invested in the city. 2. Supergeneration, when a place has been gentrified at two different historical moments. 3. Gentrification of new areas, industrial areas or ports where there is no gentrification by direct expulsion, but through all the indirect processes that occur around these neighborhoods. 4. New geographies of gentrification: spaces that have not previously been identified as spaces of gentrification, rural and suburban neighborhoods. 5. Symbolic gentrification: virtual sale and placement of new economies. 6. Resistance to gentrification: the congregation of the community to prevent the inflow of foreign capital [28]. Thus, the integration of different urban processes affects constructions and make up, renews and transforms the city and affects the dynamics, practices and design of urban spaces, which is a witness of the reinvention of the city in smaller scales.

### 4.4. Operationalization of variables

With the history of the importance of public space in the City and the influence that urban interventions for luxury housing have had in recent years, as well as the recent public space programs, a Model is created to evaluate the quality of public space in terms of inclusion or exclusion, measured using the following variables and instruments applied in the area of study of Granada, Ampliación Granada and Polanco (see **Table 4** and **5**).

For clear representation, the results are shown in a graph in a model of six concentric axes, forming two hexagons on the same axes. The perimeter of the hexagon is the coordinate zero, while the perimeter of the external hexagon is the coordinate +2 (a very inclusive space). The center of either of the two hexagons shall therefore represent a very high exclusion. In other words, the more covered the area of the hexagon is, the more inclusive that public space will be. The model was applied in all spaces of Polanco and Granadas (see **Table 5** and **Figure 6**).

| Variable | Importance of the variable in case of disaster | Instrument to collect the information |
|---|---|---|
| **Accessibility:** The degree or measure in which all people can use a public space | **Accessibility** to public space is crucial in all phases of disaster: at the emergency phase (for evacuation purposes), search and rescue activities (for organization of activities) and reconstruction (for temporary shelter) | Plan or lines of public transportation (metro, bus, collective transport), plan for taxi sites, plan for bicycle sites, plan of virtual accessibility and crosstab plan |
| **Balanced residential adjacency:** The housing around public spaces must be balanced with the rest of the services | It has been observed that **residential adjacency** to public space permits people to be close to the collapsed buildings instead of going to official temporary shelters | Land use plan (diversity of uses), residential land use plan, adjacent housing plan with real heights (2 levels, 3 levels, 5 levels, etc.), closed neighborhoods plan and Aerial Photography |
| **Lighting, temperature and humidity:** The characteristic of lighting in public spaces can determine their stay in them and their daily hours of life | A good level of **lighting, temperature and humidity** are fundamental for the use of public spaces at all phases of disaster | Heights of buildings, luminaries, terrestrial photography, aerial photography, lux meter and thermometer |
| **Urban furniture and infrastructure:** The tangible and quantitative elements that are in the public space | **Urban furniture** may enable or impede the rapid installation of emergency facilities such as tents for the reception of food or medical attention. **Infrastructure** such as water or a flood safe public space may facilitate temporary shelters | Plans of the public spaces chosen with details of furniture, urban infrastructure plan of the space, adjacent urban infrastructure plan, terrestrial photography and aerial photography |
| **Perception of the urban space:** How the resident feels about the place. In other words, if it is safe, if they feel included or excluded | A positive **perception of the urban space** (temporary used in the different phases of disaster) may be helpful to the emotional wellbeing of victims | Photography, interviews, graphs and charts |
| **Control:** Physical elements of security that control the space, such as cameras, police, surveillance modules and neighborhood watch | Physical elements of **control** and private security (as physical barriers) may impede partial or total accessibility to the public space | Security camera record plan, security module record plan, photographic record of human elements of security and interviews |

Source: Own elaboration.

**Table 4.** Variables and instruments for analyzing the quality of public space.

The main result was that in the neighborhood Granadas, although they had the determinants for their space to be recomposed through public spaces as the base of the project, this was done in an isolated manner, causing for the new pocket public spaces and linear parks determined by the economic tendencies to be places of exclusion, due to the fact that, for example:

- A lack of accessibility is seen as there are no free internet networks in the public space and there is no bicycle parking as opposed to Polanco, in which there are. Although there is public transportation near the place, it has become exclusive due to the saturation of its use.
- The residential adjacency is not balanced; although the land use is variable, the residential complexes in the area are very high and gated neighborhoods are dominant.

| No. | Type / Name of public space | Variables to analyze the quality of public space (quantitative value from 0 to 2) | | | | | |
|---|---|---|---|---|---|---|---|
| | | Accessibility | Balanced residential adjacency | Lighting, temperature and humidity | Urban furniture and infrastructure | Perception of urban space | Control |
| Neighborhood Polanco | | | | | | | |
| 1 | Lincoln Park | 2 | 1 | 2 | 2 | 1.5 | 0.5 |
| 2 | Las Américas Park | 1.5 | 2 | 2 | 1.5 | 2 | 2 |
| 3 | Uruguay Park | 2 | 0.5 | 1 | 1.5 | 1 | 1 |
| 4 | Antonio Machado Park | 1.5 | 1.5 | 2 | 1.5 | 1.5 | 1 |
| 5 | Lineal walking space Horacio | 1.5 | 1.5 | 2 | 1.5 | 2 | 1.5 |
| 6 | Lineal Park Ferrocarril de Cuernavaca | 1.5 | 0.5 | 0.5 | 0 | 1 | 0.5 |
| 7 | Pocket Park Juan Vázquez Mella | 0 | 0.5 | 1 | 0.5 | 0 | 0 |
| 8 | Pocket Park Masaryk and Mariano E. | 1 | 1 | 1 | 0 | 1 | 0.5 |
| 9 | Underbridge Masaryk and Periferico | 0 | 0 | 0 | 0 | 0 | 0 |
| Neighborhoods Granada and Ampliación Granada | | | | | | | |
| 10 | Pocket Park Ferrocarril de Cuernavaca | 1 | 1.5 | 0 | 1.5 | 0.5 | 1 |
| 11 | Pocket Park Moliere and Ferrocarril de C. | 1 | 0 | 0 | 0 | 0.5 | 1 |
| 12 | Lineal Park Ferrocarril de Cuernavaca | 0 | 0 | 0 | 1 | 0.5 | 0.5 |
| 13 | Pocket Park Ferrocarril (Rio San Joaquín) | 1 | 0.5 | 0 | 1.5 | 0.5 | 1 |

**Table 5.** Analysis of the quality of public space in the Neighborhoods Polanco, Granada and Ampliación Granada. Source: Own elaboration.

- They show records of temperature, humidity and lighting that are not comfortable in shade, since in some cases they have little exposure to the sun, and the sun directly in others. All of them are highly humid and the records go from the lower to the upper limit, due to their low vegetation and the material of their environment.

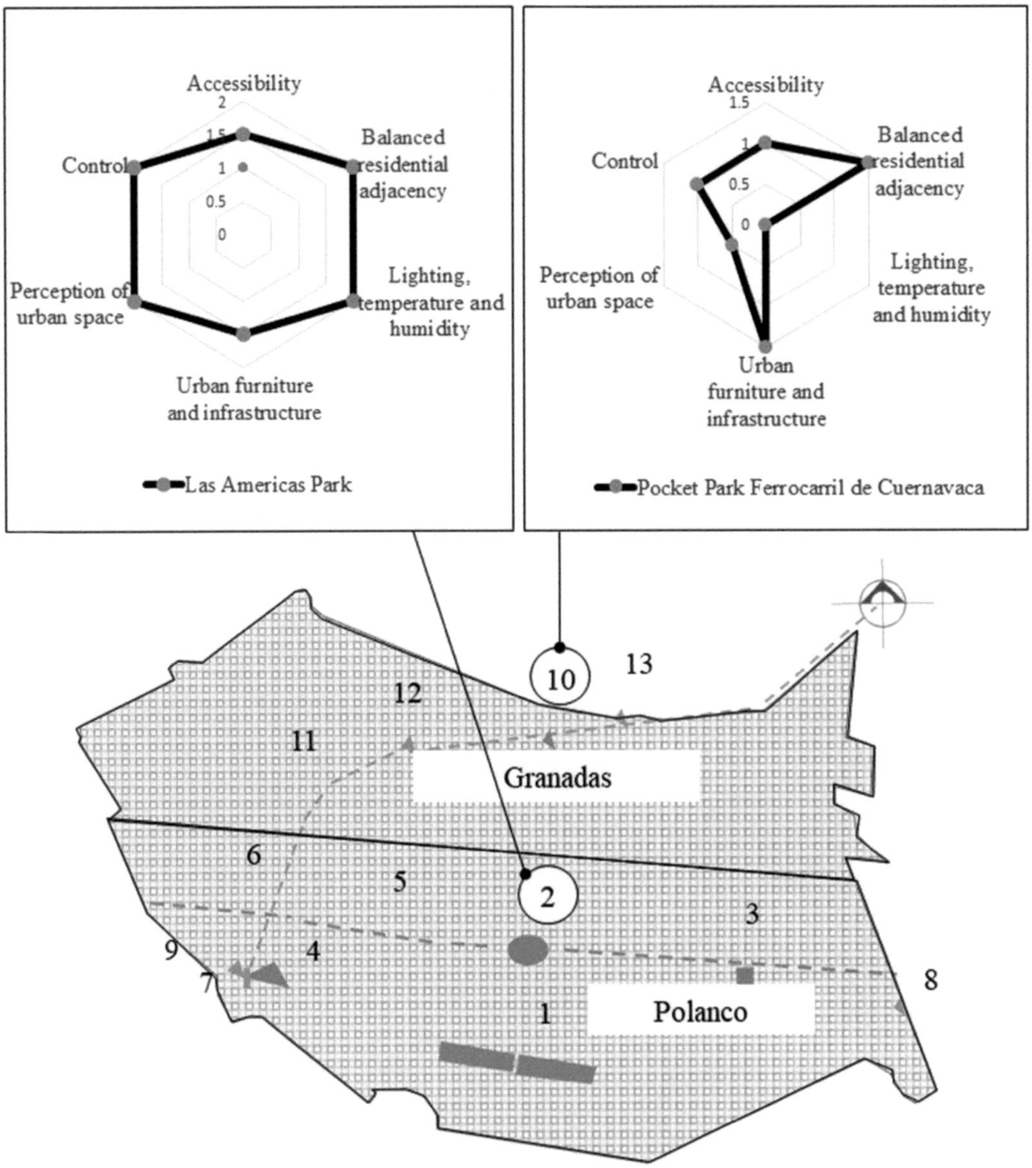


**Figure 6.** Comparative analysis of inclusion and exclusion characteristics of public spaces of the twentieth and twenty-first centuries in the Polanco and Granadas neighborhoods. Source: Own elaboration.

- In general, their urban furniture and infrastructure are normal, for they have benches. However, they do not have trash bins, much less fountains, sculptures or playgrounds. However, although they do not have their own luminaries, they have exercise machines.

http://dx.doi.org/10.5772/intechopen.72499

- An urban space is perceived in low conditions for use and enjoyment due to the previous determinants. They feel that it is unsafe and feel excluded from some parts by the physical barriers that are in the place, such as the cyclone wire fencing that divides it.
- In general, personal control systems and surveillance cameras are excessive in some parts.

## 5. Conclusions and outlook

The implementation of urban phenomenon, such as redensification and gentrification, must be treated with more care and with plans of action for all. Failure to do so may cause:

A change of identity after a short time, the loss of neighborhood values, the displacement of neighbors, an abstract public space, a collective trademark image, insufficient urban equipment, vanishing of traditional trade, a lack of roads, scarce and exclusive public spaces, change of land use and excessive trash, among others.

The need to produce and intervene in the public space is going to be determined based on the type of urban growth of the city. In other words, if it is a disordered growth, the functioning of the public space will be directly affected and it will be socially weakened.

Interventions in the city in an unplanned manner can cause problems, for example, of communication in the social and spatial sense, of urban infrastructure and of insufficient public spaces.

The creation and intervention of public spaces in Mexico City of the twenty-first century have been governed by economic, political and social determinants immersed in a global world in search of publicly owned spaces that have inclusive characteristics.

On the other hand, the production of public spaces in this century has been resulting in residual or nook spaces that have undermined spaces that make them have a struggle between the inclusion-exclusion duality.

The results show that on the one hand, there is no quantitative similarity in the characteristics of public spaces, since they are dramatically reduced as a consequence of the lack of urban planning and the lack of political intention to create habitable public spaces for any case, but especially in case of disasters. That is to say, there is no urban design. On the other hand, qualitatively, we have not seen the concern that the spaces of new creation are inclusive and open to the general population with the intention of integration; urban projects predominate not to favor urban fabric, but to delimit territories.

In this sense, if we look at the role of public space in the earthquake of 1985 and the use in 2017, we can have the minimum indicators required to be taken into account in the adaptation of existing spaces or, where appropriate, in new spaces and that should be revalued in the institutional way of thinking and deciding on public space, mainly in the developing territories within Mexico City (see **Figure 7**).

**Figure 7.** Public spaces in Mexico City used to collect food and other emergency supplies right after the earthquake of September 19, 2017. Photos by the authors.


## Acknowledgements

The authors would like to thank the reviewer for his valuable comments and suggestions of this manuscript. This article is a product of the research project IPN-SIP 20172147, funded by the Secretaría de Investigación y Posgrado del Instituto Politécnico Nacional, México.



## Author details

Milton Montejano-Castillo* and Mildred Moreno-Villanueva

*Address all correspondence to: mmontejanoc@ipn.mx

Escuela Superior de Ingeniería y Arquitectura Unidad Tecamachalco del Instituto Politécnico Nacional, Estado de México, Mexico


## References


[1] Meli R. El sismo de 1985 en México. In: Lugo Hubp J, Inbar M, editors. Desastres naturales en América Latina. 1st ed. México, DF: Fondo de Cultura Económica; 2002. pp. 125-146

[2] Guevara Pérez T. Arquitectura moderna en zonas sísmicas. 1st ed. Barcelona: Gustavo Gili; 2009. 207 p

[3] Cruz Atienza VM, Krishna Singh S, Ordaz Schroeder M. Qué ocurrió el 19 de septiembre de 2017 en México. Nexos [Internet]. September 23, 2017. Available from: https://www.nexos.com.mx/?p=33830 [Accessed: November 7, 2017]

[4] Gobierno de la Ciudad de México. Reconstrucción Ciudad de México [Internet]. October 2017. Available from: http://www.reconstruccion.cdmx.gob.mx/inmuebles-demoler [Accessed: November 7, 2017]

[5] Gobierno del Distrito Federal. Bando número 2. 2000. Available from: http://www.invi.df.gob.mx/portal/transparencia/pdf/LEYES/Bando_informativo_2.pdf

[6] Comisión Nacional de Vivienda. Guía para la Redensificación Habitacional en la Ciudad Interior. 1st ed. México, DF: Gobierno Federal, SEDESOL, CONAVI; 2010. 74 p. Available from: https://www.gob.mx/cms/uploads/attachment/file/214894/Gui_a_para_la_Redensificacio_n.pdf

[7] Alba González M. Representaciones y prácticas sociales entorno a políticas urbanas: La movilización Nimby frente a la Redensificación de las zonas centrales de la ciudad de México. Cultura y representaciones sociales. Revista electrónica de ciencias sociales. 2009;**3**(6):43-72. Available from: http://www.culturayrs.org.mx/revista/num6/deAlba.pdf

[8] Asuad Sanén NE. La desindustrialización relativa de la ciudad de México de 1980-2003. In: XX Seminario de Economía urbana y regional; June 23. México, DF: Instituto de Investigaciones Económicas de la UNAM; 2010

[9] Sobrino LJ. Desindustrialización y movilidad cotidiana en la Ciudad de México. In: de Demografía SM, editor. X Reunión Nacional de Investigación Demográfica en México. Escenarios Demográficos y Política de Población en el siglo XXI; November 3-6. México, DF: El Colegio de México, AC; 2010

[10] Palacios Monzón E. Del paisaje urbano difuso, a la ciudad compacta, Nuevas Centralidades, nuevos paisajes… "nuevos ciudadanos". 1st ed. México, D.F: Instituto Politécnico Nacional; 2010. 223 p

[11] Chain D. Espacio y Cultura. In: Ministerio de Desarrollo Urbano de la Ciudad de Buenos Aires, editor. La Humanización del espacio público. 1st ed. Buenos Aires, Argentina: Editorial Grafica Pinter S.A.; 2010. pp. 3-3. Available from: http://cdn2.buenosaires.gob.ar/desarrollourbano/publicaciones/la-humanizacion-del-espacio-publico-2010.pdf

[12] Mués Zepeda AM. Habitabilidad y Desarrollo Urbano Sostenible [thesis]. Estado de México: Instituto Tecnológico y de Estudios Superiores de Monterrey; 2011. 136 p. Available from: http://infonavit.janium.com/janium/Documentos/035079.pdf

[13] Blachére G. Saber construir habitabilidad, durabilidad, economía de los edificios. 3rd ed. Barcelona: Editores Técnicos Asociados, S.A.; 1978. 312 p

[14] Saldarriaga Roa A. Habitabilidad. Bogotá: Escala Fondo Editorial; 1981. 136 p

[15] Juárez Neri VM. Condiciones de la vivienda en la zona metropolitana del Valle de México en el año 2000. Scripta Nova: Revista electrónica de Geografía y Ciencias Sociales. 2003;**VII**(146(040)):n.p. Available from: http://www.ub.es/geocrit/sn/sn-146(040).htm

[16] Velázquez Mejía O. Habitabilidad en el parque habitacional de la AGEB 010-9 del Fraccionamiento Ojo de Agua, Tecámac, Estado de México. Revista Académica de Investigación Tlatemoani. 2011;**2**(6):1-13. Available from: http://www.eumed.net/rev/tlatemoani/06/ovm.pdf

[17] Carrión F. Espacio público: punto de partida para la alteridad. In: Segovia O, editor. Espacios públicos urbanos y construcción social. Hacia un ejercicio de ciudadanía. Santiago de Chile: Ediciones SUR; 2007. pp. 79-97. Available from: http://www.elagora.org.ar/site/documentos/Espacios_publicos_y_construccion_social.pdf

[18] Reyes Pérez R. Habitabilidad y espacio público en Barrios históricos de Mérida, Yucatán al inicio del siglo XXI [dissertation]. Mexico City: Doctorado en Arquitectura, Facultad de Arquitectura de la UNAM; 2012. 486 p. Available from: http://132.248.9.195/ptd2012/mayo/0679703/Index.html

[19] Valladares Anguiano R, Chávez Martha E, Moreno Olmos S. Elementos de la Habitabilidad Urbana. In: Seminario internacional de arquitectura y vivienda; 2008. Available from: http://insumisos.com/LecturasGratis/elementos%20de%20la%20habitabilidad%20-%20valladares%20chavez%20y%20moreno.pdf

[20] Akkar M. Questioning 'inclusivity' of public spaces in post-industrial cities: The case of Haymarket Bus Station, Newcastle upon Tyne. METU Journal of Faculty of Architecture. 2005;**22**(2):1-24. Available from: http://jfa.arch.metu.edu.tr/archive/0258-5316/2005/cilt22/sayi_2/1-24.pdf

[21] Acuña C, de Souza L, Leicht E, Musso C, Vainer D, Varela A. Aglomeración Maldonado – Punta del Este – San Carlos. Enfoques y propuestas. Hacia un modelo transformador. 1st ed. Montevideo: Facultad de Arquitectura/Instituto de Teoría de la Arquitectura y Urbanismo. Universidad de la República. Uruguay; 2013. 122 p. Available from: http://www.fadu.edu.uy/itu/files/2014/12/ITU-AGLOMERACI%C3%93N-MALDONADO-PUNTA-DEL-ESTE-SAN-CARLOS.pdf

[22] Ramírez Kuri P, Ziccardi A. Pobreza urbana, desigualdad y exclusión social en la ciudad del siglo XXI. In: Cordera R, Ramírez Kuri P, Ziccardi A, editors. Pobreza, desigualdad y exclusión social en la ciudad del siglo XXI. 1st ed. México, DF: Siglo XXI Editores/UNAM Instituto de Investigaciones Sociales; 2008. pp. 23-50

[23] Zermeño S. La centralidad de los excluidos. In: Cordera R, Ramírez Kuri P, Ziccardi A, editors. Pobreza desigualdad y exclusión social en la ciudad del siglo XXI. 1st ed. México, DF: Siglo XXI Editores/UNAM Instituto de Investigaciones Sociales; 2008. pp. 135-152

[24] Remedi G. La ciudad Latinoamericana S.A. (o el asalto al espacio público). In: Subsecretaría de Planeamiento, editor. Las dimensiones del espacio público: problemas y proyectos. Buenos Aires: Gobierno de la Ciudad de Buenos Aires; 2004. pp. 15-25

[25] Glass RL, Westergaard J. London's Housing Needs: Statement of Evidence to the Committee on Housing in Greater London. London: Centre for Urban Studies, University College; 1965. 97 p

[26] London B, Palen J, editors. Gentrification, Displacement, and Neighborhood Revitalization. Albany: State University of New York Press; 1984. 271 p

[27] Smith N, Taylor & Francis, Ltd. Gentrification and the rent gap. Annals of the Association of American Geographers. 1987;**77**(3):462-465

[28] Janoschka M. Geografías urbanas en la era del Neoliberalismo. Una conceptualización de la resistencia local a través de la participación y la ciudadanía urbana. Investigaciones Geográficas, Boletín del Instituto de Geografía, UNAM. 2011;**76**:118-132. Available from: http://www.investigacionesgeograficas.unam.mx/index.php/rig/article/view/29879/27778

Chapter 5

# On Risk and Reliability Studies of Climate-Related Building Performance

Krystyna Pietrzyk and Ireneusz Czmoch

Additional information is available at the end of the chapter

http://dx.doi.org/10.5772/intechopen.71684

**Abstract**

A design strategy based on integration of the building form and structure with its external environment in order to take advantage of natural forces (wind and buoyancy effects) has been evaluated in terms of risk and reliability measures. Tools for the probabilistic analysis (First-Order Reliability Method (FORM), Monte Carlo) have been presented and applied in the probabilistic modelling and sensitivity analysis of the response function of the studied building physics problem. Sensitivity analysis of the influence of basic random variables on the probability distribution of a response function is straightforward in FORM methodology. The case-based studies of probabilistic modelling of uncertainties coupled to wind speed and temperature difference through the specified building/environment system have been presented (i.e., the distribution models of the air change rate *ACH* and the dynamic *U* value characterising thermal performance of dynamic insulation). Sensitivities of the probability model of *ACH* to the parameters of wind speed and temperature distributions have been estimated for the consecutive values of the air change rate using FORM methodology. Reliability of *ACH* turned out to be most sensitive to the shape parameter of the wind speed distribution (in two-parameter Weibull model). The probabilistic risk analysis along with the effective tools for sensitivity analysis can be used to support design decisions and also to develop better models for evaluation of building performance.

**Keywords:** building performance, environment, risk, reliability, probabilistic approximation, FORM, sensitivity, climate, climate mitigation, wind, air infiltration, *ACH*, dynamic *U* value

## 1. Introduction

Developing tools to support decision-making to ensure comfort and safety in built environment while taking into account climate change challenges becomes important. 'Energy is the dominant contributor to climate change, accounting for around 60% of total global greenhouse gas emissions' [1]. Reducing the carbon intensity of energy is a key objective in long-term climate goals. Hence, choosing a strategy based on integrating the building form and structure with its external environment in order to take advantage of natural forces (for natural ventilation, solar heating, etc.) is an example of design decisions leading towards mitigation of climate change.

Decisions concerning the choice of the design solutions for the particular project have to be taken under uncertainties related to the unknown and variable conditions, i.e. random climatic conditions, uncertain material performance, uncertain user behaviour, etc. Confronted with significant uncertainty, deterministic modelling supporting design process has been proved to be insufficient for decision-making. However, as it is said in [2] 'Existing engineering-based models are unable to propagate uncertainties through the model, and are therefore limited in their ability to display the impact of uncertainties to decision makers'. Facing that challenge, the chapter includes the discussion of the models and the tools applied by the authors for the probabilistic transformation of uncertainties of climatic parameters through a building/environment system for the predictive modelling of building performance.

The method for the quantification of building performance in terms of probability of poor performance (failure) and satisfactory performance (safe behaviour, in general meaning) is presented. Next, the tools for the probabilistic analysis are described (FORM, Monte Carlo) in relation to probabilistic modelling and possible applications of sensitivity analysis. One of the important results of analysis is the probability distribution functions of different performances as the responses of building/environment systems to the environmental loads. Such analysis requires estimation of some climatic parameters in terms of frequency of occurrence and appropriate statistics.

The chapter includes the case-based studies of probabilistic transformation of uncertainties coupled to wind and temperature through the specified building/environment system to show the effect on the distribution model of the air change rate and further on the distribution model of the dynamic thermal transmittance (dynamic $U$ value) of the building envelope. Furthermore, the estimated distribution models could be included in risk/reliability calculations, carried out with FORM tools. The analysis of the sensitivity of the distribution of $ACH$ with respect to the randomness of wind speed and outdoor temperature exemplifies the potential of the FORM tools, which can be effectively used to find out the probabilistic characteristics typical for the combination of the important variables influencing climate-structure interaction.

## 2. Risk perspective on design for sustainable development

Design for sustainable development can be approached using risk analysis tools. To minimalise the risk of undesired consequences while increasing the chance to enhance the quality of life becomes the basic design objective. The design goal can be expressed in other terms—how to secure reliability of design under risk constraints [3]. To be clear about the terminology used further, some definitions are given below.

- **Risk**—a state of **uncertainty** where some possible outcomes have an undesired effect or significant loss [4]. It can be expressed in terms of adverse consequences scaled by the probabilities of undesired outcomes.

- **Qualitative risk**—Relative measure of risk based on ranking or separation into descriptive categories such as low, medium and high or on a scale, e.g., from 1 to 10. The example is the risk matrix of the failure mode [5] where failure severity is described by categories: minor, major, critical and catastrophic, whereas failure frequency is described in terms of very unlikely, remote, occasional, probably and frequent. A combination of both gives a qualitative assessment of risk as low, moderate or high.
- **Quantitative risk**—The most common quantification of risk is the product of likelihood of occurrence and the effect of the hazardous event [6]. Risk is treated as convolution of hazard and vulnerability, and it refers to mutual conditioning of two phenomena [7].
- **Risk assessment** is the determination of quantitative or qualitative estimate of risk related to the well-defined situation and the recognised hazard.
- **Probabilistic risk assessment** denotes the methods dealing with computation both the likelihood of undesired event and severity of possible negative consequences due to occurrence of such event.
- **Risk management process**—'the systematic process of identification, assessment, monitoring and control of risk' [8].
- **Reliability**—Ability of a system to satisfactorily perform under the specified conditions of use over an intended period of time. It is quantified by the probability of satisfactory (or safe) performance [9]. **Unreliability** is measured by the **probability of failure** (undesired performance).

### 2.1. Risk perspective on climate change challenges

Climate change threatens life on our planet. In view of high uncertainty, qualitative or semi-quantitative risk analysis based on the different scenarios is often applied. Following the quantitative definition of risk, one can write

$$Risk = P[hazard] * Consequences \tag{1}$$

$P[hazard]$ is the probability of occurrence of undesired events leading to possible *Consequences* like loss, injury, or discomfort.

Risk reduction could be accomplished by decreasing the probability of undesired event as well as diminishing the scale adverse consequences. Risk reduction of climate change and its consequences can be accomplished by climate change mitigation (decrease of the probability of occurrence of adverse events) or climate change adaptation (decrease of the adverse consequences) described as follows:

**Climate change mitigation**—'it consists of actions to limit the magnitude and/or rate of long-term climate change' [10]. 'It generally involves reductions in human (anthropogenic) emissions of greenhouse gases' [11].

**Climate change adaptation**—'anticipating the adverse effects of climate change and taking appropriate action to prevent or minimize the damage they can cause, or taking advantage of opportunities that may arise' [12].

### 2.2. Risk assessment as a tool supporting design of buildings

Designing for the integration of the building form and structure with its external environment in order to use natural forces to secure comfort (passive strategies) is an example of activities towards mitigation of climate change. If it is supported by probabilistic prediction of a local climate changes, it can be viewed from the climate adaptation perspective too.

The needs for risk reduction related to the hazards induced by climate change become an important boundary condition in the modelling of building/environment system to support building design. Following the definition of risk (Eq. (1)), adverse consequences are indicated by the set {yes = 1, no = 0}, and as a result, the probability $P[hazard]$ becomes the discriminating factor for comparison of different design solutions. It means that certain design could be chosen on the basis of comparison of probability of unsatisfactory performance or reliability, evaluated by a set of alternative design proposals. In that way risk assessment becomes a tool supporting design of buildings.

## 3. Transformation of uncertainties: probabilistic approach

Probability is a measure of uncertainty about future events. Probability of a performance of a building/environment system depends on the theoretical model used and the randomness of the influencing parameters. The epistemic uncertainty about the theoretical models applied together with the aleatory uncertainty coupled to the randomness of important phenomena contributes to the final uncertain outcome, as the result of transformation of uncertainties throughout the model. Methods and tools for probabilistic reliability analysis can be used to estimate the probabilistic response of the structure to the random climatic load. They could be an important part of risk-based design process built upon the framework of risk management methodology as proposed in [13].

### 3.1. Probabilistic analysis with FORM

Development of reliability methods resulted in variety of powerful algorithms to estimate probability of failure for complicated physical and mathematical models of building systems incorporating random variables (i.e. properties or actions). FORM denotes 'First-Order Reliability Method', which has been developed by many researchers in about 40 years ago. Short description of the FORM basics as well as sensitivity tools is presented below. For details, check [14], and for application in building physics, look in [15]. First-order reliability method (FORM) is the most popular approach applied in practice.

Once the response of a system characterised by a set of basic random variables and a mathematical model describing the relationship among them has been established, the probability density function of the response can be estimated with help of FORM tools.

In general, the performance of a system is analysed in the space $\Omega_X = \{X \in R^n\}$ of basic random variables $\boldsymbol{X}$. For a given failure mode or serviceability requirement, represented by the limit state surface $g(\boldsymbol{X}) = 0$, the space $\Omega_X$ is divided into the safe subset, i.e. satisfactory performance, $\Omega^S = \{\boldsymbol{X} \in R^n; g(\boldsymbol{X}) > 0\}$, and the failure subset, $\Omega^F = \{\boldsymbol{X} \in R^n; g(\boldsymbol{X}) \leq 0\}$, i.e., unsatisfactory performance. If all random variables are continuous with the multivariate joint probability density function (PDF) $f_{\boldsymbol{X}}(\boldsymbol{x})$, the failure probability is given by the integral

$$P_f = \int_{\Omega^F} f_x(\boldsymbol{x}) d\boldsymbol{x} \tag{2}$$

The integral (Eq. (2)) can be evaluated exactly for a few cases with the most important one: the linear limit state surface and multidimensional normal (Gaussian) distribution function of variables $\boldsymbol{X}$.

FORM algorithm starts with the non-linear transformation. Non-normal random vector $\boldsymbol{X}$ is transformed into a standard normal (Gaussian) vector $\boldsymbol{Y}$ with zero mean and unit covariance matrix $\boldsymbol{C}_{YY} = \boldsymbol{I}$. The limit state surface $g(\boldsymbol{x}) = 0$ is mapped into a limit state surface $G(\boldsymbol{y}) = 0$. Next, the design point $\boldsymbol{y}^*$, i.e. the point on the limit state surface with the minimum distance to the origin of the $\boldsymbol{Y}$ space, is determined by solving the non-linear optimisation problem with a non-linear constrain $G(\boldsymbol{y}) = 0$:

$$\beta = min\sqrt{\boldsymbol{y}^T\boldsymbol{y}} \quad for \quad \boldsymbol{y} \quad on \quad G(\boldsymbol{y}) = 0 \tag{3}$$

The hyperplane tangential to the limit state surface at the point $\boldsymbol{y}^*$ is given by formula

$$\beta - \boldsymbol{\alpha}^T\boldsymbol{y} = 0 \tag{4}$$

where $\boldsymbol{\alpha}$ is a unit outward normal vector to the hyperplane and $\beta$ is the distance between the hyperplane and the origin. Since the random vector $\boldsymbol{Y} = \boldsymbol{Y}(\boldsymbol{X})$ has standard normal distribution, the first-order approximation of the failure probability is given by

$$P_f \cong P\left[\beta - \boldsymbol{\alpha}^T\boldsymbol{Y} \leq 0\right] = \Phi(-\beta) \tag{5}$$

where $\Phi(\ldots)$ is the Laplace function.

The non-linear constrained optimisation problem (Eq. (3)) can be solved with many standard procedures as well as algorithms developed especially for this purpose, e.g. algorithm for the case of independent, non-normal random variables [16] and algorithm for problems with incomplete probability information [17].

All such solvers are iterative: for the assumed value of design point $\boldsymbol{x}^*_{(k)}$, the values of limit state function $g\left(\boldsymbol{x}^*_{(k)}\right)$ and its gradient $\nabla g\left(\boldsymbol{x}^*_{(k)}\right)$ are determined. Next, a new position of design point $\boldsymbol{x}^*_{(k+1)}$ is derived, and the process continues until the convergence criteria are fulfilled. If the state of the analysed system is described by the performance function defined by analytical formula, then the gradient can be evaluated easily, and one of algorithms solving the optimisation problem (Eq. (3)) can be applied directly. Otherwise the stochastic finite element method should be applied in order to calculate the value of the limit state function and its gradient vector at following values of design points.

The first-order reliability index $\beta$ and the failure probability $P_f \cong \Phi(-\beta)$ depend on:

- Parameters $\boldsymbol{p} = (p_1, \ldots, p_d)$ of the probability distributions of basic random variables, e.g. mean value, standard deviation, skewness or location, scale and shape parameters
- Any deterministic parameters $\boldsymbol{\Theta} = (\theta_1, \ldots, \theta_g)$ defining the form of the performance function $g(\boldsymbol{x}, \theta_1, \ldots, \theta_g)$

Practical experience shows that the failure probability is usually a strongly non-linear function of the parameter $\boldsymbol{\theta}$, whereas the reliability index $\beta$ is a rather linear function of the parameter $\theta$. Thus the change in the failure probability due to the change of the parameter $\theta$ can be approximated as follows:

$$P_f(\theta + \Delta\theta) \cong \Phi(-\beta - \Delta\beta) \cong \Phi\left(-\beta - \frac{d\beta}{d\theta}\Delta\theta\right) \tag{6}$$

The sensitivity measures of the first-order reliability index do not depend on the curvature of the limit state surface $g(\boldsymbol{x}) = 0$ at the design point. Therefore, the application of sensitivity measures is limited to small changes of the values of the parameters.

The sensitivity of the first-order approximation of the failure probability $P_f \cong \Phi(-\beta)$ is directly related to the sensitivity of the reliability index $\beta$, since

$$\frac{dP_f}{d\theta} = -\varphi(\beta)\frac{d\beta}{d\theta} \tag{7}$$

If $\theta$ is a parameter of the limit state function $g(\boldsymbol{x}, \theta)$, then derivative of the reliability index with respect to the parameter $\theta$ is equal to

$$\frac{d\beta}{d\theta} = \frac{1}{|\nabla G(\boldsymbol{y}^*, \theta)|}\frac{\partial}{\partial\theta}G(\boldsymbol{y}^*, \theta) \tag{8}$$

where vector $\boldsymbol{Y}$ contains independent standard normal variables related to the vector of basic random variables by transformation $\boldsymbol{Y} = \boldsymbol{Y}(\boldsymbol{X})$, and the limit state surface $g(\boldsymbol{x}, \theta) = 0$ defined in the space $\boldsymbol{X}$ has been mapped into the surface $G(\boldsymbol{y}, \theta) = 0$. Since the FORM index $\beta$ is equal to the minimum distance between the origin of the $\boldsymbol{Y}$ space and the limit state surface $G(\boldsymbol{y}, \theta) = 0$, thus the design point $\boldsymbol{y}^*$ is laying on the limit state surface; see **Figure 1**:

$$\beta = -\frac{\nabla G(\boldsymbol{y}^*, \theta)}{|\nabla G(\boldsymbol{y}^*, \theta)|}\boldsymbol{y}^* \tag{9}$$

The limit state surface in the $\boldsymbol{X}$ space of basic random variables $g(\boldsymbol{x}, \theta) = 0$ does not depend on any parameter $p_{ik}$ of a random variable $X_i$ with the distribution function $F_i(x_i, p_{ik})$. However, the limit state surface $G(\boldsymbol{y}, \theta) = 0$ depends on parameter $p_{ik}$ due to the transformation $\boldsymbol{Y} = \boldsymbol{Y}(\boldsymbol{X})$.

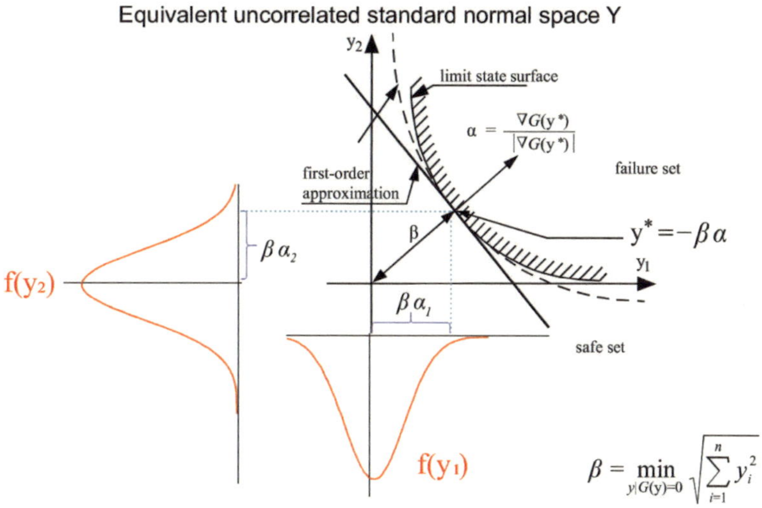


**Figure 1.** Illustration of sensitivity indices $\alpha_i$ (modified from [14]).

The derivative of the reliability index with respect to the parameter $p_{ik}$ is given by relation

$$\frac{\partial \beta}{\partial p_{ik}} = \frac{1}{\beta}(y^*)^T \frac{\partial}{\partial p_{ik}} y^* \tag{10}$$

The derivative of the vector $\boldsymbol{y}^*$ with respect to parameter $p_{ik}$ have to be evaluated for each specific transformation $\boldsymbol{Y} = \boldsymbol{Y}(\boldsymbol{X})$. For details, see [14].

Sensitivity analysis shows how the uncertainty in the output response function of a system can be allocated to the different uncertainties in the basic variables. Sensitivity analysis is straightforward for FORM methodology. The influence of the basic random variable $y_i$ on the statistics of the response can be quantified by the sensitivity indices $\alpha_i$ [18]:

$$\alpha_i = -\frac{d\beta}{dy_i} \quad \textit{for } y_i = y_i^* \tag{11}$$

where $\boldsymbol{y}^*$ is the design point in the space of normalised reduced random variables.

For uncorrelated random variables, the sensitivity vector $\alpha$ coincides with the direction cosines vector of the random variables [18]. Illustration of sensitivity indices is given in **Figure 1**.

### 3.2. FORM versus Monte Carlo simulation

An alternative technique applied for probability estimation of risk or reliability is Monte Carlo simulation (MCS). For the purpose of the zero–one indicator-based MCS, Eq. (2) defining the failure probability is given as follows:

$$P_f = \int_{\Omega^F} f_X(\boldsymbol{x}) d\boldsymbol{x} = \int_{\boldsymbol{k} \in R^n} \boldsymbol{k} h_{\boldsymbol{K}}(\boldsymbol{k}) d\boldsymbol{k} \quad (12)$$

where the random vector **K** has the non-negative sampling density function $h_K(\boldsymbol{k})$ and is defined by the transformation

$$\boldsymbol{k} = I\left(g(\boldsymbol{k}) \leq 0\right) \frac{f_X(\boldsymbol{k})}{h_{\boldsymbol{K}}(\boldsymbol{k})} \quad (13)$$

and $I(u)$ is an indicator function:

$$I(u) = \begin{cases} 1 & if \quad u \leq 0 \\ 0 & if \quad u > 0 \end{cases} \quad (14)$$

In this way the failure probability is equal to the expectation of random vector with the non-negative sampling density function $h_{\boldsymbol{K}}(\boldsymbol{k})$:

$$P_f = E[\boldsymbol{K}] \quad (15)$$

The average of $N$ simulated values of the random vector $\boldsymbol{K}$ is the estimator of the failure probability, which variance is equal to

$$Var\left[\widehat{P_f}\right] = \frac{1}{N(N-1)} \sum_{i=1}^{N} (\boldsymbol{k}_i - E[\boldsymbol{K}_i])^2 \quad (16)$$

Monte Carlo simulation technique is a powerful tool to calculate the probability of failure for the system described by non-continuous performance function as well as discrete random variables. However, the basic drawback of the MCS is long CPU time calculation, if the failure probability is of the orders $10^{-2} - 10^{-6}$, since the sample size must be very large in order to obtain estimation of failure probability with low variance and narrow confidence interval. Various variance reduction techniques have been suggested to increase the efficiency of MCS. The basic idea is to assume a sampling density function $h_Y(\boldsymbol{y})$ that reduces the variance of the estimator $\widehat{P_f}$. In the case of highly complicated systems, when time-consuming method must be applied to evaluate a single value of the limit state function, the MCS with the variance reduction technique is still an approach demanding a lot of computer time. Another drawback of the MCS, especially important, in the context of the chapter, is lack of the sensitivity analysis tools. It is simply impossible to run billions of simulation in order to study sensitivity of the system with respect to specific parameters.

## 4. Case-based risk/reliability studies of climate-related building performance

### 4.1. Building/Environment system performance

In the context of the ventilation design, air infiltration constitutes an important complement to air exchange. Furthermore, air infiltration can influence on the properties (thermal and structural)

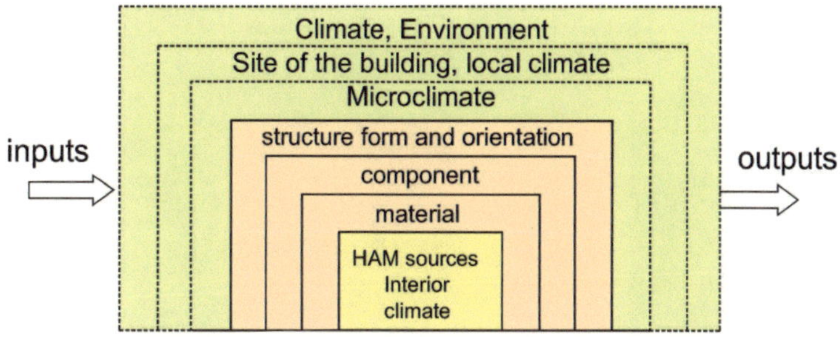


**Figure 2.** Building/Environment system applied in a traditional building physics analysis [19].

of building components. For some building technologies, e.g. lightweight timber frame with mineral wool filling, and loose mineral wool layers for roof insulation, the dependence of the thermal properties of building components on air infiltration can be observed; thus the interaction between, e.g. thermal transmittance and air infiltration should be taken into account. Therefore, it is important to apply a systemic approach to building/environment system performance taking into account different aspects of building physics. Due to the random natural driving forces governing the rate of air infiltration, the approach based on probabilistic methodology seems to be very well suited to handle these phenomena.

A building can be seen as a system transforming as well as resisting different loads (static and dynamic loads—caused by flow of air, heat and moisture) which is designed to ensure safe and comfortable living conditions inside the enclosure. The structure has to be designed in such a way that the possibilities of adverse consequences of this transformation, for example, loss of stability of the structure, inadequate ventilation or mould growth inside a building, have been minimalised. This systemic approach provides a proper theoretical tool for the analysis of the interrelations between the structure, its environment and its performance. An example of systemic model of a building, applicable in building physics studies, is shown in **Figure 2** [19].

The local environmental conditions interact with building structure to form a microclimate around a building. Sources of heat, air and moisture, including the products of HVAC systems as well as user behaviour, build up the internal load. Physical boundary conditions define the level of integration of the structure with the environment.

The output of the system can be described by the performance of the building (structure and enclosure). The performance can be considered in terms of safety, comfort and energy consumption and described by various parameters depending on physical conditions of the building structure and inside air. Those parameters should fulfil the performance requirements in order to prevent undesired performance (failure state) occurrence.

### 4.2. Case description

#### *4.2.1. Description of the test house*

The object of the study is a timber-framed low-rise naturally ventilated building with aspect ratio 2 and slope of the roof of 45° [20]. The building site in the district of Gothenburg has been considered and can be described as a semi-urban area with the surface roughness equal

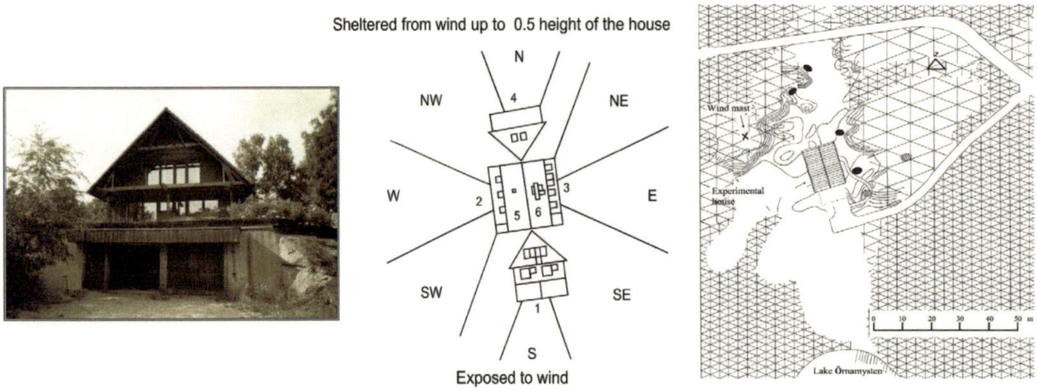


**Figure 3.** Object of the study – the Building/Environment system [20].

to 0.3 m. Example has been worked out for wind blowing from the west. It is assumed that the building is surrounded by other obstructions (other buildings, topography, vegetation, trees etc.) equivalent to half of its height. The following input data are used: volume of the house $V = 486\ m^3$, area of the building envelope $A = 336\ m^2$ and internal temperature $T_{int} = 20^{\circ}C$.

The house was constructed in 1979 with the intention of using it for experimental studies in building physics with focus on ventilation and energy saving. The garage with doors facing south is located in the extended south part of the concrete cellar as shown in **Figure 3**.

#### *4.2.2. Measurement programme*

The following parameters have been measured, as is shown in **Figure 4**: (1) leakage characteristics of the house using blower door tests, (2) mean value of pressure difference across the six building components with Validyne pressure transducers, (3) wind speed and wind direction with the anemometer located on a small hill about 25 m from the house, (4) internal and external temperatures and (5) limited number of tracer gas measurements of *ACH*. The measurement programme has been carried out during 8 months. As a result, hourly mean data have been registered.

The results of the pressure drop measurements have been used to validate the air infiltration through the envelope. An opening under the garage door has been treated separately in the calculation model for air change rate [20, 21].

### 4.3. Modelling of air change rate

The applied infiltration model takes into account the contribution of wind and stack effect to the total air change rate (*ACH*) in the following form [22]:

$$ACH = \sqrt{ACH_s^2 + ACH_w^2} \tag{17}$$

where $ACH_s$ is the air change rate caused by stack effect and $ACH_w$ is the air change rate caused by wind.

The model refers to low-rise building with light-weight construction, single ventilation zone, single temperature zone and steady-state conditions of air flow.

The infiltration model developed by Pietrzyk [20] indicates the air change rate *ACH* as a random function of three basic random variables: **temperature difference, wind speed** and **wind direction**. Wind direction is divided into eight sectors and is treated as a uniformly distributed within each sector. Finally, the air change rate conditioned by the wind direction sector is given by the following expression:

$$ACH_d = \sqrt{s_1 \Delta T^2 + s_2|\Delta T| + s_3|\Delta T|^{1,5} + w_{d,1}v_d^4 + w_{d,2}v_d^2 + w_{d,3}v_d^3} \tag{18}$$

where $d$ is a wind direction sector; $s_1, s_2, s_3, w_{d,1}, w_{d,2}$ *and* $w_{d,3}$ are the deterministic coefficients related to the house dimensions, position of neutral pressure layer, level of external and internal pressure coefficients; $\Delta T$ is an ext.-int. temperature difference; and $v$ denotes wind speed.

Distributions of air change rate averaged over one hour (1-h) periods at a randomly chosen time in the year have been estimated with the help of the model described by Eq. (18). One-hour mean data ensure steady-state conditions of the airflow through the building envelope. Wind is the most important source of variations in the process of air exchange. However, according to wind energy spectrum presented in [23] for the frequency range 0.00014–0.0033 cycles/hour related to time interval from 5 min to 2 h, the wind speed varies slightly. This range is called spectral gap. Measurements carried out for periods of that duration can be regarded as representing the steady-state conditions [24].

Performance criteria in terms of *ACH* should take into account the minimum threshold evaluated with respect to unhygienic conditions. Then, probability of unsatisfactory performance is equal to $P[ACH < threshold]$.

**Figure 4** presents how the building response such as *ACH* depends on the uncertain environmental conditions. The wind speed is traced from the meteorological station to the site and eventually to the building envelope which in turn influences the microclimatic conditions near to structure. The zone of wind-structure interaction is included in the model of designed system (see boundary conditions of the system presented by the solid lines). Serviceability performance due to wind action can be evaluated in terms of probability of undesired performance (failure). It is worth noticing that measurement data have been used to model the building performance as well as to validate the results of analysis carried out with the help of the established model.

The probability density function for air change rate as a function of basic random variables 1-h mean wind speed and 1-h mean temperature difference at time points chosen randomly during the year has been estimated with the help of the FORM sensitivity analysis for the performance function

$$g(x_1, x_2; a) = ACH(x_1, x_2) - a \tag{19}$$

where $ACH(x_1, x_2)$ is given by Eq. (18), $x_1 = \Delta T$ and $x_2 = v_d$.

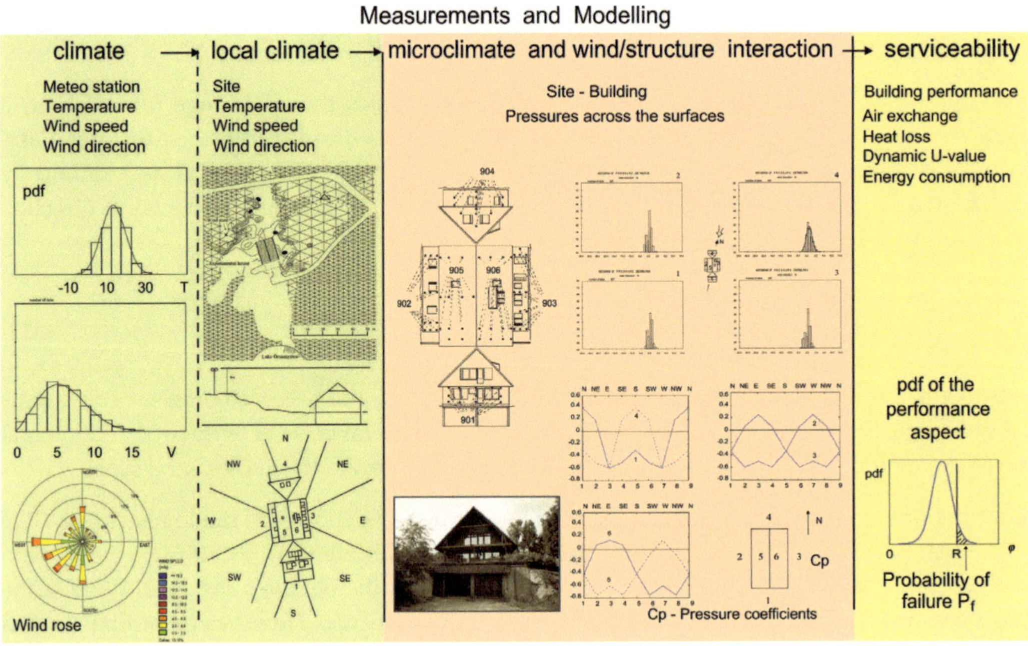


**Figure 4.** Transformation of uncertainty within the modelling of building performance.

The parametric sensitivity analysis applied to FORM measures (reliability index or failure probability) is used in order to determine the probability density function for the random response $ACH = ACH(x_1, x_2)$. The cumulative distribution function of random function $ACH = ACH(x_1, x_2)$ is actually equivalent to the probability of failure defined for the performance function Eq. (19):

$$F_{ACH}(a) = P[ACH(x_1, x_2) \leq a] = P[g(x_1, x_2; a) \leq 0] \tag{20}$$

Thus, the cumulative probability function can be estimated with the help of the FORM analysis:

$$F_{ACH}(a) \approx \Phi(-\beta_a) \tag{21}$$

where $\Phi(u)$ is the Laplace function and the reliability index $\beta_a$ has been determined for the limit state surface $g(x_1, x_2) = ACH(x_1, x_2) - a = 0$ defined for a given value of parameter $a$.

Following the sensitivity measures presented earlier in the chapter, the probability density function of the random response $ACH$ can be estimated with the help of formula:

$$f_{ACH}(a) \approx -\varphi(\beta_a)\frac{d\beta_a}{da} = \frac{\varphi(\beta_a)}{|\nabla G(\boldsymbol{y}, a)|_{for\ \boldsymbol{y}=\boldsymbol{y}^*}} \tag{22}$$

where $\varphi(u)$ is the probability density function of the standard Gaussian distribution, $G(\boldsymbol{y}, a)$ is the limit state function in the space $\boldsymbol{Y} = \boldsymbol{Y}(\boldsymbol{X})$ of normalised random variables and $\boldsymbol{y}^*$ is the design point, i.e. the point on the surface $G(\boldsymbol{y}, a) =$ at the shortest distance to the origin of the coordinate system. The value of probability density function of random response function $f_{ACH}(ACH)$ can be obtained by means of FORM sensitivity analysis for consecutive values of parameter $a$; for details, see [14].

#### 4.3.1. *Wind transformation – climate/local climate/microclimate*

The input basic random variable for the infiltration model is wind speed in the vicinity of the building envelope. Wind speed and direction are usually measured at the meteorological stations. The mean value of 1-h mean wind speed can be evaluated from the mean value of 10-min mean wind speed obtained from meteorological station using the principle that the mean velocity increases by 5% when the averaging period is reduced from 1 h to 10 min. The transformation of the abovementioned data to the site of the building is often needed especially for wind that changes drastically due to the roughness of the ground surface.

The hourly mean wind speed $v$ is assumed to follow the two-parameter Weibull distribution with probability density function as follows:

$$f(v; c, \lambda) = \frac{\lambda}{c}\left(\frac{v}{c}\right)^{\lambda-1} exp\left\{-\left(\frac{v}{c}\right)^{\lambda}\right\} \tag{23}$$

where $\lambda$ is a shape parameter and $c$ is a scale parameter.

In general, parameters $\lambda, c$ for wind speed averaged over 1-h should be estimated for different wind direction sectors that shall result in quite different Weibull distributions due to two reasons: directional variability in the terrain surrounding the house and the predominance of certain wind directions. In particular, most building sites are subjected to sheltering effects from topography, trees and buildings. The roughness of the ground surface changes the mean wind speed and its turbulent characteristics and is described by the surface roughness height (aerodynamic roughness length) denoted $z_0$. Roughness height depends on the mean element height of the roughness field. The results of laboratory measurements show that the value of $z_0$ is approximately equal to 1/30 of the height of the roughness elements. **Table 1** presents the classification of roughness height for different types of surfaces with reference to the categories of terrain roughness used in Swedish Code [24].

Transformation of the wind speed between terrains of different surface roughness is possible due to the similarity theory [25], based on the equilibrium boundary layer height, which is according to [26] equal to 1200 m. The wind flows with the gradient velocity $v_g$ along the isobars:

$$v_g - \frac{u_*}{\kappa}\left[ln\left(\frac{u_*}{fz_0}\right) - A_u\right] \tag{24}$$

where $z_0$ is the surface roughness height (m); $\kappa$ is Karman's constant, $\kappa = 0,4$; $A_u$ is const.; assumed $-1$; $f$ is Coriolis parameter (1/s), $f = 1,12{*}10^{-4}$ for latitude of order of 50° [25]; and $v_g$ is gradient velocity (m/s).

$u_*$ is friction velocity depending on the surface shear stress $\tau_0$ as given in Eq. (25):

$$u_* = \sqrt{\tau_0/\rho} \tag{25}$$

where $\tau_0$ is surface shear stress (kg/ms$^2$) and $\rho$ is air density (kg/m$^3$).

The mean velocity profile $u(z)$ near the ground, where $z$ is the height above the ground ($z < 100\ m$), can be expressed by the log-law model described by Eq. (24) assuming ideal conditions, i.e. the uniform height of roughness field and the neutrally stable atmosphere when thermal gradient is weak or absent.

$$u(z) = \frac{u_*}{\kappa} ln\left(\frac{z}{z_0}\right) \tag{26}$$

Eq. (26) is used for wind speeds greater than 10 m/s. For low wind speeds, the influence of thermal gradient both for unstable and stable atmosphere should be taken into account. The modified logarithmic formula can be found in [27].

The 10-min mean wind velocity measured at a meteorological station (usually at the level of $z = 10\ m$ above the ground) for upwind surface roughness $z_{0m}$ can be transformed to any other location described by upwind surface roughness height $z_{0s}$ through similarity of the wind speed at the gradient height for all terrain types [27]. Hence, the gradient velocity takes the same value for both locations and can be expressed by Eq. (27):

$$v_g = \frac{u_{*m}}{\kappa}\left[ln\left(\frac{u_{*m}}{fz_{0m}}\right) - A_u\right] = \frac{u_{*s}}{\kappa}\left[ln\left(\frac{u_{*s}}{fz_{0s}}\right) - A_u\right] \tag{27}$$

where $u_{*s}$ is friction velocity at the building site (m/s) and $u_{*m}$ is friction velocity at the meteorological station (m/s).

| Types of surface roughness | Height $z_0$ (m) | Category |
|---|---|---|
| Calm open sea, water area | 0.0001 | I |
| Sand surface (smooth) | 0.001 | I |
| Snow surface | 0.003 | I |
| Bare soil | 0.005 | I |
| Airport runway area, mown grass | 0.01 | I |
| Farmland with very few buildings, trees, etc. | 0.03 | I |
| Farmland with open appearance | 0.05 | R |
| Farmland with closed appearance | 0.1 | II |
| Many trees and bushes | 0.2 | II |
| Shelter belts | 0.3 | II |
| Suburbs | 0.5 | II |
| City, forest | 1.0 | III |

**Table 1.** Roughness height for different types and categories of surfaces, acc. to Swedish Code [24].

The friction velocity at the meteorological station $u_{*m}$ is computed from Eq. (28), which has been derived on the basis of Eq. (27), by substituting the friction velocity at the site with friction velocity at the meteorological station.

$$u_{*m} = \frac{u_m(z)\kappa}{ln\left(\frac{z}{z_{0m}}\right)} \tag{28}$$

where $u_m(z)$ is 10-min mean wind speed measured at the meteorological station at the height $z$. The mean wind velocity $u_s(z)$ at the site and at the height $z$ characterised by upwind surface roughness $z_{0s}$ can be estimated from Eq. (26):

$$u_s(z) = \frac{u_{*s}}{\kappa} ln\left(\frac{z}{z_{0s}}\right) \tag{29}$$

The ratio between wind velocity at the site and the wind velocity measured at the meteorological station denoted as $\eta$ is a function of the surface roughness $z_{0m}$ and $z_{0s}$:

$$\eta = \frac{u_s(z, z_{0s})}{u_m(z, z_{0m})} \tag{30}$$

It can be shown that non-linear relationship $\eta(u_m)$ can be approximated with errors of order of 7% or less by a constant factor $\eta$ for specified surface roughness at the building site. As the surface roughness appears in an implicit form in the expression for wind velocity (Eq. (29)), an analytical expression is not available. Instead, values of the factor $\eta$ have been computed for different combinations of the surface roughness at the site and at the meteorological station (**Table 2**).

Simple wind transformation between categories of roughness is possible for $z < 20z_0$ [28]. Thus, for $z = 10\ m$, the transformation is valid for $z_0 < 0{,}5\ m$. In the case of non-homogeneous upwind terrain, implementation for multiple roughness changes is required [26].

Values of wind speed measured at the meteorological station can be transformed using Eqs. (27)–(29). The statistical parameters of wind speed evaluated for the site, i.e. the mean value $\mu_{u_s}$ and the standard deviation $\sigma_{u_s}$, can be easily evaluated on the basis of the mean wind speed measured at the meteorological station $\mu_{u_m}$ and the standard deviation of the wind speed measured at the meteorological station $\sigma_{u_m}$, since the wind speed at site is related the wind speed measured at the meteorological station by simple Eq. (30). Hence:

| | $z_{0m} = 0.01$ | $z_{0m} = 0.05$ | $z_{0m} = 0.3$ |
|---|---|---|---|
| $z_{0s} = 0.01$ | 1.00 | 1.16 | 1.53 |
| $z_{0s} = 0.05$ | 0.86 | 1.00 | 1.31 |
| $z_{0s} = 0.3$ | 0.66 | 0.76 | 1.00 |

**Table 2.** Values of the ratio η corresponding to different roughness conditions.

$$\mu_{u_s} = \eta\mu_{u_m} \text{ and } \sigma_{u_s} = \eta\sigma_{u_m} \tag{31}$$

Concluding, the shape parameter $\lambda$ characterising the distribution of wind speed at the site of the building remains the same as the shape parameter $\lambda$ for the meteorological station. The scale parameter for the distribution of wind speed at the site is equal to $\eta c$ (Eq. (23)).

Hence, probability density function of wind speed transformed to building location is given by 2p–Weibull probability model:

$$f(v; \eta c, \lambda) = \frac{\lambda}{\eta c}\left(\frac{v}{\eta c}\right)^{\lambda-1} \exp\left\{-\left(\frac{v}{\eta c}\right)^{\lambda}\right\} \tag{32}$$

where $c$ is a scale parameter and $\lambda$ is a shape parameter of the PDF of wind speed measured at the nearest meteorological station.

Modelling of microclimate around the structure takes into account the influence of structure form, orientation and the quality of the surrounding. Usually, the effect of wind pressure on the façade is estimated with the help of the tabulated values of wind pressure coefficients. In the analysed case pressure differences across the six building components on the structure were measured.

### *4.3.2. Air flow through the building envelope (influence of wind and temperature)*

Some building performance aspects are dependent on the wind-structure interaction. Wind together with temperature difference causes airflow through building envelope.

**The probability distribution model of external temperature** depends on the specific geographical region. For temperate regions characterised by four seasons evenly distributed over the year, the normal (Gaussian) model with probability density function $\varphi(T; \mu_T, \sigma_T)$, given by Eq. (33), can be used for 1-h mean external temperature at "a random time" [29, 30]:

$$f(T; \mu_T, \sigma_T) = \frac{1}{\sigma_T\sqrt{2\pi}} exp\left\{-\frac{1}{2}\left(\frac{T-\mu_T}{\sigma_T}\right)^2\right\} \tag{33}$$

Also the full-scale measurements carried out near Gothenburg indicate [20] that the outdoor temperature can be approximated by the normal distribution.

Climatic data consist of 40-year record of observations made on meteorological stations at the airport in Säve, near Gothenburg. External temperature at the building site has been assumed to be equal to the temperature measured at the meteorological station, and its randomness is modelled by the normal distribution with the mean value of 11.1 and the standard deviation of 6.1 as shown in **Figure 5**.

Temperature difference across the building envelope is also described by the normal PDF but shifted towards positive values by the average value of internal temperature.

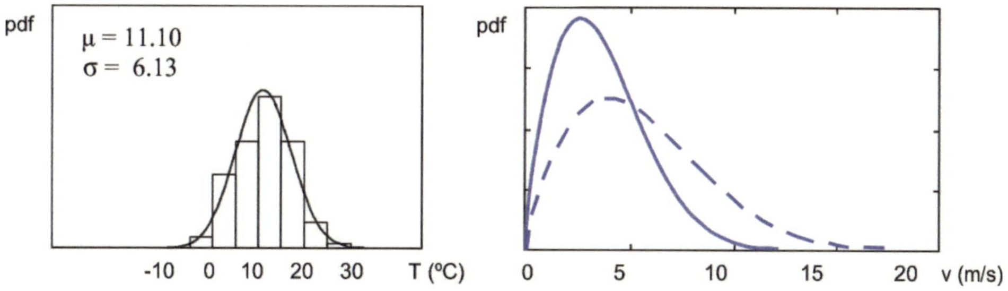


**Figure 5.** Normal PDF of ext. temperature T (*C) (left) and PDF of wind speed (m/s) for data coming from the Säve meteorological station (dashed line) and for local wind (solid line).

**The Weibull probability density function** for a local wind speed has been evaluated on the basis of 10-min mean values of wind speed measured at meteorological station (see **Figure 5**). The meteorological station is located at the airport with assumed surface roughness 0.01. The ratio between wind velocity at the site and the velocity measured at the meteorological station has been calculated and is equal to 0.66 (**Table 2**). Probabilistic models of local wind speed together with wind speed measured at the meteorological station are given in **Table 3**.

The probability density function of the random function *ACH* (**Figure 6**) has been evaluated using FORM approach (Eq. (22)). Probabilistic inference leads to the conclusion that the randomness of *ACH* is best described by the log-normal distribution with the mean value of 0.73 and the standard deviation of 0.38. Mean value and standard deviation are denoted, respectively, by $\mu$ and $\sigma$. The PDF of the air change rate due to stack effect $ACH_s$ and the PDF of air change rate due to wind $ACH_w$ are also shown in **Figure 6**. Randomness of air change

| | Mean value | Standard dev. | Scale parameter | Shape parameter |
|---|---|---|---|---|
| Meteo | 5.65 | 3.30 | 6.35 | 1.77 |
| Local | 3.73 | 2.18 | 4.19 | 1.77 |

**Table 3.** Stochastic parameters of the wind speed.

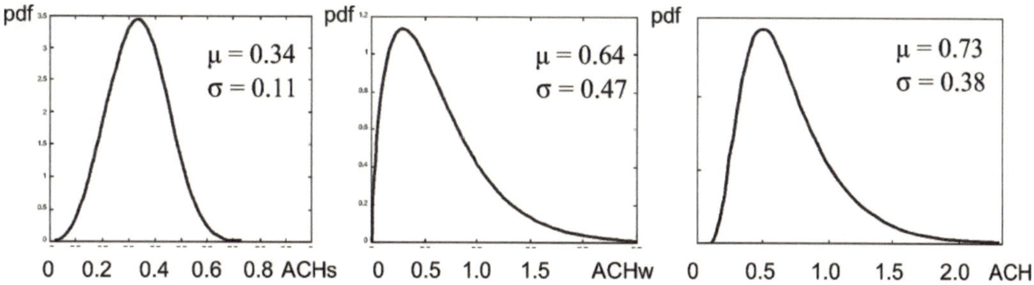


**Figure 6.** The probability density function for $ACH_s$ (left), $ACH_w$ (middle) and *ACH* (right) established with the help of FORM analysis.

rate due to stack effect can be described by the normal distribution whereas due to wind by the Weibull distribution skewed to the right.

*4.3.3. Sensitivity analysis of the probabilistic variability of air change rate with respect to the variability of wind and temperature*

Dependence of *ACH* on the mean values of input variables follows the trends showed by sensitivity indices for individual variables (see **Figures 7–9**) [31]. For the values of *ACH* above 1.0, where $-\alpha_{\Delta T}$ approaches 0 and $-\alpha_v$ is equal to 1, the changes of reliability indices are dependent almost only on the changes of wind speed. Concluding, the wind velocity and temperature difference contribute significantly to the variability of the air change rate with sensitivity indices up to 0.8 for ΔT (for lower *ACH*) and up to near to 1 for wind speed (for higher *ACH* (**Table 4**)).

Sensitivity of *ACH* distribution with respect to mean values and standard deviations of input variables leads to the following conclusions: (1) strong dependence on wind variation, (2) temperature difference variations affect only low values of *ACH* (up to 0.4), (3) variations of $\Delta T$ affect the lowest values of the *ACH* distribution, and (4) variations of the wind speed are significant for performance studies of *ACH* within the whole range of wind speed values.

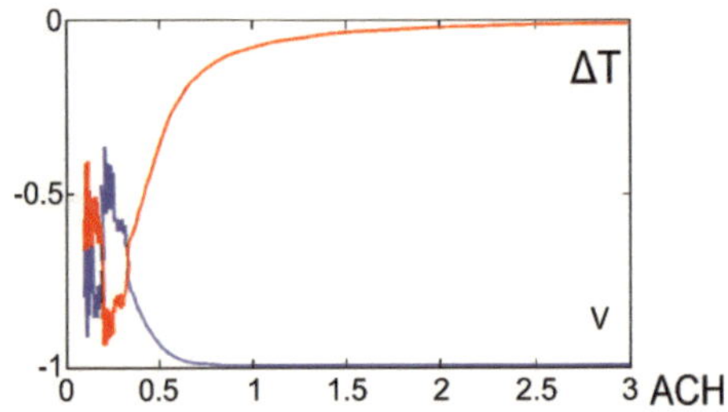


**Figure 7.** Course of sensitivity index $\alpha$, for variables $\Delta T$ and $v$.

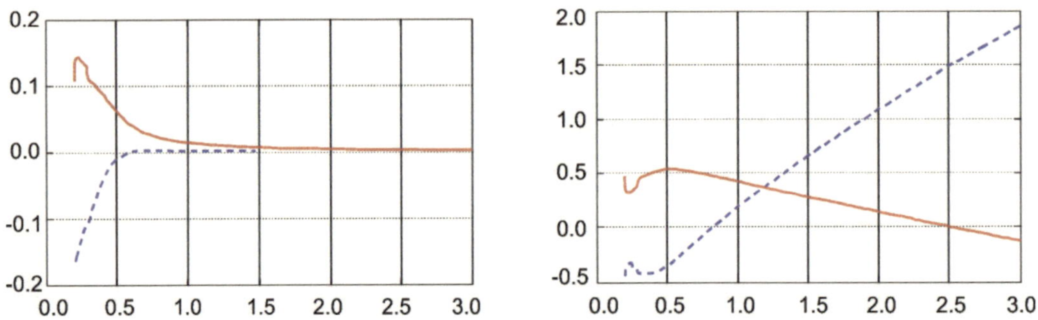


**Figure 8.** *ACH* sensitivity to the $\mu_{\Delta T}$ (solid) and the $\sigma_{\Delta T}$ (dashed) (left) and *ACH* sensitivity to the $\mu_v$ (solid) and the $\sigma_v$ (dashed) (right).

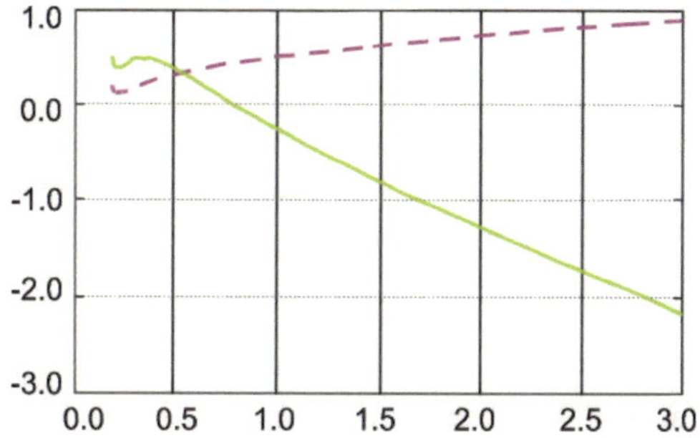


**Figure 9.** *ACH* sensitivity to scale (dashed) and shape (solid) parameter of Weibull distribution of wind speed.

| | ACH = 0.32 | ACH = 0.64 | ACH = 3.0 |
|---|---|---|---|
| $\alpha_{\Delta T}$ | 0.8 | 0.4 | 0.0 |
| $\alpha_v$ | 0.6 | 0.8 | 1.0 |

**Table 4.** Some results from **Figure 7**.

**Figure 9** shows the measure of sensitivity of the PDF of *ACH* with respect to scale or shape parameter of the Weibull distribution of wind speed for consecutive values of air change rate. The changes of shape parameter are the most important for the distribution of air change rate, especially for the threshold values close to the tail of distribution.

### 4.4. Probabilistic modelling of airflow-dependent thermal transmittance

For lightweight timber frame with mineral wool filling, the dependence of the thermal properties of building components on air infiltration is well acknowledged. An example is so-called dynamic wall [32], specially designed to save energy. In such a wall, the ventilation air passes through the insulation exchanging heat with a porous material reducing its conduction heat loss. The air entering the building is preheated by the conduction heat of the insulation (infiltration case), or the air leaving the building heats up the insulating material (exfiltration case) [33]. In the case of dynamic wall, the thermal transmittance becomes the most interesting parameter that can vary with the climatic data. Dynamic wall as a natural heat exchanger is a future of high-performance housing. The interaction between thermal transmittance and airflow through the components should be considered while calculating heat loss through a building envelope. A modelling approach based on probabilistic methods is proposed in [34].

Probabilistic model of **dynamic *U* value** takes into account only some of the uncertainties related to the properties of the thermal insulation described by the thermal transmittance $U^0$, the climatic load and the internal load coming from the building installations and occupants' behaviour (ventilation strategy). The model described by Eq. (34) can be used to estimate a probability distribution of the dynamic *U* value of the building envelope consisting of *i-th* elements with total area of $A_{tot}$:

$$U = \frac{1}{4}\sum_{i}^{n} Nu_i U_i^0 A_i \tag{34}$$

The Nusselt number $Nu_i$ is equal to 1 for the element without convection flow. In general, the value of the Nusselt number depends on the velocity of the airflow through the insulation, the direction of the flow and the thickness and the density of the insulation.

The example of approximation of the probability density functions of a dynamic $U$ value has been carried out with the help of FORM techniques. PDF of dynamic $U$ value has been evaluated using FORM sensitivity analysis (see Section 3.1.1). It depends on statistical parameters of the joint distribution of two random variables: thermal transmittance $U^0$ (varying with the temperature) and wind as well as buoyancy-driven airflow in terms of air change rate $ACH$ (see **Figure 10**). It has been assumed that stochastic information is limited to the parameters of marginal probability density functions of those variables and the correlation coefficient between them.

Probability density functions of thermal transmittance depend on the direction of the airflow through the envelope as well as on the probability model of the air change rate. Respectively, to the contribution of the natural forces (wind, temperature) and mechanical forces, different probability distributions (normal, log-normal, Weibull and gamma) can be fitted to model randomness of the air change rate [35]. In general, the probability density functions of the dynamic $U$ value are skewed to the left—in the case of infiltration—and are skewed to the right, in case of exfiltration. The specific character of the relationship between Nusselt number and air change rate may explain these results. For the case of infiltration, the best fit according to the Kolmogorov-Smirnov test has been obtained for the Weibull distribution, while for the exfiltration case, the three-parameter gamma (or alternatively Gumbel) distribution has been obtained (see **Figure 10**).

The model could be further developed to include uncertainties due to other mechanisms and factors, e.g. influence of wind or radiation on external heat transfer coefficient or the influence of non-homogeneity of the material characteristics.

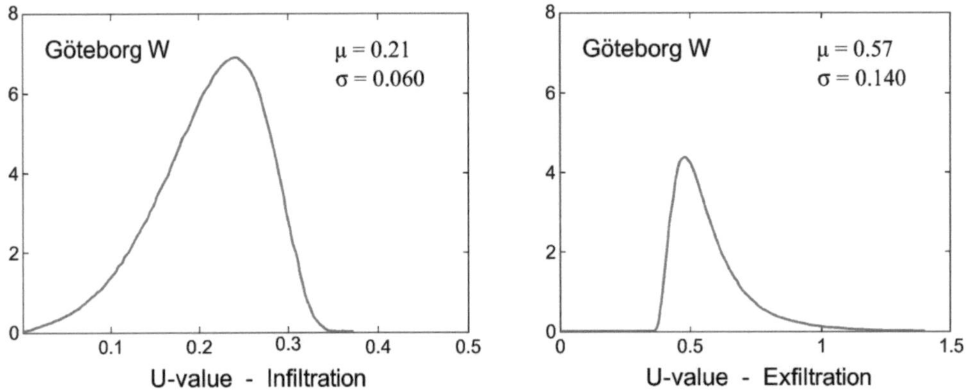


**Figure 10.** Probability density functions of dynamic U value for the cases of infiltration (left) and exfiltration (right) approximated for the building located near to Gothenburg for western winds.

The probabilistic model for estimation of **heat loss** accounting for interactions between ventilation and transmission heat losses has been presented in [20, 33]. The model predicts the probability density function of the heat loss distribution over a specified period of time (e.g. a heating season) on the basis of the design parameters of the house, temperature characteristics of the site as well as the air change rate due to mechanical or natural ventilation. The probability of heat loss exceeding certain number of kW can be compared for different design options concerning various ventilation strategies (natural or/and mechanical ventilation) and various transmittance properties (tight envelope contra dynamic wall) of the building envelope. Hence, rational engineering decisions promoting low-energy solution contributing to climate change mitigation can be taken into account in the design process.

## 5. Conclusions

Risk analysis together with appropriate tools can support building design strategies concerning climate change mitigation and adaptation. Lower levels of uncertainty can be handled by means of risk analysis based on system's risk or reliability estimations. In the case of higher order of uncertainties [36], other strategies could be developed based on the concept of resilience.

Risk analysis of building performance enables the selection of the best design based on comparison of probabilities of undesired performance estimated for alternative design solutions. Systemic approach gives opportunity to identify important relationships between variables. For example, air infiltration as a result of climate/structure interaction may be a significant variable in the thermal performance of building envelope. However, in order to handle the whole complexity of the real system, multivariable decision models for different design solutions should be further developed.

The examples of dynamic *U* values resulting in the different characters of distribution models for the cases of infiltration and exfiltration show that the probabilistic methods and tools can be effectively used to establish the probabilistic characteristics typical for the combination of the important variables influencing climate-structure interaction.

The sensitivity measures are important in the case of risk or reliability-based design. Sensitivity analysis of the distribution of a response variable (random function) with respect to the basic random variables and its parameters is straightforward for FORM methodology, whereas it is not easy in the case of the Monte Carlo simulation. The results of case studies show that the air change rate distribution depends on the temperature difference $\Delta T$ and the wind velocity significantly. Dependence of PDF model of *ACH* on the mean value of input variables is similar in the trends for both studied variables: temperature and wind speed. Sensitivity analysis of *ACH* probability distribution model to standard deviations of input variables shows the high contribution of wind speed and limited to low values of *ACH* (up to 0.4) influence of temperature difference.

Approximate transformation of wind speed data from the meteorological station to a specific location, where analysed building is situated, can be carried out by multiplying wind speed by a constant factor $\eta$ (with 7% error or less), established for a specific ranges of roughness conditions. The transformation of the probabilistic model of 10-min mean wind speed from meteorological station to the probabilistic model of hourly mean speed for the site of the building results in change of the scale parameter, while the shape parameter remains the same. The form of PDF for *ACH* as well as reliability index is sensitive to the value of the shape parameter of the Weibull distribution of wind speed and much less sensitive to the scale parameter. Hence, the transformation of probabilistic model of wind speed to the local site seems to be robust for the analysed case.

As it was shown, the sensitivity analysis has helped to understand the relationships between model inputs. It can also help to test the model outcome in terms of its robustness in the presence of uncertainty. The probabilistic risk analysis along with the effective tools for sensitivity analysis can be used to support design decisions and also to develop better models for evaluation of building performance.

## Author details

Krystyna Pietrzyk[1*] and Ireneusz Czmoch[2]

*Address all correspondence to: krystyna.pietrzyk@chalmers.se

1 Department of Architecture and Civil Engineering, Chalmers University of Technology, Gothenburg, Sweden

2 Faculty of Civil Engineering, Warsaw University of Technology, Warsaw, Poland

## References

[1] United Nations. Sustainable Development Goals. Available from: http://www.un.org/sustainabledevelopment/energy/

[2] Booth AT, Choudhary R, Spiegelhalter DJ. Handling uncertainty in housing stock models. Building and Environment. 2012;**49**:35-47

[3] Pietrzyk K. Quantification of building/environment system performance. In: The World Sustainable Building Conference (SB08), Vol. 2; Melbourne, Australia; 2008. pp. 501-506

[4] Bedford T, Cooke R. Probabilistic risk analysis. Foundations and Methods. United Kingdom: Cambridge University Press; 2001

[5] Rausand M, Høyland A. System Reliability Theory: Models, Statistical methods, and Applications. 2nd ed. Hoboken: Wiley; 2004

[6] Corotis RB. Risk and risk perception for low probability, high consequence events in the built environment. In: Haldar A, editor. Recent Developments in Reliability-Based Civil Engineering. Singapore: World Scientific Publishing Co; 2006

[7] Cardona OD. The need for rethinking the concepts of vulnerability and risk from a holistic perspective: A necessary review and criticism for effective risk management. In: Bankoff G, Frerks G, Hilhorst D, editors. Mapping Vulnerability: Disasters, Development and People. London: Earthscan Publications; 2004. pp. 37-51

[8] CUP (Central Unit on Procurement). No. 41 Managing Risk and Contingency for Construction Work. C.U.P. Guidance. London: HM Treasury; 1997

[9] Misra KB. Reliability engineering: A perspective. In: Misra KB, editor. Handbook of Performability Engineering. London: Springer; 2008

[10] Fisher BS et al. Ch. 3: Issues related to mitigation in the long-term context. In: Contribution of Working Group III to the Fourth Assessment Report of the Intergovernmental Panel on Climate Change (IPCC AR4 WG3); 2007 (3.5 interaction between mitigation and adaptation, in the light of climate change impacts and decision-making under long-term uncertainty)

[11] IPCC. Summary for policymakers. Climate Change 2007: Working Group III: Mitigation of Climate Change, Table SPM.3, C. Mitigation in the Short and Medium Term (until 2030) (IPCC AR4 WG3); 2007

[12] Climate Action. European Commission https://ec.europa.eu/clima/policies/adaptation_en

[13] Pietrzyk K. Risk management of windows performance. Energy Procedia. 2015;**78**:2476-2481

[14] Czmoch I. Influence of Structural Timber Variability on Reliability and Damage Tolerance of Timber Beams [Doctoral Thesis]. Sweden: Division of Structural Mechanics of Luleå University of Technology; 1998

[15] Pietrzyk K, Hagentoft C-E. Reliability analysis in building physics design. Building and Environment. 2008;**43**(4):558-568

[16] Rackwitz R, Fiessler B. Structural reliability under combined load sequences. Computers and Structures. 1978;**9**:489-494

[17] Der Kiureghian A, Liu P-L. Structural reliability under incomplete probability information. Journal of Engineering Mechanics Division. 1986;**112**:85-104

[18] Haldar A, Mahadevan S. Probability, Reliability and Statistical Methods in Engineering Design. John Wiley & Sons; 2000

[19] Pietrzyk K. A systemic approach to moisture problems in buildings for mould safety modelling. Building and Environment. 2015;**86**:50-60

[20] Pietrzyk K. Probabilistic modelling of air infiltration and heat loss in low rise buildings [PhD thesis]. Gothenburg, Sweden: School of Architecture, Chalmers University of Technology; 2000

[21] Pietrzyk K, Hagentoft C-E. Probabilistic analysis of air infiltration in low-rise buildings. Building and Environment. 2008;**43**(4):537-549

[22] Etheridge D, Sandberg M. Building Ventilation: Theory and Measurements. John Wiley & Sons; 1996

[23] Van der Hoven I. Power spectrum of horizontal wind speed in the frequency range from 0.0007 to 900 cycles per hour. Journal of Meteorology. 1957;**14**:160-164

[24] Handa K. Estimation of mean wind speeds for different surface roughness. Series on Designing for Wind. Report No. 1. Sweden: Department of Architecture, Chalmers University of Technology; 1996

[25] Simiu E, Scanlan RH. Wind effect on structures fundamentals and applications to design. Hoboken: John Wiley & Sons (Wiley-Interscience Publication); 1996

[26] Cook N. The Deaves and Harris ABL model applied to heterogeneous terrain. Journal of Wind Engineering and Industrial Aerodynamics. 1997;**66**:197-214

[27] Azad RS, editor. The Atmospheric Boundary Layer for Engineers. Berlin: Springer Verlag; 1993

[28] Wieringa J. Updating the davenport roughness classification. Journal of Wind Engineering and Industrial Aerodynamics. 1992;**41-44**:357-368

[29] Kaczmarek Z. Statistical Methods in Hydrology and Meteorology (in Polish). Warszawa: Wydawnictwa Komunikacji i Lacznosci; 1970

[30] Guyot G. Physics of the Environment and Climate. Paris: John Wiley & Sons; 1998

[31] Pietrzyk K, Czmoch I. Sensitivity analysis of air infiltration through the building envelope to the stochastic characteristics of wind speed and air temperature. In: Proceedings of the COBEE (2nd International Conference on Building Energy & Environment); USA. 2012

[32] Anderlind G, Johansson B. Dynamic insulation. In: A Theoretical Analysis of Thermal Insulation through which a Gas or Fluid Flows. Stockholm: The Swedish Council for Building Research; 1983

[33] Pietrzyk K. Thermal performance of a building envelope—A probabilistic approach. Journal of Building Physics. 2010;**34**(1):77-96

[34] Pietrzyk K, Hagentoft C-E. Probabilistic modelling of dynamic U-value. In: Proceedings of the Thermal Performance of the Exterior Envelopes of Buildings IX; American Society of Heating, Refrigerating and Air-Conditioning Engineers, Inc. (ASHRAE); Florida, USA. 2004

[35] Pietrzyk K. Probability-based design in ventilation. The International Journal of Ventilation. 2005;**4**(2):143-156

[36] Walker WE, Lempert RJ, Kwakkel JH. Deep uncertainty. In: Gass SI, Fu MC, editors. Encyclopedia of Operations Research and Management Science. USA: Springer; 2013. pp. 395-402

# Machinery Safety Requirements as an Effective Tools for Operational Safety Management

Hana Pacaiova

Additional information is available at the end of the chapter

http://dx.doi.org/10.5772/intechopen.71152

**Abstract**

Free circulation of goods is the major pillar of the single united market of European Union's (EU) member states and main motivated power of competitiveness and economical acceleration in EU. By using the legislative were defined base requirements on goods and also high level of authorized interest protection of goods users. The main changes in H&S (Health and Safety) management and also in Machinery safety started after implementation the Framework Directive 89/391/EEC and Directive 89/392/EEC in 1989 year. Directive 89/391/EEC implements systematical tools in H&S management as: H&S politics, Risk Management, education requirements, review activities, employee's involvement in H&S procedures. Directive 89/392/EEC, on the present time 2006/42/EC directive, implements for machinery producer or its contractor duties to access risk for each stage of machinery life cycle and implement adequate measures. These legislative requirements had changed all procedures and rules, which were used in H&S area.

**Keywords:** safety management, risk assessment, equipment criticality, maintenance, safety integrity

## 1. Introduction

Legislation defines a framework of organization operations and their fulfillment is a part of organization's business policy. Accepting customer's requirements also means fulfillment of legal requirements (e.g. laws, standards, regulations) [1, 2].

Risks management is a basic tool for demonstration of meeting the requirements in different areas of organization management (e.g. occupational health and safety, accidents prevention, critical infrastructure, dangerous substances transportation, environmental or financial requirements). Management of the organization often times is kind of "lost" while determining effective economical actions to be able to follow the required legal frameworks of a business environment or achieve its own and more difficult goals. Benefits of the decision-making process, their reliability and efficiency are determined by a risk assessment analysis, which is increased by improper applied processes, risk assessment methods and a measurements selection for their management [1, 3, 4].

ISO 31000:2009 standard allows us to globally understand the risks assessment versus "unwanted losses" on an integrated level of all the management activities. However, this integrated management requires specific processes and methods of the risks management, which is derived from the system/object properties, risks assessment goals and processes management level in the organization [3].

## 2. Safety legislative requirements

Risks assessment is a basic requirement of technical systems safety and occupational health and safety (OHS) management.

### 2.1. Occupational health and safety management

Human legal requirement is a basic for assessing the safety level at work by meeting the minimal requirements which is defined by Council directive 89/391/EEC ("Framework Directive"); on the introduction of measures to encourage improvements in the safety and health of workers at work.

The scope of this Directive is defined for employers and employees in all sectors of the productive and non-productive sphere [1].

The Directive defines general requirements for prevention for an employer, who is obliged to apply the general principles of the prevention when implementing the measures necessary to ensure the safety and health protection at work, including the information, education and organization of work and tools. These principles include [1]:

- Exclusion of the hazard and the possible resulting risk.
- Risk assessment that cannot be excluded, especially while selecting or using working tools, materials, substances and methods.
- Implementation of the measures to eliminate hazards at the site of their occurrence.
- Prioritization of collective protective measures against individual protective measures.
- Planning and implementing a policy of the prevention through the safety working tools, technologies. methods, improvement of the working conditions with regard to working environment factors and through social measures.

The Directive application can be defined as follows:

- It is applied to all public and private areas, such as industry, agriculture, commerce, services, education, culture, leisure, and so on.

- It does not concern areas, where specific public services and activities are involved, e.g. military and police activities or civil protection activities that may conflict with its requirements.

A number of specific directives (see **Figure 1**) have been adopted to implement the requirements of this Directive as individual directives within the meaning of Article 16 (1) of Directive 89/391/EEC in the following order (1. e.g. first individual Council Directive, etc.):

1. 89/654/EEC of 30th of November 1989 concerning the minimum safety and health requirements for the workplace.
2. 2009/104/EC (original 89/655/EEC) of 16th of September 2009 on the minimum health and safety requirements regarding the use of work equipment by workers at work.
3. 89/656/EEC of 30th of November 1989 concerning the minimum health and safety requirements for the use of personal protective equipment by workers at work.
4. 90/269/EEC of 29th of May 1990 on the minimum health and safety requirements for the manual handling of loads, especially where is a risk of injury to the lumbar spine of workers.
5. 90/270/EEC on minimum safety and health requirements for work with displaying units.

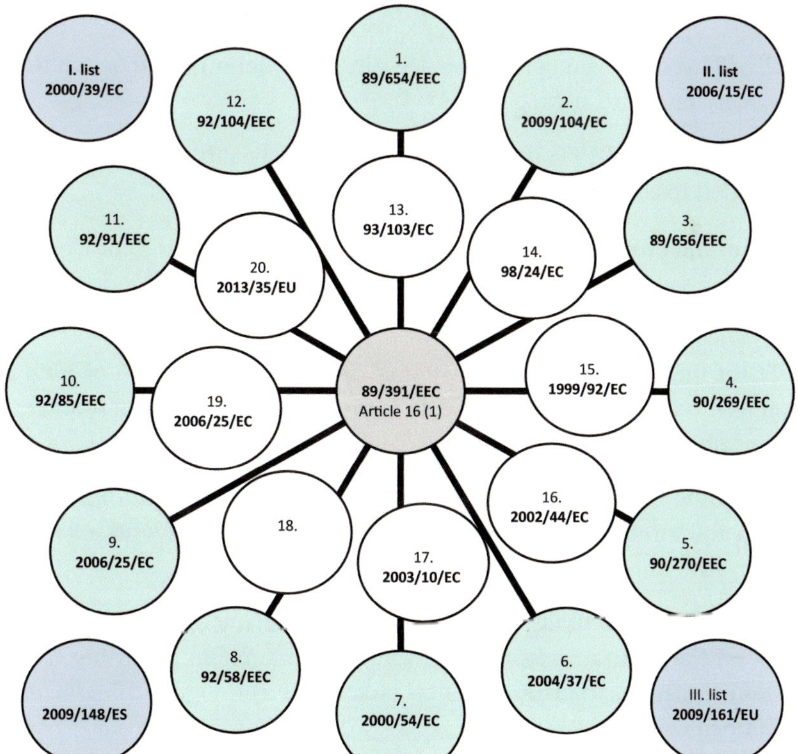


**Figure 1.** Relation between commission directive 89/391/EEC and individual council directives in accordance with Article 16 (1) [1].

6. 2004/37/EC of the European Parliament and of the Council of 29th of April 2004 on the protection of workers from the risks related to exposure to carcinogens and mutagens at work.

7. 2000/54/EC of the European Parliament and of the Council of 18th of September 2000 on the protection of workers from risks related to exposure to biological hazards at work, which is a consolidated directive of the previous Directives.

8. 92/57/EEC of 24th of June 1992 on the introduction of minimum safety and health requirements for temporary or site-changing buildings.

9. 92/58/EEC of 24th of June 1992 on the minimum requirements for the provision of safety and health signs at work.

10. 92/85/EEC of 19th of October 1992 on the introduction of measures to encourage improvements in the safety and health at work of pregnant workers and workers who have recently given birth or are breastfeeding.

11. 92/91/EEC of 3rd of November 1992 on the minimum requirements for improving the safety and health protection of workers in the extractive industry.

12. 92/104/EEC of 3rd December 1992 on the minimum requirements for improving the safety and health protection of workers on the ground and underground mining.

13. 93/103/EC of 23rd of November 1993 concerning the minimum safety and health requirements for work on board fishing vessels.

14. 98/24/EC of 7th of April 1998 on the protection of the health and safety of workers from the risks related to chemical factors at work.

15. 1999/92/EC of the European Parliament and of the Council of 16th of December 1999 on minimum requirements for improving the safety and health protection of workers potentially at risk from explosive environment.

16. 2002/44/EC of the European Parliament and of the Council of 25th of June 2002 on the minimum health and safety requirements regarding the exposure of workers to the risks arising from physical factors (vibration).

17. 2003/10/EC of the European Parliament and of the Council of 6th of February on minimum health and safety requirements regarding the exposure of workers to the risks arising from physical nuisance (noise).

18. Canceled Directive 2004/40/EC of the European Parliament and of the Council of 29th of April 2004 on the minimum health and safety requirements regarding the exposure of workers to the risks arising from physical factors (electromagnetic fields), note: replaced by 20th Council Directive.

19. 2006/25/EC of the European Parliament and of the Council of 5th of April 2006 on the minimum health and safety requirements regarding the exposure of workers to the risks arising from physical factors (artificial optical radiation).

**20.** 2004/35/EC of the European Parliament and of the Council of 18th of June 2013 on the minimum health and safety requirements regarding the exposure of workers to the risks arising from physical factors (electromagnetic fields).

Other significant directives on OHS, but not issued as individual directives under Article 16 (1) The Health and Safety Directives can be classified as:

- Commission Directive 2000/39/EC of 8th of June 2000, which establishes the first list of indicative occupational exposure limit values for the implementation of Council Directive 98/24/EC on the protection of the health and safety of workers from the risks related to chemical factors at work.
- Commission Directive 2006/15/EC of 7th of February 2006, which establishes the second list of indicative occupational exposure limit values for the implementation of Council Directive 98/24/EC and amending Directives 91/322/EEC and 2000/39/EC.
- Commission Directive 2009/161/EU of 17th of December 2009, which establishes the third list of indicative occupational exposure limit values for the implementation of Council Directive 98/24/EC and amending Commission Directive 2000/39/EC.
- Directive 2009/148/EC of the European Parliament and of the Council of 30th of November 2009 on the protection of workers from the risks related to exposure to asbestos at work.

### 2.2. Machinery safety

The Machinery Directive 2006/42/EC on the approximation of the laws for the Member States relating to machinery is intended especially for machinery suppliers. For the machinery operators, the rules required in accordance with Directive 2009/104/EC, which replaced Directive 89/655/EEC, the so-called "The second individual Directive within the meaning of Article 16 (1) of Directive 89/391/EEC on the minimum safety and health requirements for the use of working equipment by workers at work (**Figure 2**).

The Machinery Directive covers the use of all technical equipment, including mobile and lifting equipment. These devices must be regularly inspected and maintained to ensure their readiness and security.

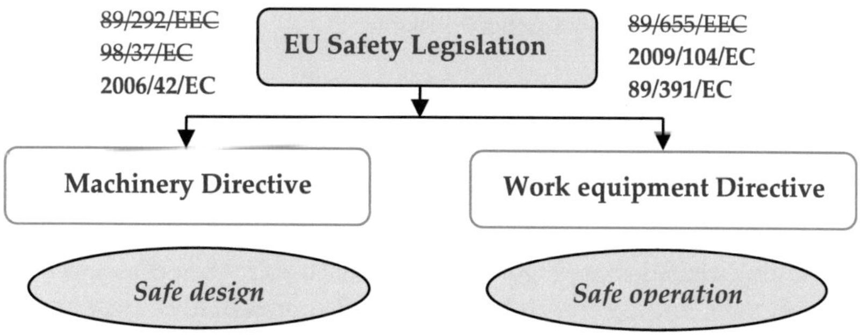


**Figure 2.** History of legislation on machinery safety and OHS.

The objective of this directive is to increase the level of equipment safety, with an emphasis on the analysis of possible risks in their intended operation and maintenance at the design and construction stage of the equipment. Emphasis is also placed on the creation of metering points for monitoring the status of equipment by the methods of the technical diagnostics already at the stage of the project. This creates conditions to prevent the occurrence of breakdowns and possible accidents by defining a real technical state, the impact of which would have an obvious effect on the safety and health of the operator or public. An important aspect is a detailed description of the operational regulations requirements in the native language of the country, where the technical equipment is in operation. Then, there are created standardized procedures for informing the operator of the existing hazards and residual risks arising from the operation of these devices [1, 5].

In accordance with Annex I, 1 Machinery Directive is a manufacturer of a machine or a complex technical device obliged to define the hazards, that arise during the operation of the machine, to estimate the consequences of possible injury or damage to health as well as the probability of their occurrence, and then determine and asses risks in order to take measures to minimize them. This is also connected with an obligation to provide machine user relevant information on residual risks.

These requirements make activities of machinery designers, engineers, manufacturers and users of machinery (including maintenance requirements—item 1.6) conditional upon them. Relevant activities must be conducted in accordance with risk management rules.

### *2.2.1. Integrated safety principle*

The Integrated Safety Principle is defined in Annex I to the Machinery Directive in five steps, as follows:

**I.** Devices must be designed and constructed in such a way that they are adapted to their function and can be operated, set and maintained, so people who use them are not exposed to the risks under the foreseeable conditions, also taking into account their reasonably foreseeable wrong use (e.g. operator error).

**II.** When selecting the most appropriate solutions, the manufacturer or his authorized representative, must apply the following principles in the following order:

- Eliminate or reduce risks as much as possible,
- Take the necessary measures to protect against risks,
- Inform users of the residual risks caused by the various shortcomings in the protective measures taken, notifying whether special training is required and determining any need to provide personal protective equipment.

**III.** When designing and constructing a machine device and when drawing up the instructions for use, the manufacturer or his authorized representative must assume not only the intended use of the machinery but also his reasonably foreseeable misuse.

http://dx.doi.org/10.5772/intechopen.71152

**IV.** Machine devices must be designed and constructed in such a way that include taking into account the limitations to which the operator is exposed as a result of the necessary or foreseeable use of personal protective equipment (PPE).

**V.** Machine devices must be supplied with all the necessary special equipment and accessories to enable them to be safely adjusted, maintained and used.

The objective of the taken measures must be exclusion of any risk for the machine life cycle, including the phases of transport, assembly, disassembly, decommissioning and disposal [5–7]!

The instruction manual must inform about residual risks, meaning that informs a user of the ways, in which the machine devices should not be used.

A **risk** (under the Machinery Directive) is defined as a combination of probability and severity of injury or injury to health that may result from a dangerous situation!

## 3. Basic principles of a risk assessment

In the risk assessment in the field of OHS, there are usually used simple methods based on the causal model of the accident (hazard → hazard situation → initiation → harm → loss). These methods are usually combined according to their use in the individual steps of the risk assessment algorithm (Brainstorming, Check-list, Risk matrix [1–3]).

The basic risk assessment algorithm is a structured logical sequence of steps (**Figure 3**) [1]. It does not matter whether it is a project, process, technology, device or a provided service. The analyzed system/object must be broken down into individual elements as is required to fulfill a defined task (activity, function). Each element is evaluated separately in terms of the possibility of endangering the target role (function). The probability or frequency, with which this hazard situation may occur at the time considered (duration of action), is the basis for the risk assessment together with the assessment of the severity (consequence) for the target function. In the financial sector, the risk is also declared positively (as ISO 31000 also accepts the concepts of opportunities), while the analysis of technologies and work activities is assessed only in relation to negative consequences [3].

Measures derived from the assessed risks are defined either by legislation (relevant directives for specific areas—hazards, such as work with display units, noise protection, vibration, etc.), or/and requirements resulting from the overall culture and the advancement of the organization's management to reduce the risk value to the lowest level possible (risk acceptability level) [1, 8].

Normally, risk assessment for OHS in organizations uses a "Risk matrix" (see **Table 1**), which is based on an assessment of the probability and consequence of the analyzed hazard that is determined from work activities [1, 9]. Emphasis is placed on a simple form of an evaluation and risk assessment and its understanding by all involved parties (employees, third parties, etc.).

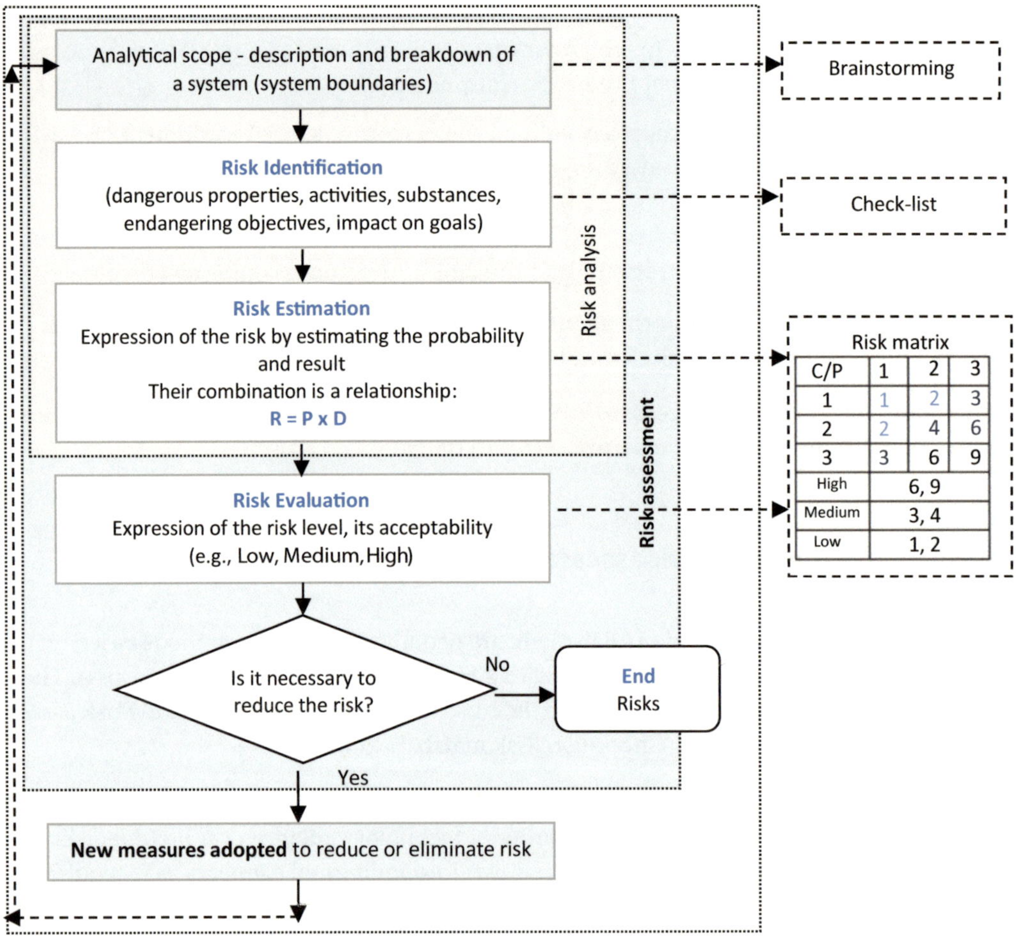


**Figure 3.** Basic algorithm for risk assessment and management.

| Consequence | Probability | | |
|---|---|---|---|
| | Low | Medium | Almost certain |
| Insignificant | L | L | M |
| Significant | L | M | H |
| Catastrophic | M | H | H |
| Risk level | L – low | M – medium | H – high |

**Table 1.** "Risk matrix" example.

It is essential to apply appropriate tools and procedures to meet legislative requirements. These are evolving and changing in terms of requirements for risk assessment, risk management and health and safety management at work. They can be broken down as follows:

### 3.1. Risk assessment

- Canceled standard EN 1050 Machines Safety: Principles of risk assessment (1998- replaced by EN ISO 14121-1: 2007, nowadays canceled too).
- IEC 60300-3-9 Reliability Management: Part 3, Section 9: Risk Analysis of Technical Systems (1995).
- EN ISO 12100 Machines Safety: General principles of machine design; Risk Assessment and Reduction (2010), Consolidation of ISO 12100-1 and ISO 12000-2 requirements.
- Canceled standard ISO 14121-1: Machines Safety: Risk Assessment. Part 1: Principles; TNI/ ISO/TR 14121-2: Practical Guides and Examples (2007).

### 3.2. Risk management

- ISO 31000 Risk management—principles and implementation guides (2009).
- ISO 31010 Risk management—risk assessment techniques (2009).

### 3.3. Health and safety management systems

- OHSAS 18001:2007: Occupational Health and Safety Management System—Requirements.
- OHSAS 18002:2000: Occupational Health and Safety Management System—Implementation guide OHSAS 18001.
- New standard ISO 45001:2017, which is used to transform and complement the requirements of OHSAS 18001 internationally.

## 4. Machinery risk assessment standards and tools

The safety issues of machines and machine devices are devoted to a number of harmonized standards which have their hierarchy [1] (see **Figure 4**).

**Type A standards**: safety standards, providing basic concepts and principles for design, construction and general considerations that can be applied to all machine devices. Basic safety standards of Type A include for example EN ISO 12100.

**Type B standards**: safety standards that mostly take care of only one safety aspect or one type of safety device that can be used for a larger amount of machines. They are divided into: Type B1 standards, which are related to individual safety aspects (e.g. safety distances, surface temperatures, noise, etc.) and Type B2 standards for the relevant safety devices (e.g. different shields, pressure sensitive devices, two-hand control device, locking device, etc.).

**Type C standards**: safety standards for machines that define detailed safety requirements for a particular machine type or group of machines. They refer to related Type A and B

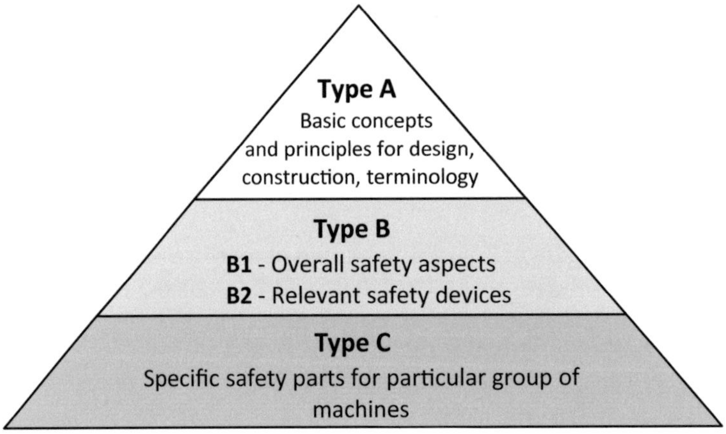


**Figure 4.** Hierarchy of standards for the machines and equipment safety.

standards or, if possible, also to other Type C standards and define safety requirements and identify the risks and priorities that are required. The principle is that the Type B and C standards cannot be repeated or verbally describe the text of the other standards to which they refer.

From a legal point of view, any product is safe, which is meeting the requirements of the relevant regulation or where no prescription for this product is meeting the standards requirements or corresponding to the state of scientific and technical knowledge known at the time of its placing on the market.

In general, the safety assessment rules for health and safety at work are based on the basic principle of meeting the requirements of the technical regulations and standards.

### 4.1. Requirements of EN ISO 12100

The requirements of directive 2006/42/EC support EN ISO 12100—defines the terminology and methodology used to achieve machine device safety. The purpose of this standard is to provide to the constructors a basic framework for designing safe machines. This standard has a historical development from the basic standard EN 292-1, 2, through EN 1050. Now it is replaced also EN ISO 14121-1 standard [1, 10].

The standard is principally structured to:

- Risk assessment, i.e. basic principle and hazards identification.
- Risk reduction, i.e. three-step method and measures.

This standard has a list of the potential hazards to be taken into account when designing a machine device (examples of hazards are part of Annex B, which is taken from the canceled standard ISO 14121-1. Hazards analysis must take into account the entire life-cycle of the machine—from its design, construction, manufacture, installation, operation and maintenance to its disposal.

### 4.2. Risk assessment and reduction strategy

This safety strategy—the risk assessment and risk management steps are defined as follows [1, 11, 12]:

**Step 1**: determination of machine boundaries, including intended use of the machine and consideration of its foreseeable misuse (e.g. operator errors),

**Step 2**: identification of hazard sources and hazardous situations,

**Step 3**: The risk estimation for each hazard and the resulting hazard situation,

**Step 4**: the risk evaluation and consideration of the necessary reduction by introducing measures,

**Step 5**: elimination of a hazard or risk reduction associated with hazards by applying appropriate measures (application of the so-called three-step risk reduction method).

The first three steps represent a process of **risk analysis**—a combination of specifications for determining the machine boundaries, hazard and hazard situations identification and risk estimation.

The overall procedure includes risk analysis and **risk evaluation** (fourth step). It is the process of **risk assessment**.

The **risk management** process is based on a risk assessment and proposes to take appropriate measures to implement and monitor their effectiveness.

*Note*: Risk assessment does not include a step of taking measures, only the steps of consideration—the classification of the estimated size of the risk according to the pre-selected scheme (e.g. the risk matrix) and the decision-making process "what to do with that now" based on the risk acceptance rate.

This risk management process is a basic and unchangeable process, and represents an iterative approach (ALARP—As Low As Reasonably Practicable), which the designer or constructor must observe in designing the machine, but also the user in managing workplace safety [1, 12].

The designer must consider designing the machine, all anticipated activities (even not expected ones during normal use of the machine), production must take into account possible risks in a machine manufacturing, the user (or the employer) must ensure the safety of the machine in the working environment.

#### *4.2.1. Step 1: determination of machine boundaries*

The purpose of this step is to understand the principles of machine operation, the conditions and the way it is used. Determining machine boundaries serves to identify sources of hazards, a description of possible hazard scenarios while performing the required activities (e.g. machine operator and maintenance, visit, or third-party activities performed at the working site), or predictable behavior when using the machine by unskilled workers. Also an appropriate procedure is to define the so-called functional machine structures for identifying dangerous elements on the machine. It can be, for example a control function, safety function, stability function, etc.

Procedures to determine the machine boundaries according to EN ISO 12100 standard [1]:

Usage limits (intended use and foreseeable misuse)

- Operating modes and preventive procedures, including manipulation with the machine when misused,
- The way and the place for the machine use (household, industry) by persons, their skills and the ability to use the machine,
- Expected level of qualification, experiences, education and capabilities of the concerned persons (a maintenance worker, an attendant, an apprentice or public),
- Other persons who may be at risk from the machine (other machines operation, administrative staff, visits).

**A.** Layout: range of motion, operating and maintenance area, relationship between the machine and power source.

**B.** Time limit: machine lifespan (parts), maintenance intervals.

**C.** Other boundaries: properties of the processed material, purity, environment (temperature, external conditions, etc.).

#### *4.2.2. Step 2: identification of hazards*

After determining machine boundaries, the basic step of the risk assessment is to identify the types of hazard situation depending on the hazard properties of the machine, taking into account each stage of the machine's life-cycle [1, 4, 8, 9, 11].

Account is also taken of the behavior of the operator [1, 11, 13], e.g.:

- Loss of control by the operator (e.g. manual or mobile machines),
- Improper behavior of the person in the event of failure of the machine, in the event of a breakdown or accident,
- Behavior resulting from lack of a concentration or inattention,
- Behavior resulting from the search for options beyond the prescribed procedure (instruction manual), the "least resistance way,"
- Behavior resulting from the effort to keep the machine running at all costs,
- Behavior of another group of people (children, people with disabilities).

EN ISO 12100 provides a description of 10 types of potential hazards (e.g. mechanical, electrical, thermal, noise, vibration, radiation, ergonomics, etc.), their potential sources and possible consequences. It is based on the requirements of the Machinery Directive.

Similarly, it is possible to proceed with identifying hazards in relation to the work being done at the workplace in order to assign appropriate personal protective equipment.

#### 4.2.3. *Step 3: risk estimation*

This is one of the most important risk assessment steps [1, 3, 9, 11]. The level of the risk reflects the severity of a hazardous situation and is dependent on the following parameters:

**a.** Consequence *C* or the severity of the hazard situation: the impact on health, property or environment,

**b.** Probability *P* of the damage occurrence that depends on:

- Exposure of the person to the hazard situation, Exposure time: *E*,
- Probability (or frequency) of occurrence of a hazard situation: $P_H$,
- Technical and human possibilities to prevent or limit the range of possible damage: *M* (measure).

The level of risk can be calculated as function of these parameters, using this formula:

$$R = f(E, P_H, M, C) \tag{1}$$

The risk assessment uses simple methods based on the expression of probabilities and consequences and on the risk evaluation, so-called "Risk matrix" (risk rating tool) [...].

Usually the level of risk is defined as combination of these parameters:

$$R = P \times C \tag{2}$$

Creating a Risk matrix as a tool for analysis and risk assessment requires establishing criteria for estimating probabilities and consequences (**Tables 2** and **3**).

For the risk assessor, the "common sense" principle must be applied to determine the range of the level of the assessed parameter (e.g. from 1 to 3).

#### 4.2.4. *Step 4: risk evaluation*

The risk matrix (see **Table 4**) can be created by the "ordinary" multiplication of the individual levels assigned to the probability and consequence. The number of levels of the estimated

| Probability | Level description of a probability | Level |
|---|---|---|
| Low | Low probability of event occurrence | 1 |
| Medium | An event can be expected with a higher probability | 2 |
| High | The probability of occurrence is almost certain | 3 |

**Table 2.** Description of the probability of occurrence of a hazardous event: *P*.

| Consequence | Level description of a severity/consequence | Level |
|---|---|---|
| Negligible | Small event impact range, minimal or no consequence, near-miss | 1 |
| Serious | Medium range of an event consequence - serious consequence, injury—occupational accident (e.g. from 3 days off work) | 2 |
| Very serious | Large range of an event consequence—very serious consequence, death or mass injury | 3 |

**Table 3.** Description of the consequence *C* or the severity of the hazard situation: *C*.

| Probability | Consequence | | |
|---|---|---|---|
| | Negligible | Serious | Very serious |
| Low | 1 | 2 | 3 |
| Medium | 2 | 4 | 6 |
| High | 3 | 6 | 9 |

**Table 4.** Risk matrix 3 × 3.

parameters determines the type of matrix, for example, 3 × 3, 4 × 5, 6 × 4, etc. Determining the number of levels depends on the depth, to which the risk assessor intends to specify the probability and consequence of a negative effect.

As can be seen from **Table 4**, the estimated risk sizes range from 1 to 9. In the next step, the risk (risk evaluation) needs to be evaluated, so for the assessor which level is high, medium, and low in severity level (e.g. acceptability) of the risk.

Values: 1–2 can be assigned to a low level, meaning small or low risk: L; from 3 to 4: medium level: M; from 6 to 9: high level: H.

For a better illustration, the Risk matrix can be adjusted more clearly, where the principle of so-called "traffic light" effect is applied, **Table 5** [1, 9, 11].

There is no binding rule to determine the level of a risk (e.g. H: high risk, M: medium risk, L: small or low risk), whether in qualitative, quantitative or semi-quantitative form. The applied methodology depends on the area of investigation (e.g. machine failure and its consequences) and data availability (e.g. monitoring machine failures) [6, 11].

| Probability | Consequence | | |
|---|---|---|---|
| | Negligible 1 | Serious 2 | Very serious 3 |
| Low 1 | L(1) | L(2) | M(3) |
| Medium 2 | L(2) | M(4) | **H(6)** |
| High 3 | M(3) | **H(6)** | **H(9)** |

**Table 5.** Risk matrix 3 × 3 "traffic light".

Important at this stage of the risk assessment is to ensure sufficient information, e.g. historical data about machine failures, near-misses, injuries, accidents, as well as opinions of the experts and practitioners in the investigated area or system.

Risk analysis can be done principally in two ways, applied in specific methods [1, 11]:

a. **Up-bottom approach** (deductive methods)—lead off from information based on statistics of accidents and other undesirable events, analysis of their causes and consequences. So it is based on the events that have already occurred.

b. **Bottom-up approach** (inductive methods)—proceeds from the examination of all hazards and consideration of ways in which damage can occur, meaning: from predicting the probabilities and consequences of a possible undesirable event.

The choice of these methods depends on the experience and knowledge of the team that deals with the assessment process. The inductive methods may have the advantage over deductive methods in a more advanced analysis of all possible hazards and hazard situations but on the other hand they may be more time consuming.

#### *4.2.5. Step 5: risk reducing measures*

Reducing the risk to the residual level is conditional on machines by following the **three-step method**—constructional measures excluding or limiting the risks; by installing the necessary protective systems and additional protective measure for those risks that could not be reduced or eliminated in the first step; by providing information on residual risks to the machine user (by providing the instructions for use) [1, 9, 11].

1. Step: **Custom Construction Safety**—this phase of the risk reduction is the most important. Even the most reliable protective systems and additional protective measures can fail during the life cycle of the machine.

2. Step: **Safety protection and additional protection measures**. Protective covers and protective devices must be used when their own construction safety has not adequately eliminated hazard and so does not sufficiently reduce the risk. In this case, another additional protective measure may be applied. Typical examples of protective measures include locking covers, light curtains, safety mats, two-hand control and activation switches. Additional protective measures are the devices which perform emergency stops, escape routes, equipment for manual start of certain parts during emergency stops, communication devices to make emergency calls, disconnection from the power source, handling devices (load lifting), and so on.

3. Step: **User manual**—information on safe use of the machine must be provided in the required quality, understandable language and to the extent that all information on the machine uses and its operating modes. They must inform and warn about the residual risk.

**Residual risk**—is a risk that remained after the adoption of the implemented measures (protective measures) so it can be manageable. It can describe protective or safety measures taken at the design stage or other additional measures taken by the user of the device, at the stage of its operation [1].

## 5. Process safety

Machine safety under the Machinery Directive requires an integrated approach to the safety. However, the machine is a complex construction that is not only mechanical or electrical, but often times it is a complex control unit whose reliable function affects not only machine safety but also the whole process [6, 12, 14]. For this reason, safety integration is understood as a requirement not only for the safety of the machine itself, but also for the safety of the whole process (IEC 61511). Standard IEC 61511 defines requirements for safety control systems of continuous technological processes and on the other hand IEC 61508 defines functional safety requirements for electrical/electronic/programmable electronic safety systems (**Figure 5**) [1, 5, 10].

The objective of ISO 13849-1 (Type B-1) is to provide guidance on the design and construction of control (safety) systems so the requirement of integrated security is ensured.

A designer—constructor while reducing a risk considers applying safety measures that contain one or more safety features. The parts of machine control systems that provide a safety function are called safety-related parts of the control system and labeled as SRP/CP SRP—Safety related parts; CP—control system). They may consist of hardware and software, but may not be part of the machine's control system.

### 5.1. Safety control systems

Safety control systems are designed to perform a safety function. It's the part of the control system (or the control system itself) that prevents the hazards. It could be said that it creates a barrier between hazards and hazards situation (e.g. shields). For these reasons, the safety system must work reliably, under all foreseeable circumstances.

The safety function is implemented by the machine components of the machine control system in such a way so it maintains the device (or bring it into a state) in a safe state with respect to the specific risk circumstances.

According to ISO 13849-1 standard, this is the function of a machine, whose failure can lead to an immediate increase of a risk.

The main task of the designer of the safety system is to avoid hazardous conditions and to prevent the possibility of an unintentional machine start.

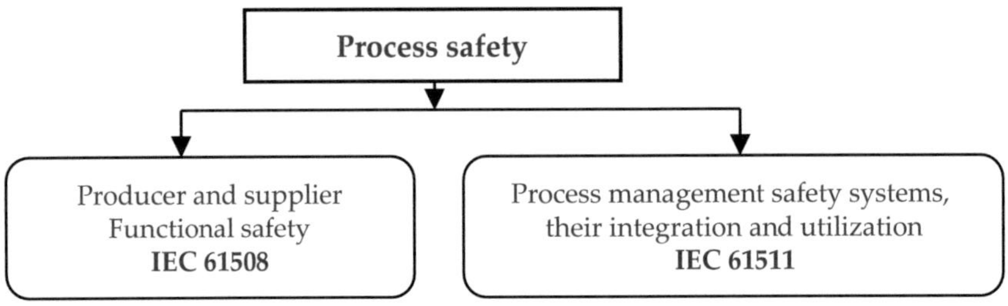


**Figure 5.** Relation between IEC 61508 and IEC 61511.

The safety feature may have several parts, e.g. for a protective cover it is possible to define it in three steps:

- When the cover is closed, the risks (for which the cover has been constructed) can not endanger the person,
- By opening the cover, the exposure to an operational risk must be excluded,
- Closing the cover does not restore the risk.

For safety systems, the use of "safety requirement or after safety requirement" are used as a result of their mode. An example of the requirement for the safety function is the interruption of the light curtain, the opening of the cover, where the operator may require to stop the machine parts or leave them without power if they had already been stopped after the safety function triggered [1, 5, 10].

The safety function is performed by the machine control system safety parts (elements). The safety function begins by sending a command and ends with a response (by doing an activity).

The safety system must be designed with a level of integrity that corresponds to the machine's risk level. Higher risks require a higher level of integrity so the safety performance is ensured. The machine safety system can be partitioned to the performance level of the capability to ensure the performance of the safety function or otherwise, the functional level of safety integrity.

### 5.2. Functional safety of control systems

Functional safety—it is a part of the overall safety that depends on the correct functioning of the systems or devices in responding to their inputs (stimulus) [1, 10, 15].

Functional safety is the identification of potential hazards based on the activation of protective or corrective devices or mechanisms to prevent a dangerous event or to reduce the level of its consequence.

According to IEC 61508 standard, an example of functional safety is, for example, an overheat protection device that uses a thermal sensor in the motor winding to disconnect the voltage before it is overheated and subsequently could occur a destruction. But, e.g. a special insulation, resistant to high temperatures, is not an example of functional safety, even though it provides protection against the same hazards as a thermal sensor. Similarly, the fixed door as an intrusion barrier does not have a characteristics of the functional safety feature, like on the other hand, door locked housing.

In order to achieve the functional safety, it is necessary to meet the requirements for:

- Safety function,
- Safety integrity.

Risk assessment is the basis for creating the functional safety requirements: risk analysis provides the basis for the safety function requirements and risk evaluation forms the basis for specifying the safety integrity, i.e. the levels of system properties!

### 5.3. Standards for the functional safety of control systems

The basic standards for the functional safety of the machine control systems are [1, 5, 10, 15]:

**a.** IEC/EN 61508: Functional Safety of Electrical, Electronic/Programmable Electronic Safety Systems (Part 1, 2 and 3). This standard is general, not limited to the field of a machinery and contains requirements that apply to the design of complex electronic and programmable control systems.

**b.** IEC/EN 62061: Machine Safety—Functional Safety of Safety-Related Electrical/Electronic/Programmable electronic control systems that are connected with the safety. This is, in fact, a specific implementation of the IEC/EN 61508 standard for the machinery. The requirements of this standard can be applied to system level design for all types of electrical control systems as well as to not very complex subsystems and devices.

**c.** EN ISO 13849-1: Machine Safety—Safety Parts of Control Systems. This standard provides requirements and guidance for designing, constructing and integrating safety parts of the safety control systems (safety-related parts), including the software design. For these components, there are specific characteristics that include the power level required to ensure the safety function.

**d.** IEC 61511: Functional Safety. Safety Control Systems of Continuous Technological Processes. This standard was developed in accordance with the introduction of IEC/EN 61508 for industrial processes.

The application of standards has its own possibilities and limitations, e.g. IEC/EN 62061 and EN ISO 13849-1 standards, which deal with an electrical safety management systems (later on they should be unified), they use different methods to achieve their results, and the user can choose them as they are both harmonized under the EU Machinery Directive. The difference between them is in a use in different technologies. IEC/EN 62061 standard is restricted to electrical systems only, while the second ISO 13849-1 standard deals with pneumatic, hydraulic, mechanical and electrical systems.

### 5.4. SIL and IEC/EN 62061

This standard describes an extent of the risk, which needs to be reduced but also the capability of the control system to reduce this risk with the Safety Integrity Level (SIL) [5, 15]. In the field of machinery, there are three levels from SIL1 to SIL3 (highest level of integrity) are used.

Since the risks may also occur in a different industry, such as the petrochemical, energetic or a rail sector, e.g. in the manufacturing industry (applies specific standard IEC 61511), this standard also offers another category of the safety integrity level, SIL4.

The SIL category refers to the safety function. The subsystems or elements of the system, into which the safety function is implemented, must have the appropriate capability to be assigned to a particular SIL category. This capability is called SIL Claim Limit.

| PL | $PFD_{avg}$ (average probability of a dangerous failure per hour) | SIL (safety integrity level) |
|---|---|---|
| a | $\geq 10^{-5}$ to $<10^{-4}$ | No special safety requirements |
| b | $\geq 3 \times 10^{-6}$ to $<10^{-5}$ | 1 |
| c | $\geq 10^{-6}$ to $<3 \times 10^{-6}$ | 2 |
| d | $\geq 10^{-7}$ to $<10^{-6}$ | 3 |
| e | $\geq 10^{-8}$ to $<10^{-7}$ | 4 |

**Table 6.** Relation between SIL and PL.

### 5.5. PL and EN ISO 13849-1

This standard does not use SIL, instead of that; it uses Performance Level (PL), properties level or performance. It defines five levels, where PLa is the lowest and PLe is the highest (**Table 6**) [1, 5, 15].

Application of adequate methods of determining SIL requirements depend on organization's risk criteria.

Some standards offer similar methods, e.g. EN 61508 offers three methods: quantitative, risk graph, risk matrix and IEC 61511 offers more as semi-quantitative methods, Safety layer matrix and Layer protection analysis (LOPA) [5, 10].

## 6. Risk management

The risk is an occurrence of a random event that can happen with a certain probability, when it occurs, it may have a negative impact on the organization's business objectives.

In the process of a risk management, it is necessary to accept three principles [1, 3]:

1. The outcome of the assessment is uncertain. If it is certain, it is not possible to talk about the risk.

2. It is possible to assign at least one of the types of estimated negative consequences (high loss of property, death, environmental damage, financial loss, etc.).

3. Assessment depends on time and changing circumstances, that's why it must be systematic and repeated.

These three principles are often ignored in practice. The manager in the company expects the results of the risk assessment to produce a clear result: "what is wrong and how to fix it" or "I did everything I could and we have no risks at all."

Risk management, particularly in terms of social acceptability, is expressed through the ALARP principle (As Low As Reasonable Practicable). Its priority is to reduce the level of a risk "to such an extent as is reasonably practical," while working with the level of risk between an unacceptable and fully acceptable (tolerable) level.

**Acceptable Risk**: represents a risk that is reduced to a level that can be tolerated in an organization, but at least it must respect provided requirements of binding regulations and the organization's own policy [1, 3, 11].

ALARP was defined by health and safety executive (HSE) organization in Great Britain. The goal is to manage the residual risk to the extent that it is practical (bearable) for the organization. In Great Britain and New Zealand, this model is also described as SFAIRP (So Far As It Is Reasonably Practicable) in the USA by ALARA (As Low As Reasonably Achievable) [1, 12].

When implementing the ISO 31000 standard for considering the so-called "positive risk"— an assessment of opportunities, it would be possible to apply a new approach for assessing an effectiveness, that is AHARP (As High As Reasonably Practicable) [1, 12].

The OHS management system is part of the organization's overall management system that creates and implements the OHS concept and manages health and safety risks. So it represents a set of mutually beneficial elements to make a policy and achieve the set goals.

The OHSAS 18001 standard required the most time to obtain the "standard" status. Since the safety requirements were different in each country and are strongly supported by the country's legislation, the transition to standard was relatively slow.

After accepting the British BS 8800 standard, the ISO organization had issued for the first time the OHSAS 18001 standard in 1999, which was first revised in 2007. In 2017, transition to the HLS structure (High Level Structure) is expected, as well as the standard also gets a new definition by ISO 45001 standard [1, 3].

To understand connection between Machinery Safety and OHS management is important for organization maturity and its competitiveness. Management system requirements are coming from context description of organization operating (external and internal relationships). This context is a base for risk assessment process coming from organizational business activities and is defined as Risk-based Thinking principles (RBT) [1, 4]. Newly prepared standard ISO 45001:2017 requires proactively approach in Risk Management processes. RBT distinguishes term risk and term opportunities, and also is linked with principles of ISO 31000. This brings a natural pressure for assuming methods and tools for risk assessment in relation with organization objectives on all management level.

## Acknowledgements

This work was developed within the projects APVV-15-0351 "Development and Application of a Risk Management Model in the Setting of Technological Systems in Compliance with Industry 4.0 Strategy" and 7FP entitled "iNTegRisk," no. CP-IP213345-2 and co-financed by APVV based contract No. DO7RP-0019-08.

## Author details

Hana Pacaiova

Address all correspondence to: hana.pacaiova@tuke.sk

Safety and Quality Production Department, Faculty of Mechanical Engineering, Technical University of Kosice, Slovakia

## References

[1] Pačaiová H, Markulik Š, Nagyová A. The Importance of Risk in Management Systems. Kosice: Beki Design; 2016. 276 p ISBN 978-80-553-2618-4

[2] Ferjencik M, Slovackova I. Trust managers and respect workmen: What does it mean to be competent in caring about safety? Journal of Loss Prevention in the Process Industries, Elsevier; pp. 95-104. DOI: 10.1016/j.jlp.2014.08.001

[3] ISO 31000: Enterprise Risk Management: CERM Academy Series on Enterprise Risk Management. USA. 235 p. Greg Hutchins PE CERM; ISBN 9780965466578

[4] Pacaiova H, Sinay J, Nagyova A. Development of GRAM – A risk measurement tool using risk based thinking principles. Elsevier: Measurement; pp. 288-292. DOI: 10.1016/j.measurement.2017.01.004

[5] Blecha P, Blecha R, Bradáč F. Integration of Risk Management into the Machinery Design Process. Mechatronics Recent Technological and Scientific Advances. Springer-Verlag Berlin Heidelberg. 9th International Conference Mechatronics 2011; Varšava; 2011; pp. 473-482

[6] Krajček K, Nikolić D, Domitrović A. Aircraft performance monitoring from flight data. Tehnicki Vjesnik; pp. 1337-1340. DOI: 10.17559/TV-20131220145918

[7] Starr A, Al-Najjar B, Holmberg K, Jantunen E, Bellew J, Albarbar, A. Maintenance today and future trends. In: E-Maintenance. London: Springer-Verlag London; 2010. pp. 5-10. DOI: 10.1007/978-1-84996-205-6_2

[8] Balazikova M, Tomaskova M, Dulebova M. Study of non-auditory effects of noise MM Science Journal; pp. 912-913. DOI: 10.17973/MMSJ.2016_06_201609

[9] Baybutt P. Calibration of risk matrices for process safety. Journal of Loss Prevention in the Process Industries. Elsevier; pp. 164-166. DOI: 10.1016/j.jlp.2015.09.010

[10] Gulland W.G. Methods of determining safety integrity level (SIL) requirements—pros and cons. In: Redmill F, Anderson T, editors. Practical Elements of Safety. Springer, London. p. 105-122. DOI: 10.1007/978-0-85729-408-1_6

[11] Haimes Y Y, Sage P A. Risk modelling, assessment and management, 4nd ed. Hobocen: Wiley; pp. 56-117. ISBN: 978-1-119-01798-1

[12] Tablot J. ALARP (As Low As Reasonably Practicable). Available from: http://www.jakeman.com.au/media/alarp-as-low-as-reasonably-practicable [Accessed: 24-09-2016]

[13] Pacaiova H. Human reliability in maintenance task. Frontiers of Mechanical Engineering in China; pp. 184-187. DOI: 10.1007/s11465-010-0002-4

[14] Maletic D, Maletic M, Al-Najjar B, Gomiscek B. The role of maintenance in improving company's competitiveness and profitability: a case study in a textile company. Journal of Manufacturing Technology Management; pp. 441-452. DOI: 10.1108/JMTM-04-2013-0033

[15] Kingsley J. Safety Integrity Level (SIL)—Explained Simply. 2017. Available from: https://www.linkedin.com/pulse/safety-integrity-level-sil-explained-simply-john-kingsley [Accessed: 20-07-2017]

Chapter 7

# Integrated Risk Assessment of Safety, Security, and Safeguards

Mitsutoshi Suzuki

Additional information is available at the end of the chapter

http://dx.doi.org/10.5772/intechopen.71522

**Abstract**

The peaceful use of nuclear energy has been pursued for more than half a century, and even after the catastrophic disaster at the Fukushima nuclear power plant in 2011, a new market in East Asia has been growing from the viewpoint of stable supply of nuclear energy. Countermeasures against malicious aircraft attacks have been introduced worldwide after the 9/11 terrorist attack in 2001 as synergies between safety and security. Although safeguards and security communities have different histories and technical aspects compared to safety, not only the mitigation plans as emergency preparedness but also a risk assessment as a supplement to the current requirements could be developed to promote synergism between safety, security, and safeguards (3S). The optimal installment of 3S countermeasures could be encouraged by a risk assessment to enhance reliability, robustness, and transparency of those facilities. One of the synergies of the integrated 3S risk assessment is a 3S by Design (3SBD) approach for new nuclear facilities. An introduction of 3SBD into the conceptual design stage increases regulatory effectiveness as well as operational efficiency and also reduces expensive and time-consuming retrofitting.

**Keywords:** probabilistic risk assessment (PRA), safety, security, and safeguards (3S), integrated risk assessment, 3S by design (3SBD), proliferation risk, sabotage risk

## 1. Introduction

After the Fukushima accident, the Nuclear Regulation Authority (NRA) in Japan developed the new safety standard, and soon after many utility companies submitted revised license applications to restart their nuclear power plants as soon as possible. Malicious aircraft attacks are considered in the standard, and mitigation plans are required to minimize possible consequences as synergies between safety and security in [1]. This time-consuming installment of

3S countermeasures could be encouraged by a risk assessment to enhance reliability, robustness, and transparency of the facilities as in [2].

The 3S initiative launched in 2008 emphasized on three (safety, security, and safeguards) of the 19 infrastructure elements of the Milestones in the Development of a National Infrastructure for Nuclear Power by IAEA as in [3]. One of the most apparent synergies is an adoption of 3SBD approach for new nuclear facilities. This benefit of the concept of 3SBD was pointed out recently in [4], and the Safeguards by Design (SBD) approach has been discussed extensively. The international safeguards has been implemented by the IAEA, and the IAEA has clear responsibility for the verification activities for the state's compliance with the NPT treaties and agreements, while their role in safety and security is limited on regulatory standardization. This means that there is not an obligatory authority governing safety and security regulations worldwide. Therefore, in order to achieve the 3SBD synergy, the realization of SBD approach with the IAEA and member states is challenging to achieve an international consensus in [5].

In addition to this, regarding institutional and technical issue for the internal and international regulators, a risk notion should be harmonized to be shared with the 3S authorities concerned. In safety, a frequency of accident is estimated from the past experienced data, and the accident sequence is analyzed with ETs/FTs, and the probabilistic assessment methodologies have been developed by the long historical trials and discussions. Because of the recent concern about nuclear security, the similar probabilistic assessment is extended to be used in the guideline against sabotage in nuclear security in [6]. The conventional vulnerability assessment in physical security has been well developed on a deterministic and prescriptive basis, on the

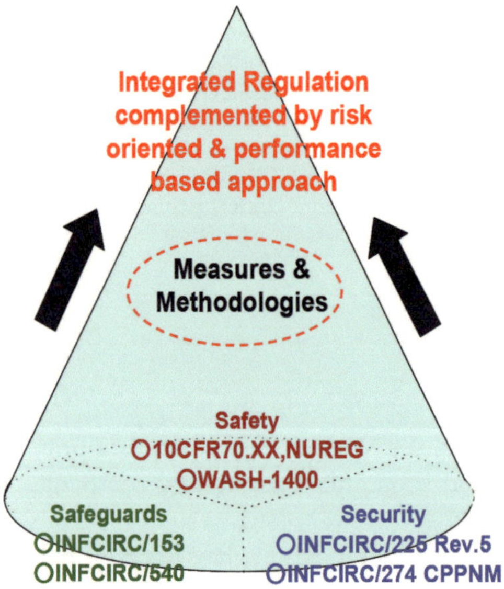


**Figure 1.** Integrated regulation by risk-informed and performance-based approach.

other hand, safeguards effectiveness is involved in proliferation resistance (PR) evaluation as extrinsic barriers, and a diversion pathway analysis is used to investigate a proliferation risk of nuclear cycles. Initial efforts for harmonization between reliability and safety and PR and physical protection (PP) are initiated under the Generation IV (GEN IV) international framework in [7]. In this regard, integrated measures and methodologies could be developed to evaluate an optimized balance between the 3S performance quantitatively as shown in **Figure 1**.

## 2. Mathematical model for 3S risk assessment

A major difficulty encountered in applying the probabilistic methods to safeguards is how to determine an initiation of diversion and misuse. In safeguards, the diversion of nuclear material and misuse of technology are induced by motivation of states and intentional acts of facility operator, so that estimating from historical incidences and predicting intentional human acts are generally very difficult. In comparison with the security, event sequence is analyzed probabilistically on a basis of the plant layout, system design, and structural robustness. Candidate tools for proliferation assessment were broadly investigated in [8]. As well-known international PR methodologies, Innovative Nuclear Reactors and Fuel Cycles (INPRO) program that was IAEA-led international project has developed the checklist approach, while the GEN IV's Proliferation Resistance and Physical Protection Working Group (PR&PP WG) has developed a risk-informed methodology in the qualitative and quantitative manner in [9]. To assess 3S risks, several mathematical tools are categorized to consider incident frequency and governing law resulting in the incidence shown in **Figure 2**.

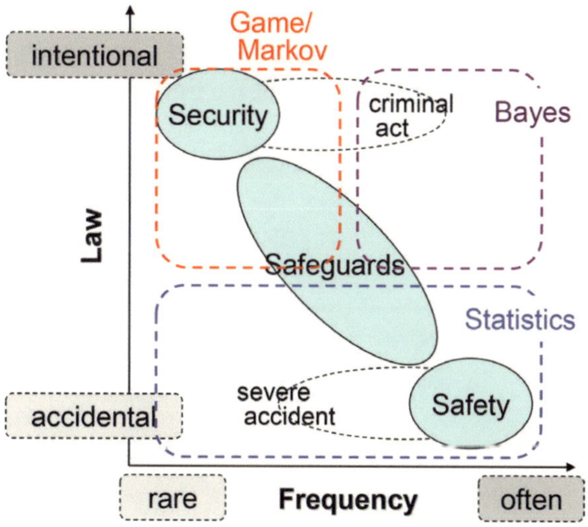


**Figure 2.** Mathematical models and assessment methodologies applied to safety, security, and safeguards (3S). The governing law and incidence frequency are selected to classify the inherent nature among the 3S incidences. The mapping of individual 3S region is drawn heuristically.

### 2.1. Probabilistic risk model in safeguards

Although the mathematical formalization for the international safeguards has been developed for several decades as in [10], the discussion of adopting probabilistic methodology to address nonproliferation issues was done in a different perspective in [11]. In safeguards, the estimation of intentional act leading to the diversion and misuse of nuclear material is generally very difficult. In addition to the incidence probability, there is another uncertainty related to measurement error in material accounting. This is the significant quantity (SQ) and timeliness goal that underline the basis for nuclear material accountancy (NMA). Based on the prescriptive and deterministic logic, uncertainty of NMA should be controlled under this limit as a first priority in safeguards. The IAEA determined the threshold value for nuclear material losses for each type of facility and process. However, as an amount of nuclear material increases in large-scale facilities, uncertainty due to measurement error becomes large and likely exceeds the limit. Because it is important to control the measurement error within the absolute threshold of NMA, a probability distribution of the measurement error of NMA has to be considered in conjunction with the incidence probability as shown in **Figure 3**.

The two-dimensional probability formalization is proposed as in the Eq. (1) as follows in [12, 13]:

$$R = P \times C = P(t, m) \times C = P(t) \times P(m \mid t) \times C \tag{1}$$

In the Eq. (1), the measurement error probability is defined as (*P(m)*) related to measurement uncertainty in material accounting. The measurement error probability is expressed as the probability density function in the measurement error axis in **Figure 3**. The accumulated distribution function leads to the detection probability. On the other hand, the incidence probability is defined as a Poisson density function under an assumption of random occurrence of diversion incidence. It should be noted that both probabilities are not independent and those would be closely correlated each other because of the inherent nature of intentional acts.

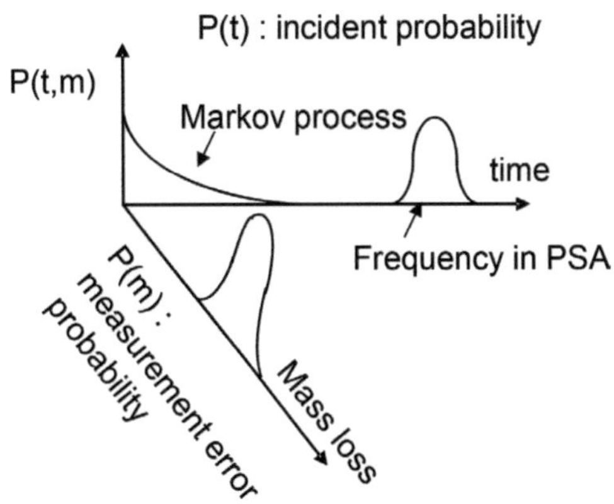


**Figure 3.** Two-dimensional probability for safeguards. The probability distribution composed of two random variables, the incidence time and the measurement error, is a characteristic feature of the proliferation risk.

### 2.2. Probabilistic risk model in security

In safety, Probabilistic Safety Analysis (PSA) has been developed by the long historical trials and discussions. This approach is to estimate the frequencies of accidents and failures from the historical data and to analyze the accident sequence with ETs/FTs based on these parameters. Because of the recent concern about nuclear security, similar probabilistic assessment was extended for use in developing guidelines for protection of nuclear power plants against sabotage in [14]. Although the conventional vulnerability assessment in physical security has been well developed on a deterministic and prescriptive basis, an inherent difficulty in determining the frequency of terrorist attack by malicious acts is undertaken by the conservative estimate. The risk formalization in security is expressed as in the Eq. (2) in [15]:

$$\mathrm{R} = P_A \times (1 - P_E) \times C = P_A \times (1 - P_I \times P_N) \times C \quad (2)$$

where ($P_A$) is the incidence probability, ($P_E$) the performance probability, ($P_I$) the interruption probability, ($P_N$) the neutralization probability, and ($C$) the consequence, respectively. Because of the difficulty of specifying the incidence probability, the security system is usually evaluated by the performance probability in which the timeline analysis is performed to identify the interruption probability and the security countermeasures; fence, sensor, camera, and so on are designed and installed into actual nuclear facilities. The neutralization probability is the unique feature of the security risk assessment and is determined by the performance of the response force. In addition to this, the deterrence effect can be estimated with a Bayesian method utilizing historical data, the game method assuming rational behavior and payoff matrix, and others, and the incidence probability could be evaluated qualitatively as in the decision process in [16].

Especially in the security risk study, sabotage risk is defined by taking the product of the frequency of sabotage incidence and the magnitude of consequences. Although it is difficult to estimate the initiation frequency, the risk can be described using the conditional probability and the magnitude of consequence as follows in [17]:

$$R_j = \pi_j p_j c \quad (3)$$

$R_j$ = The risk due to sequence $j$ leading to consequence, $\pi_j$ = The probability that an adversary will attempt to complete sequence $j$, $p_j$ = The conditional probability of success of causing consequence given attempt of sequence $j$, c = The magnitude of consequence.

For certain sabotage attacks, it is assumed that it is possible to identify well-defined sabotage sequences leading to consequence. In addition, a sequence is a cut set of a sabotage fault tree equation and does not necessarily imply a particular time order because a saboteur might attack the fault tree components in an intentional way.

Considering all sequence levels, the total risk, $R$, is expressed as follows:

$$R = \sum_{j=1}^{\mu} \pi_j p_j c = c \sum_{j=1}^{\mu} \pi_j P_{DC_j} \prod_{k=1}^{\eta_j} q_{jk} \quad (4)$$

$\mu$ = The number of sequences leading to consequence, $P_{DCj}$ = The probability of release reduction by the damage control measures, $q_{jk}$ = The probability of completion of the $k^{th}$ event in sequence $j$, $\eta_j$ = The number of discrete events in sequence $j$.

The three categories of measures, which are physical protection, damage control, and plant layout design, provide protection against radiological sabotage. The physical protection measures have been regulated; however, the other two measures are not fully discussed. In order to investigate the effect of damage control and plant layout design on the sabotage protection, the probability of release reduction by damage control measures, $P_{DCj}$, and the probability of completion of event sequence, $q_{jk}$ , are important.

## 3. Case study of individual risk analysis

### 3.1. Probabilistic risk analysis (PRA) in safety

The terminology, PSA, has been used in nuclear engineering field in Japan to introduce probabilistic risk analysis (PRA) that has been fully developed in safety assessment of nuclear power plant worldwide. According to the unique history of the introduction of PSA, the case study of PSA in safety is shown as the PRA study.

In the advanced fuel cycle project, Japan Atomic Energy Agency (JAEA) studied fast breeder reactor (FBR), advanced aqueous reprocessing, and fuel fabrication technologies. The details of these technologies have not been developed yet. However, a conceptual design of the advanced aqueous reprocessing process is used for safety risk study as shown in **Figure 4** as in [12, 13]. In this diagram, some challenging technologies and revolutionary instruments are included, and further development should be needed to proceed to engineering phase. After measuring the input plutonium amount at input accountability tank, most of uranium is removed at the crystallization process indicated as no. 9, and the remaining solution is adjusted to be treated correctly at the extraction process as nos. 13 and 14. Fission product and minor actinide are extracted at the extraction process. After the extraction process, the separated plutonium is always accompanied by uranium, and plutonium does not exist solely in the entire process. Through the evaporator output plutonium is measured at output accountability tank, and the mixture of plutonium and uranium is stored as reprocessed product. Several innovative technologies have been investigated in the FBR project. And in this feasibility study, as a typical FBR reprocessing process, the process throughput is assumed to be 200 ton-HM/year for spent fuel from FBR that is approximately corresponding to 18 ton-Pu/year, and the process is to be operated during 200 days/year.

Using PSA methodology, the risk for radioactive material release and damage to public health is estimated based on failure data of instruments and associated reference values in [6]. Severe

http://dx.doi.org/10.5772/intechopen.71522

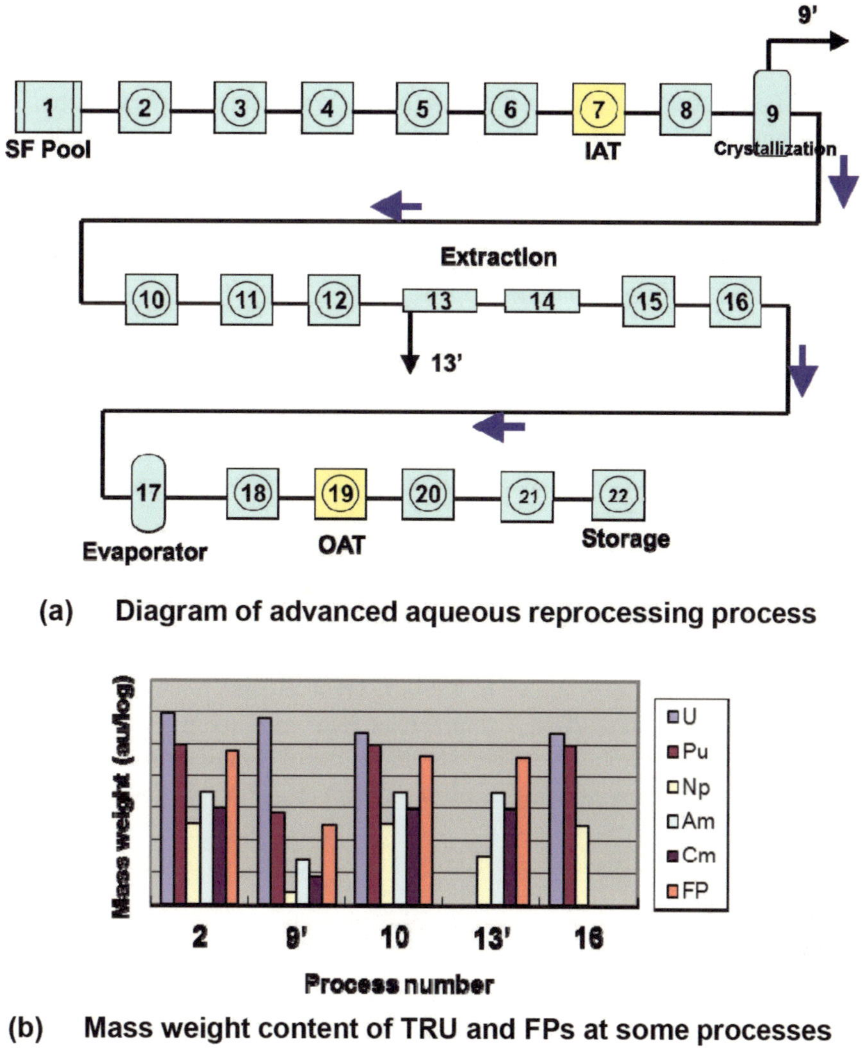


**Figure 4.** Diagram of advanced aqueous reprocessing process. Spent fuels discharged from fast breeder reactor are stripped to remove the uranium contents at the crystallization, no. 9, and extracted to remove the minor actinide contents using centrifugal extraction machines.

accident that would cause release of radioactive material is evaluated with expert judgment. With an assumption of multiple failures of instruments and human error, some important scenarios are selected as follows:

- *Scenario 1*: At the outlet piping of the extraction process, plutonium is leaked into high active liquid waste (HALW) tank, because a shape of the HALW tank is not manufactured to avoid criticality and then a critical accident is induced.
- *Scenario 2*: After removing americium (Am), curium (Cm), and fission product (FP) into raffinate component at the extraction process, Am and Cm are recovered using chromatography

technique into minor actinide (MA) tank. At that time, MA is leaked from the outlet piping of MA tank, and then the leaked waste solution is boiled.

- *Scenario 3*: The coolant system is broken down due to a pump failure, and then self-heat-generation source tanks, HALW and MA tanks, are boiled. Radioactive solution materials are evaporated. After chocking and destructing high efficiency performance air (HEPA) filter, those effluents are discharged into outside.
- *Scenario 4*: Organic solvent is very volatile and is leaked from the outlet piping of the extraction process, and then fire accident is induced.

After specifying important factors and making an ET for success and/or failure branches, an accident sequence is decided. Moreover the incident frequency is estimated with a FT and instrumental data, and the total probability for the sequence is calculated. The assessment result is shown in **Figure 5**. Even in the worst case scenario 2, the estimated risk is still two orders lower than the safety target that is a design goal in the FaCT project.

### 3.2. PRA in safeguards

As the case study of the risk analysis in safeguards, the proliferation risk assessment that is applied to reprocessing process is described in this section. General description of large aqueous and PUREX process model is shown in **Table 1**. These process parameters are assumed to represent characteristics of large PUREX commercial plant and do not contain any proprietary information and sensitive technologies. They are simply decided to perform a preliminary investigation on this study while maintaining characteristics of large commercial reprocessing plant.

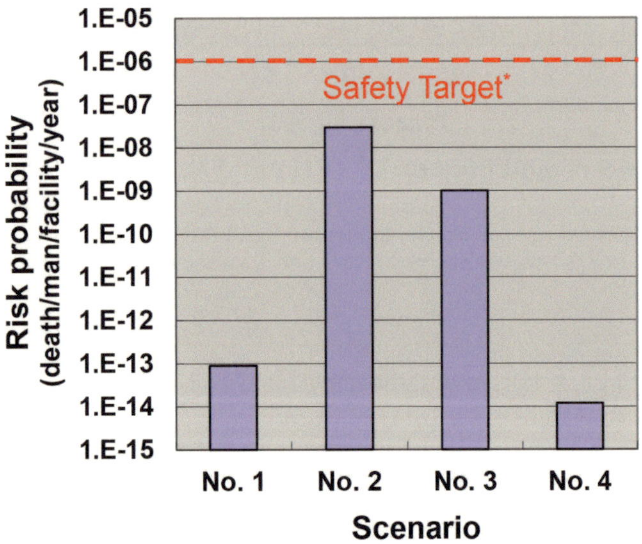


**Figure 5.** Probabilistic safety assessment for possible accident scenario. The safety target is decided to be $1{\times}10^{-6}$ (death/man/facility/year).

| Throughput (t/year) | 800 |
|---|---|
| (tPu/year) | 7 |
| Input (batch/day) | ~ 1b/day |
| (gPu/L) | ~ 3 |
| Product (batch/day) | 1b/5 days |
| (gPu/L) | 250 |
| Inventory (kg-Pu) | ~400 |
| Working days | 200 |

**Table 1.** Typical parameters in a large reprocessing plant.

The schematics of the process components are shown in **Figure 6**. These are composed of adjusting and input accountability, extraction and partition, and plutonium purification and concentration processes. The annual throughput is 800 ton-HM/y, and the working days per year are 200 days. In a steady-state operation, about 40 kg-Pu is a daily throughput, and plutonium inventory of the entire process is around 400 kg-Pu. It should be noted again that the model lacks the proprietary information. This is constructed from general PUREX specification and does not include design and performance information of dissolver, extraction process, and other sensitive technologies.

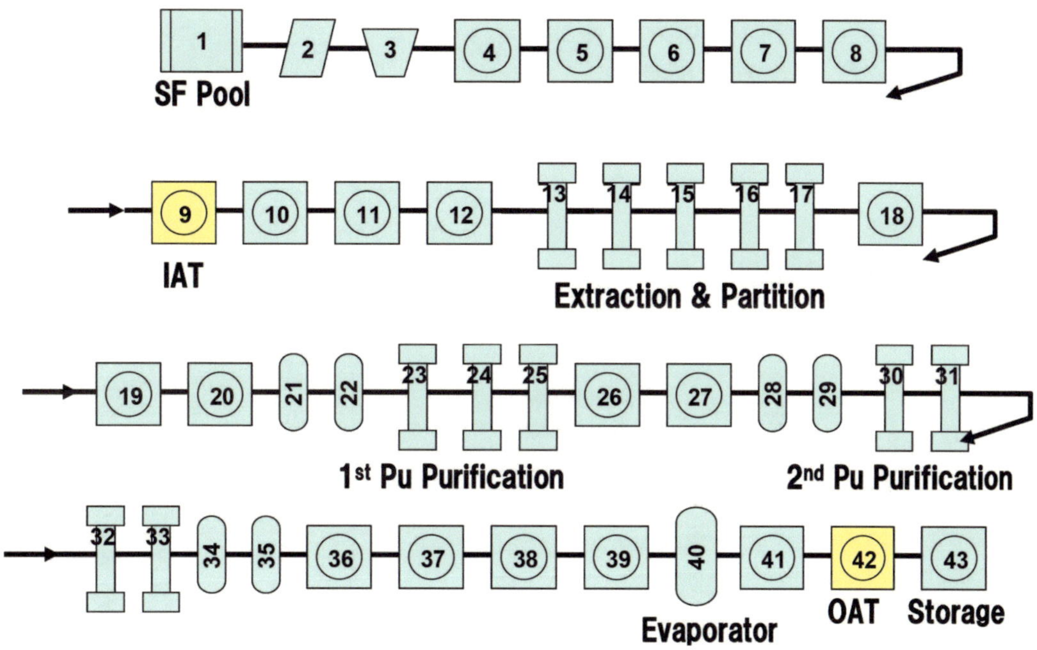


**Figure 6.** Schematics of PUREX reprocessing process.

In order to investigate adaptability of risk notion in a PUREX reprocessing process, PRA is carried out using the Markov model developed by the PR&PP WG. The PR&PP WG has also discussed the safeguard ability of an installation, defined as the degree to which a system can be put with effectiveness and efficiency under international safeguards, and the attributes have been defined for its characterizations. One of the evaluation methodologies has clearly noted for the notion of proliferation risk with assuming the Markov process model, and the proliferation risk analysis directly indicates the vulnerable diversion path instead of an expert elicitation.

As shown in **Figure 7**, the Markov process model is applied to the PUREX process to perform proliferation risk analysis, and in addition to the extrinsic effects by safeguards

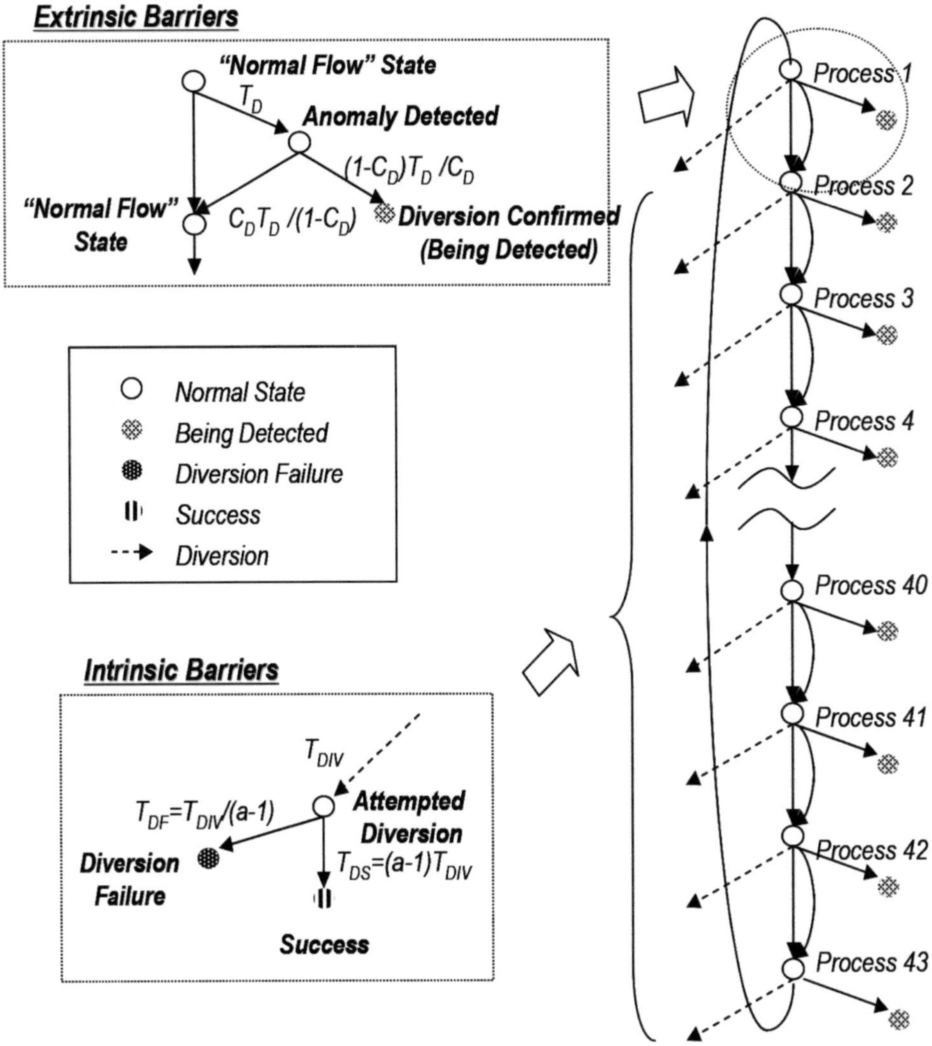


**Figure 7.** Markov model for proliferation risk analysis. Both extrinsic effects and intrinsic barriers are considered in the model. The extrinsic effect is modeled for safeguards implementation and the intrinsic barriers are for material difficulty due to radiation exposure and high temperature.

implementation, characterized by "$T_D$" and "$C_{D,}$" the intrinsic effects by radiation strength in materials, by "a," are considered as technical difficulty. Dip tubes for the solution monitoring system is installed in the 82 tanks in Rokkasho Reprocessing Plant (RRP) and can be used as in-line level and density monitoring. While the solution monitoring can generate real-time signals of the solution, the sensitivity should depend on not only performance of pattern recognition algorithm but also meaningless background by sampling, homogenization, evaporation, and so on. The solution monitoring sensitivity is considered according to the process steps with an assumption that the plutonium concentration would determine the detection capability of the solution level change corresponding to 1 SQ (= 8 kg-Pu). Both the extrinsic effects of safeguards implementation and the intrinsic barriers by radiation from residuals are also considered to evaluate a proliferation risk in the reprocessing process. As shown in **Figure 8**, the detection probability at the process number 1, spent fuel pond, shows the largest probability due to the residence time. The success probability increases clearly after the process number 34, downstream from the second purification process, and especially after the number 40, evaporator, due to high concentration of plutonium as well as low radioactivity in [18].

Although the proliferation is caused by intentional acts, it is assumed that a Poisson process, which is based on random incidence and is a theoretical background of the Markov model, could be applied to the risk analysis in the reprocessing. It is not yet applied to classical safeguards because PRA is not yet a quantitative safeguards component. In addition, measurement error probability in nuclear material accounting should be considered simultaneously with the incident probability that is a key component in the Markov model.

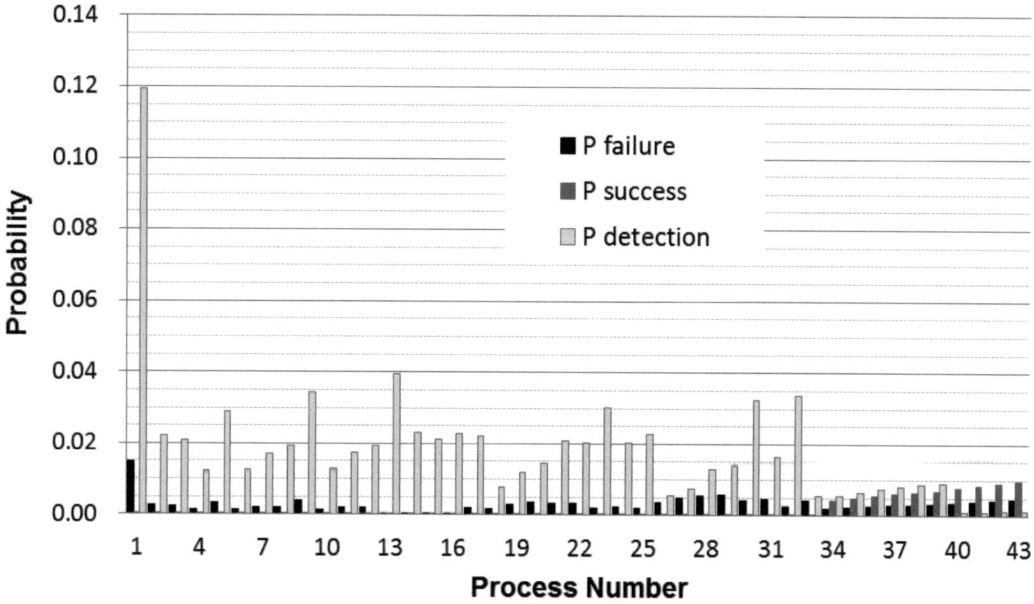


**Figure 8.** According to the individual process numbers, the detection, failure, and success probabilities are shown. A typical PUREX process is assumed to be composed of the 43 different processes: pool, dissolution, extraction, purification, evaporation, and storage.

### 3.3. PRA in security

In this case study, firstly, the Markov approach is applied to the sabotage risk assessment, and the Bayes updating is used to estimate the incident probability. And finally, the risk is considered with taking into account the sabotage sequences.

#### 3.3.1. *Markov model approach*

In **Figure 9**, the sabotage pathway of hypothetical nuclear reactor is shown. This example scenario represents a sabotage of nuclear reactor at full power operation by disabling decay heat removal function of the reactor, and the decay heat removal can be performed either by destroying coolant loop or failing the sea water circulation. It is assumed that this sabotage scenario is carried out by a conventional strategy such as unauthorized intrusion. And this unauthorized intrusion is defined as design basis threat (DBT) for physical protection system

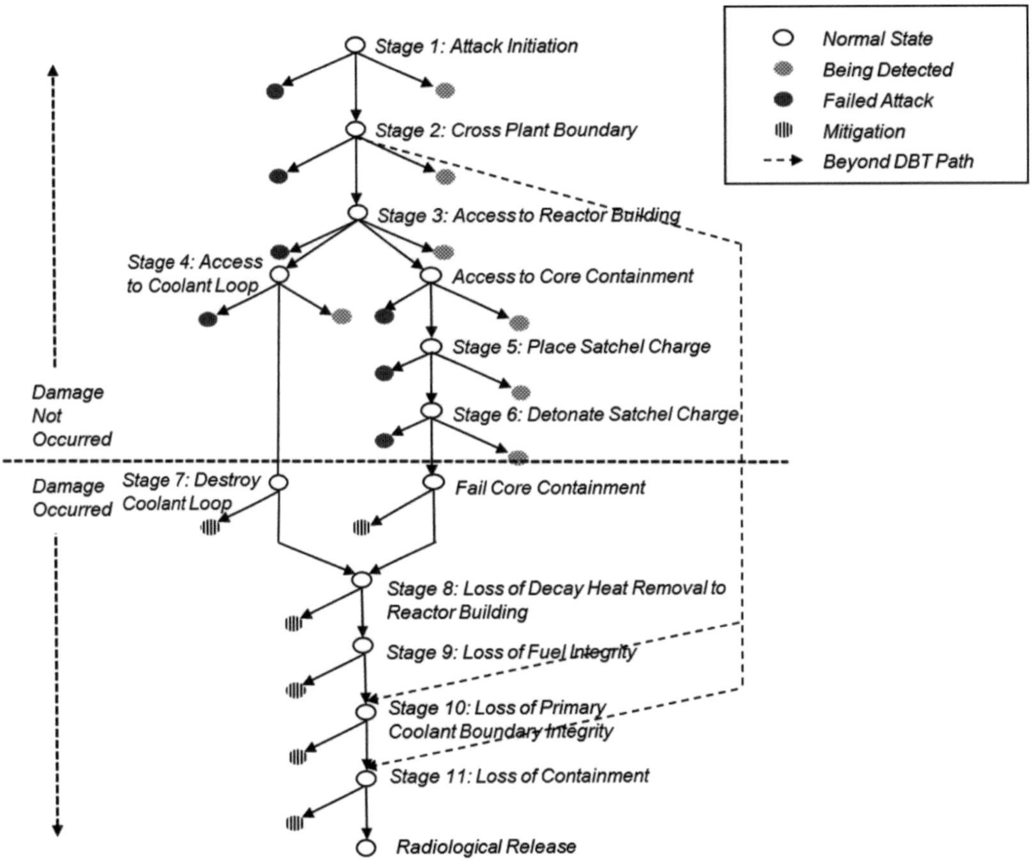


**Figure 9.** Sabotage pathway to radiological release of nuclear power reactor. The Markov model is used to model this sabotage pathway analysis. An intrusion attack is assumed as DBT, and a standoff stack is as beyond DBT.

in this hypothetical nuclear reactor. In addition, a standoff attack scenario is considered as a typical example of beyond DBT. In the beyond DBT, standoff attacks are performed to fail the primary coolant boundary integrity and containment integrity, and the serious consequences result in radiological release to the atmosphere.

In this security risk evaluation, two sabotage scenarios for nuclear power plants are considered. As the DBT, an intrusion incidence is investigated in the case (a), and terrorists attempt to overcome physical barriers and to destroy a reactor building using explosives. In the case (b), a standoff attack is modeled, and aircraft and/or missile attacks are assumed as beyond DBT. The failure, mitigation, detection, and success probabilities are calculated according to the elapsed time shown in **Figure 10**. It is understood that the detection probability is very high, and the success probability is very low in the case (a). This means that physical protection system perform efficiently for the DBT scenario. On the contrary, the mitigation and success probabilities increase gradually in accordance with fuel melting in the case (b). This indicates that the physical protection system does not work well in the case of beyond DBT and mitigation plans are important to minimize the consequence. Therefore, it is understood that a good cooperation between facility operator and national response authority is essential to mitigate the consequence against the sabotage attack.

### *3.3.2. Bayes updating*

To compare the risk representation in security with that in safety, an incidence probability is roughly estimated with the Bayes updating using past 17 data taken from global terrorism incidences against nuclear power plant in the world from 1972 to 2007 as shown in **Figure 11** as in [19].

In the Bayes updating, the prior probability distribution is assumed to be a gamma distribution, and the most updated mean value of the probability is about $4 \times 10^{-2}$ (1/the number of global NPPs/year). This is about $10^{-4}$(1/year) for individual nuclear power plant.

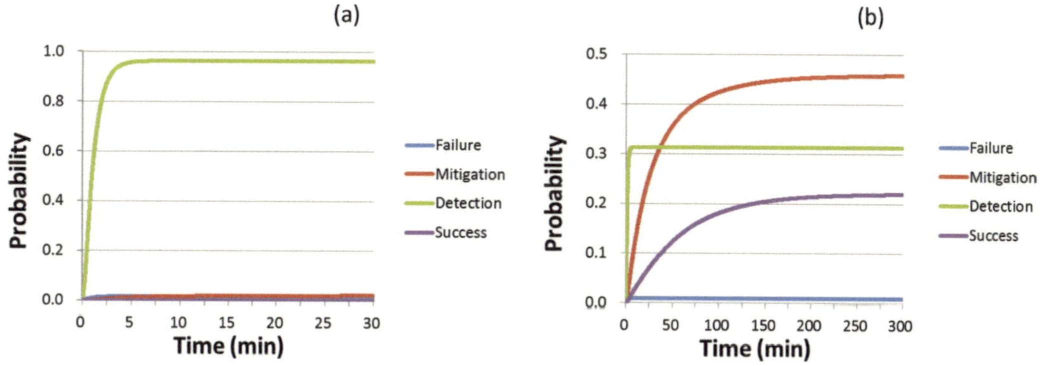


**Figure 10.** The failure, mitigation, detection, and success probabilities are calculated according to the elapsed time from an incidence. In the case (a), sabotage by an intrusion is assumed as design-based threat (DBT), and aircraft and/or missile attacks are assumed as beyond DBT in the case (b).

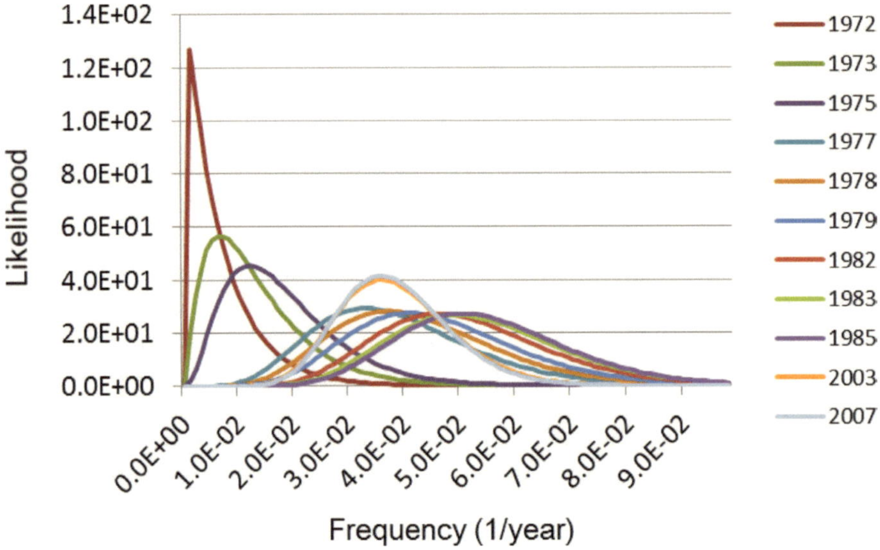


**Figure 11.** Incident probability is roughly estimated by Bayes updating. Experienced 17 data taken from past terrorism incidences in the world from 1972 to 2007 are used.

### *3.3.3. Sabotage sequence analysis*

A loss of offsite power (LOOP) is the typical case scenario in the sabotage protection study because of its vulnerability of the offsite power source and transmission line. In addition, a loss of onsite power by emergency diesel generator (EDG) is assumed because of the new safety regulation considering the Tsunami in the Fukushima accident. All these sabotage sequences including a reactor cooling by auxiliary feed water (AFW), breeding steam into containment vessel (CV), and loss of all direct current (DC) and alternate current (AC) power are shown. In order to investigate an effect of damage control design, five design changes cited from reference [20] and emergency power source are included into the event sequence as in **Figure 12**.

In order to evaluate an effect of the design change, the number of target sets is shown in bar charts as histograms in **Figure 13**. The horizontal position of each bar corresponds to the vulnerability derived from the individual target element. The vulnerability is shown as the polyline in the same figure. In the reference case, the hollow bars are seen around the high vulnerability region, the left-hand side in the figure, and the number of elements is 1 or 2. The cumulative number of the target sets is not so large; however, these target sets should be very vulnerable for the sabotage protection. On the contrary, considering the design change, the cumulative number of the target sets is large, but those vulnerabilities are very low. This does not mean that the target sets with the design change are vulnerable.

The sabotage risk in Eq. (3), which is proportional to a summation of the multiplication of the number of target set and the vulnerability, is shown as a function of the number of elements in the target set with and without the design change in **Figure 14**. The number of element

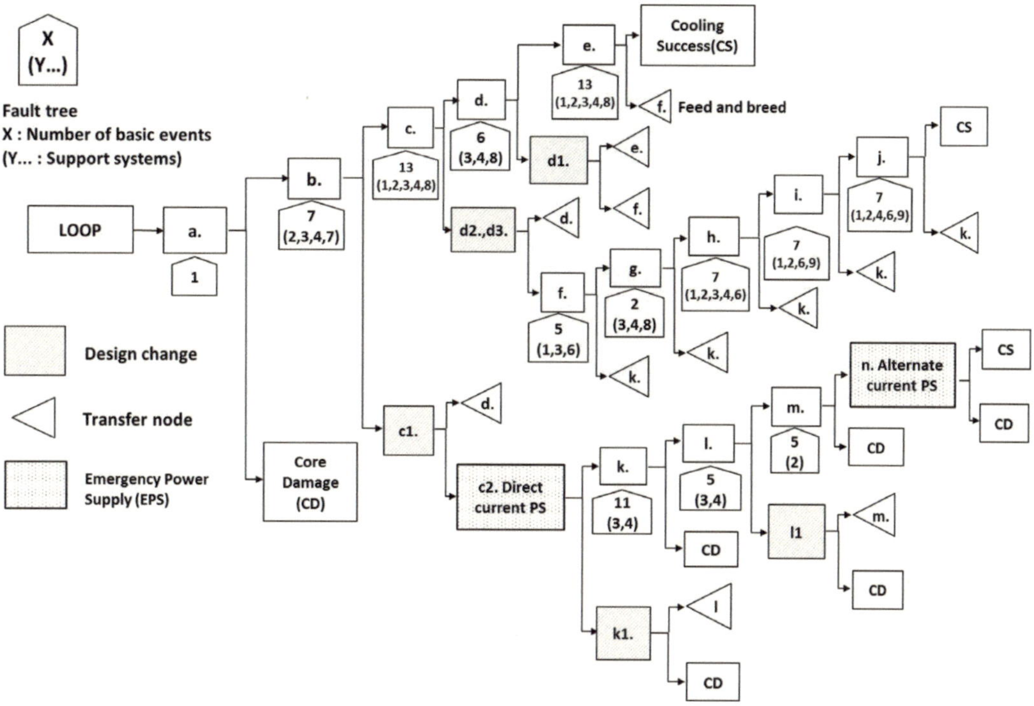


**Figure 12.** The name of heading and that of the design change in the event tree and all fault trees are abbreviated due to security concerns.

constituting the target set changes in the range of 1–3, and the effect of design change for damage control is shown. The total risk in all cases, regardless of the number of element, is reduced by considering the design change for damage control. It is verified that the built-in measures are effective and resistant compared to the emergency equipment placed outdoors. The movable equipment is flexible and resilient measure in accident management. It should be noted, however, that the equipment has to be used properly as the defense-in-depth (DiD) measures due to the possible adversary's interference.

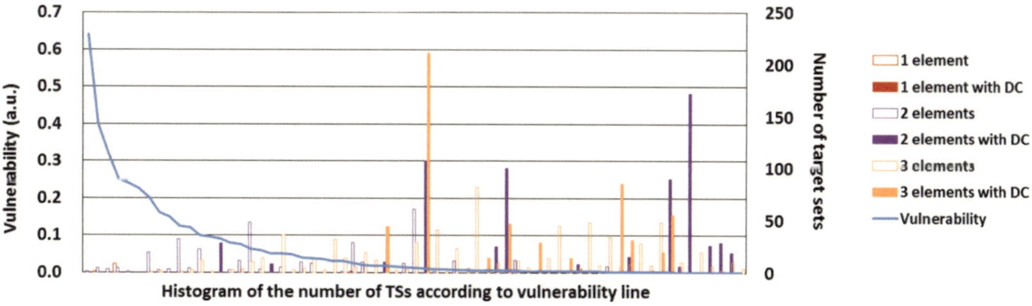


**Figure 13.** The number of target set (TS) and the vulnerability with and without the design change (DC).

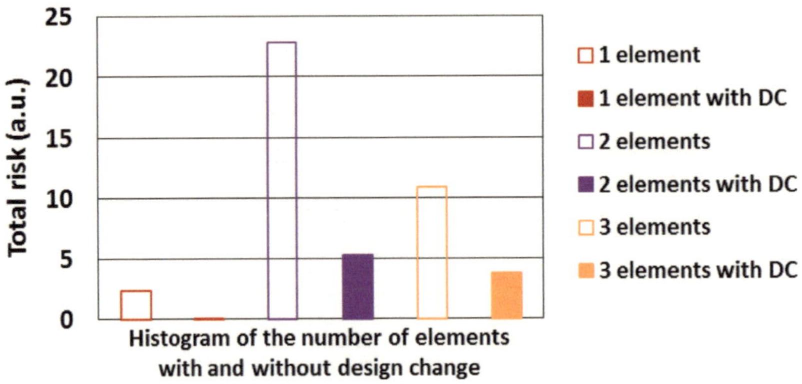


**Figure 14.** Reduction of total risk due to the design change (DC).

## 4. Prospect of integrated risk analysis

The PRA is an important method to evaluate equality in cost-effectiveness (CE) among 3S countermeasures, and quantification of risk in safeguards and security is always a challenge. The safety CE can be calculated by Eq. (5). And the frequency and damage cost in Eq. (5) have been well investigated. For the security CE, the incidence probability can be roughly estimated as shown in the previous section, and the damage cost can be evaluated according the individual scenario. On the contrary, for the safeguards CE, there is no method to estimate the incidence of diversion and/or misuse:

$$Safety\ CE = Frequency\left(\frac{1}{\frac{NPPs}{year}}\right) \times DamageCost\ (\$) \tag{5}$$

$$Safeguards\ CE = Unknown \times DamageCost\ (\$) \tag{6}$$

$$Security\ CE = Rough\ Estimation\left(\frac{1}{\frac{NPPs}{year}}\right) \times DamageCost\ (\$) \tag{7}$$

This is a current status toward the integrated 3S risk evaluation trial. However, the balanced management in resource allotment is highly appreciated to introduce nuclear energy in full compliance with international regimes as well as in a cost-effective manner. An integrated management system based on the quantitative risk evaluation would be the future research area in nuclear engineering field.

PRA in safeguards and security has been evolving and applying to the promotion of 3SBD activities. However, the theoretical basis is diverse, and the effectiveness of PRA in these areas has not been clearly demonstrated yet. Not only for an advanced instrument but also for risk-informed installation, it is shown that the Markov model approach is a good example of Safeguards by

Design activities. The model is applied to PRA with the PUREX model, and it is clearly demonstrated that the vulnerable path in the PUREX process is safeguarded by the solution monitoring (SM) originally installed based on the expert elicitation. The recent study on SM is the uncertainty analysis to optimize the safeguards measures with the trade-off relation between the safeguards performance due to measurement error and the economical consideration as increasing the throughput in the advanced reprocessing process. Both the harsh circumstances with the residual MA and FPs and the increase of measurement uncertainty due to the large throughput support more NDA installment than DA with considering the initial and running cost of those measures.

The probabilistic risk methodologies in security have been developing, and the inherent difficulties due to intentional acts are still challenges. However, the Markov model, the Bayes updating, and the sabotage sequence analysis could be applicable to the decision problems in security. In fact, the sabotage scenario analysis using vital area identification methodology has been used to increase an effectiveness of sabotage protection in nuclear power plants. And the sabotage logic trees that have been originally developed as ETs/FTs in the safety PRA are used for the security protection.

Finally, integrating the PSA in safety as the risk assessment techniques with the PRA in safeguards and security would have a potential to be fascinated by the younger generation, and the comprehensive 3S regulation based on the qualitative and quantitative risk discussion should be transparent and persuasive for a reasonable approach in the mandatory 3S implementation.

## Author details

Mitsutoshi Suzuki

Address all correspondence to: suzuki.mitsutoshi@jaea.go.jp

Integrated Support Center for Nuclear Nonproliferation and Nuclear Security, Atomic Energy Agency, Japan

## References

[1] Outline of New Regulatory Requirements For Light Water Nuclear Power Plants (Severe Accident Measures), April 3, 2013. http://www.nsr.go.jp/data/000067119.pdf

[2] Suzuki M, Izumi Y, Kimoto T, Naoi Y, Inoue T, Hoffheins B. Investigating 3S Synergies to Support Infrastructure Development and Risk-Informed Methodologies for 3S by Design. Vienna Austria: IAEA-CN-184/64, IAEA safeguards symposium; 2010. Nov. 1–Nov. 5

[3] IAEA Nuclear Energy Series, No. NG-G-3.1, Milestones in the Development of a National Infrastructure for Nuclear Power, Vienna: IAEA; 2007

[4] Stein M, *et al.* Safety, security, and safeguards by design – An industrial approach, proceedings of global 2009, Paris, France; Sep, 2009. 6-11

[5] Suzuki M, Burr T, Howell J. Risk-informed approach for safety, safeguards, and security (3S) by design, ICONE19-43154, 19th international conference on nuclear engineering. Vol. 16-19. Chiba, Japan. May, 2011

[6] Kurisaka K, Kubo S, Kamiyama K, Niwa H. Comprehensive Safety Examination of Commercialized Fast Reactor Cycle Systems – Examination of Safety Development Target and Risk Analysis of the Aqueous Fuel Cycle Systems-, JNC TN9400 2002-031, 2002

[7] Yue M, Cheng LY, Bari R. A Markov model approach to proliferation-resistance assessment of nuclear energy systems, Nuclear Technology. 2008;**162**(26):28-44

[8] Suzuki M, Terao N. Solution Monitoring Evaluated by Proliferation Riks Assessment and Fuzzy Optimization Analysis for Safeguards in a Reprocessing Process, Science and Technology of Nuclear Installations. Vol. 2013, Hindawi; 2013

[9] Gen IV International Forum, PR-PP Expert Group, Evaluation Methodology for Proliferation Resistance and Physical Protection, Rev. 5, GIF/PRPPWG/2006/005, OECD, November 30, 2006

[10] Cobb D. Sequential tests for near-real-time accounting, INMM Proceedings; 1981. pp. 62-70

[11] Avenhaus R, Canty MJ. Formal models for NPT safeguards, Journal of Nuclear Material Management. 2007;**354**:69-76

[12] Suzuki M, Burr T, Howell J. Risk-informed approach for safety, safeguards, and security (3S) by design, ICONE19-43154, proceedings of ICONE19. Chiba, Japan: 19th International Conference on Nuclear Engineering, May 16-19, 2011

[13] Suzuki M, Demuth S. Proliferation Risk Assessment for Large Reprocessing Facilities with Simulation and Modeling, Paper 399247, Proceedings of Global 2011, Chiba, Japan, December 16-11-2011

[14] "Engineering Safety Aspects of the Protection of Nuclear Power Plants against Sabotage", IAEA Nuclear Security Series No.4. Vienna: Technical Guidance, IAEA; 2007

[15] Sandia National Laboratories Security Risk Assessment Methodologies. Available from: http://www.sandia.gov/ram/RAM-CI.html

[16] Kardes E, Hall R. Survey of literature on strategic decision making in the presence of adversaries, CREATE Report, March 15; 2005

[17] Ericson DM, Bruce Varnado G. Nuclear power plant design concepts for sabotage protection, NUREG/CR-1345, SAND80-0477/1, 1981

[18] Suzuki M, Howell J, Burr T, Demuth S. Proposal of Proof-of-Principle Study on Aqueous Reprocessing Facility, IAEA CRP meeting; 2012

[19] Steinhausler F. Countering Security Risks to Nuclear Power Plants, Jeddah: International Symposium on the Peaceful Applications of Nuclear Technology in the GCC countries; 2008

[20] Lobner P. Nuclear Power Plant Damage Control Measures and Design Changes for Sabotage Protection, NUREG/CR-2585, SAND82-7011, 1982

Chapter 8

# Practical Propagation of Trust in Risk Management Systems

Kristian Helmholt, Matthijs Vonder,
Bram Van Der Waaij, Elena Lazovik and
Niels Neumann

Additional information is available at the end of the chapter

http://dx.doi.org/10.5772/intechopen.70741

**Abstract**

Using risk management systems for large-scale asset management is not without risk itself. Systems that collect measurement from a geographically diverse area, across many organisations, contain many interacting components that can fail in many different ways. In this chapter these systems are discussed from a risk assessment point of view, using practical examples. It provides suggestions how trust can propagate between interacting components of risk management systems by making information needed for risk assessment information explicit.

**Keywords:** asset management, systems architecture, uncertainty, trust, distributed systems

## 1. Introduction

This chapter is about trust propagation in risk management (related) systems used for large-scale asset management that are largely constructed using information and communication technology. An example of such a system is a smart grid monitoring system used for managing the risk of power outage. Based on measurements and failure models, it determines the probability of future failure of components of the grid and the impact of such failure. The focus will not be on the type of risk management these systems were designed to support. Instead, the focus is on assessing risks involved with these systems.

We look at large-scale distributed systems that are so complex that no individual member of the set of people involved in the life cycle (design, construction, operations, etc.) of such a system can understand the entire system. It would require more study and training from any individual than what could be reasonably expected of a human being. Assessing risk in this context requires the propagation of trust between the collaborating multidisciplinary members of the group. Members can vouch for certain aspects of the system based on their knowledge and expertise but need to rely on each other's judgement to come up with a verdict about the system itself and its output as a whole.

Special attention will be given to (1) automation of propagation of trust and (2) make the trust propagation process itself transparent, so that the system can be studied/audited/inspected by third parties that were not involved either in creating a system (of systems) or in monitoring that system. We work from the assumption that these independent third parties must be able to assure people that a system that supports (or even makes) decisions can be trusted. If these systems fail for whatever reason, it should be detected quickly. We go by this assumption, as we think, that without the ability to have people trust decision support/decision-making systems, these systems are useless in practice. Without these systems our societal toolkit to deal with a growing dependency on relatively scarce resources like energy, fresh water, skilled labour force, etc. is becoming dangerously empty. Dealing with these problems requires effective decisions that take into account analysis of many different aspects of physical reality which is above individual human capacity. We will describe several of those systems that are quintessential in making optimal decisions in societal domains where nonoptimal decisions—let alone failure—are becoming less and less of an option.

We start this chapter by describing cases where risk management (related) systems (should) provide information to decision-makers. There is risk involved in all decisions. The risk (R) is defined as the product of impact and probability of an undesired event/outcome. For example, the decision of an electricity grid operator not to replace an electricity cable might result (probability p) in power outage (undesired event), while at the same time, the decision to replace the cable might result (1-probability p) in unnecessary spending of money. What is 'worse' depends on the valuation of the undesired events, which can be difficult to compute. If, for example, the loss of power (indirectly) results in the death of a person that could not call an emergency number, the impact is huge. Valuation of such impact is not a topic of this chapter, neither is the computation of probability for each case. Instead, after providing a description of the cases, we will show an approach for propagating trust. Last but not least, this chapter is not about specific improvements to fault tree analysis (FTA) or failure mode, effects (and criticality) analysis (FME[C]A).

## 2. Examples of trust propagation in risk assessment

In this section we provide examples of risk management (related) systems. We will not provide a comprehensive risk assessment of each system as this is beyond the scope of this chapter. The function of these examples is to serve as a backdrop later on in this chapter for illustrating approaches for propagating trust within these systems and to its end-users. All examples are derived from real-life cases on which we have worked. For reasons of clarity, explanation purposes and customer confidentiality, they are not 'verbatim copies' of reality.

### 2.1. Railroad degradation: finding the cause behind the effect

This example serves as a means to show several risks involved in cause and effect relationships in complex systems.

Rail transport operations involve many risks at different levels of abstraction. People depend on the transportation of objects (e.g. people, cargo) to arrive safely within time, possibly comfortable and within budget constraints. End-users run many different risks, for example, not arriving on time, getting killed in an accident, paying too much, etc., different probabilities and different costs of impact. Railroad operators in turn run the risk of creating an accident and delay and providing uncomfortable or overpriced services, etc. In this example we focus on a specific aspect: determining the probability that a specific segment of physical track becomes unavailable due to physical degradation. This probability is a welcome ingredient to sophisticated risk assessment. For example, it can be used to determine when to perform maintenance, or it can be used to determine which tracks are available for routing trains. It is far from trivial to determine this probability, due to the many cause and effect relationships present in railroad operations. This is because of the many physical interactions of objects and forces that together influence the physical condition of the track. These interactions are the results of actions that in turn are the results of processes at different organisations and people involved in railroad operations. For example, an object that exerts influence is the train that interacts with the track through its wheels. The presence of a train is the result of planning of railroad transport carrier organisations. The interaction of the train with the track is influenced by the type of train, its weight, its length and the amount of trains. The track also interacts with its surroundings, like the geotechnical situation (e.g. 'soft or wet soil'), and the weather. The degradation of the track is also influenced by the construction materials, shape and specific construction. Last but not least, maintenance activities influence the degradation process also (i.e. 'it disrupts degradation'). All these influences interact in a way that seems (and probably is) far from trivial to understand. There is a lot of uncertainty with respect to determining the probability of degradation.

In recent years parties involved in railroad operations (i.e. railroad operators, contractors, etc.) have come up with approaches for arriving at a better estimation of the probability of (un) availability of the track. Many of these approaches are based on the idea that future states of the track can be estimated on knowing previous states. This is based on the assumption that a future state is (at least partially) determined by previous states, which can be the case in physical systems. The idea is to analyse recordings of previous states and discover a mathematical relationship between past and present states. This relationship could in turn lead to a prediction/estimation of future states by carrying out computations with previous states as input. From the viewpoint of risk assessment, roughly stated, two types of mathematical relationships can be identified: statistics based and physical model based. The first approach looks at parameters that describe the level of track degradation (e.g. height, shift, etc.) through time. It tries to fit those parameters in a mathematical function that describes the development of the parameters through time as accurate as possible (e.g. 'linear regression', 'curve fitting'). The future state is then estimated using this function. There is no real physical understanding of the system observed. The second approach also looks at parameters that might influence degradation (weather, soil type, etc.). It tries to find a mathematical relationship between influence parameters and degradation parameters through time, resulting in a physical understanding in terms of a model. If that relationship is well known, the future state can be predicted based on measurement of the previous state, influence parameters and application of the model. In practice there are combinations of both approaches. For example, Bayesian networks can be used to determine the conditional chance that a track will degrade. The idea is to determine the contribution of different influence parameters in the probability that a track will degrade. For example, an outcome of an analysis using Bayesian networks might be 'if 80% of the trains crossing the track travel at a speed of 100 km/h, the probability of severe track height degradation is 65%, where as it is reduced to 30% if only 20% of the trains travel at that particular speed'. Using this approach does not provide a complete physical model, but it does provide a deeper understanding of the observed system when compared to extrapolating on a degradation parameter like height alone. In summary, discovering/determining a physical model is far from trivial in railroad degradation. This is due to the many interactions of potential influence parameters and diversity in track construction and operations.

From the perspective of assessing risk in using an approach like this, several types of risk can be identified. There are risks involving:

1. **Measurement.** For example, the recorded measurement might not reflect the actual physical reality. This can be caused by different things: faulty measurement sensors, errors in recording, errors in converting data during data preparation, etc. Also, measurement methods might change throughout the years. A more accurate set of date might become available with different characteristics. This in turn might result in the observation that 'during the years something changed in track behaviour', while it was only a change in the measurement method.

2. **Analysis.** For example, analysts might assume that a set of measurements is more accurate than it is in reality. Analysis errors might be made due to difficulties in comparing different measurement sources (e.g. height of track, soil saturation of the underground, speed of trains, etc.). As different aspects of reality are measured at different points in time and location (in this case), mathematical interpolation of these types of data is needed for combining them. This might be a wrong conclusion as they are not tightly coupled in and come from complex multivariate and multi-organisational systems.

3. **Persistence of analysis results** (i.e. identified mathematical relationship). For example, if a mathematical relationship has been discovered, while an influence parameter did not change during statistical analysis of measurement data, the model might not take that parameter into account. As soon as this influence parameter changes, the modelled mathematical relationship will probably be no longer valid.

These risks directly translate into trust issues. Measurement experts must be able to trust sensors. Analysis experts must be able to trust measurements. Risk assessment experts must be able to trust analysis results. As all three areas are different areas of expertise, there is a need for propagation of trust. Again, this is far from trivial as railway systems contain thousands of kilometres of track in different surroundings, used in many different ways. Coming up with a methodology (or system that automates this method) that estimates the probabilistic chance of unavailability due to degradation in a uniform way is a challenge. It involves collaboration from different experts working at different organisations at different locations. Each organisation has to be trusted to provide the right information. This can be extra difficult in competitive markets, which is the case in some countries (e.g. the Netherlands). In that market cargo transportation companies compete for cargo, and contractors compete for the right to maintaining track for a period of several years. Sharing data or analysis methods might interfere with the rules of competition. For example, determining the cause of a stop in degradation requires knowing where and when a contractor performed maintenance on the track. This however is also part of the competitive edge a contractor has. This is an impediment to sharing this kind of data in general.

### 2.2. Pipeline management: trusting a computed future

This example [1–3] serves as a means to show the risks involved in computing possible future states of complex systems.

In the Netherlands distribution of drinking water and gas is largely done through underground pipelines, as the soft soil of most of the Netherlands permits for relatively easy modification of the top (1–2 m) layers of soil. Failure to provide water and gas has a severe impact on economy and society. Risk management is part and parcel of the work carried out by the organisations responsible for the management of these utility networks. They target at minimising risk within budget restraints by spending resources (time, money, etc.) in the most optimal way. Risk assessment constitutes an important part of their activities in minimising risk. In this example we focus on assessing the risk of structurally unreliable pipelines due to the influence of ground settlement. This risk is significant enough to be investigated, as (in the Netherlands) the top soft soil layer can—and does—move at different speeds at different places. This can cause strain in the pipelines, and depending on the materials and specific geometry they can rupture or break, it can result in leakage. Next to not delivering gas and water, there are other types of impact. Water leakage can result in local flooding that can destroy roads ('sinking cars'). Gas leakage might result in explosions that (at least partially) destroy houses as gas seeps into basements from underneath the roads where most pipelines are situated. Risk managers therefor want to reduce as much uncertainty as possible with respect to the probability of this type of pipeline failure occurring.

One approach that is currently being used in a project at TNO is to use physical models of reality to estimate the likelihood of possible future states of pipelines. The following models are used:

1. **Soil settlement**, as a result from the interaction of difference in forces on the ground (e.g. heavy load), geomechanical behaviour of soil types (e.g. clay, sand, etc.), etc.
2. **Forces on a pipeline**, as a result from differences in soil settlement
3. **Mechanical stress in a pipeline**, as a result from forces on the pipeline and its geometry
4. **Probability of breaking/tearing**, as a result from the mechanical stress and the material of the pipeline

The construction and maintenance of these models require specific expertise, for example, in the area of geotechnics and structural reliability. A multidisciplinary team is needed for the constructing a supermodel that integrates all of these models. This 'supermodel' requires a broad spectrum of input parameters, including:

1. Detailed geographical description of soil type
2. Forces on the ground throughout the years, as geotechnical processes can take years (e.g. 30 years for settlement due to a specific load)
3. Location, geometry and material of the pipeline

Information about these parameters is not available in a uniform way. There are drill samples of the soil, but not from all locations. This has to be derived using another model that provides an (probabilistic) estimate of what type of soil could (most likely) be located at a specific spot. Often, there is information about when an area was transformed into a 'built environment', but what exactly was built or deposited when and where is mostly forgotten in history. This has to be derived and estimated from other sources. There are geographical information systems (GIS) containing information on 'where to dig' for pipelines, but the exact height and location are often not known. Not in the least because of the fact that pipelines can move due to the softness of the soil. So, the 'supermodel' has to deal with a lot of uncertainty. This is partially done by including the actual settlement of the top soil layer, as measured by satellites. This data can be assimilated in order to keep estimations of soil settlement closer to reality. How this is done is beyond the scope of this chapter.

Because there is a lot of uncertainty with respect to input parameters, the engineers behind this approach have decided to use 'stochastic modelling'. Roughly stated, it means that not one possible future state is computed using the chain, but many different possible states, based on probability density functions. For several possible variations in a variable, a possible state is computed. Given the amount of variables involved and the different probability density functions, these results in many different possible future states. These are subjected to a statistical analysis, which then results in a 'most probable future'. Note that this approach has only become affordable recently due to developments in the automation of distributed computation. The amount of time to wait for results can now be reduced drastically by computing in parallel across clouds and combining the results later on.

The use of so much input data means that there are risks involved with measurement and (statistical) analysis, just as in the previous 'railroad degradation' example. Other risks are:

1. **Untimely arrival of data.** As the 'supermodel' requires a lot of data on a regular basis to update its estimations, it might be the case that one of the suppliers fails to deliver on time. This will impact the output of the 'supermodel' in the sense that it is an outdated advice.
2. **Inaccurate integrated model of the physical world.** For example, the models might represent a correct understanding of the physical world separately, but when integrated into a 'supermodel', they do not. For example, a pipeline itself might influence the behaviour of the soil too, which might not be taken into account by the soil settlement model.

http://dx.doi.org/10.5772/intechopen.70741

3. **Incorrect understanding of uncertainty.** Expressing uncertainty with respect to input or a result is far from trivial. If an experts states that his model provides 75% accuracy in settlement, what does that mean? Does it mean that the soil could settle a 25% extra? Or does it mean that 75% of the estimate is spot-on. This has to be communicated in very strict way for proper risk assessment.

4. **Errors in computation.** In order to provide actual numbers, a model needs to be implemented in software to be run on a (distributed) computer (system). The more software is needed to be written for carrying out the computations involved, the bigger the chance that some programmer makes an error during implementation. As the computations are complicated and many, it might be difficult to detect this error.

5. **Errors in presentation.** Finally, if the results are somehow misrepresented on screen or on paper, it is still possible to make the wrong decision.

### 2.3. Smart grid analysis: non-available data

This example [4, 5] serves as a means to show the risk of data becoming unavailable.

In the Netherlands, electricity is distributed throughout underground low-voltage utility networks, just as water and gas. That makes the cables and their connections invisible to direct visual inspection. However, as the condition of the cable isolation contributes significantly to the risk of power outage, knowledge about the actual state of the isolation is an important ingredient for assessing the risk of power outage. An approach, currently under investigation in research projects, to deal with this uncertainty is to come up with an estimate of the condition of the isolation part of the cable, using a model that takes into account the material of the cable, its construction, its surroundings and the power loads it has been subjected to. This model will be developed based on analysing measurements. Using methods from applied statistics, researchers will try to identify relationships between power outage and specific influence parameters. This approach can be compared to the one used in the 'railroad degradation' example above.

What makes this example different from the previous ones is the possible need for accessing data that is protected by privacy laws. Specific power might influence degradation of the condition of the insulation part of the cable. To determine if this is the case measurement, data is needed on the demand and supply of power to a distribution network. Smart metres (at households) might be able to provide this data. However, this data could also provide insight into what equipment is used at which times of the day. This is why smart metres have been the topic of many heated privacy debates. Smart grid operators might be allowed by special law to only use the data for grid analysis. However, in the future there might be new and tougher laws on privacy that stop the operators from having access. Keeping the system (i.e. failure models) up to date will then become a challenge.

This case shows the risk that data is not (always) available, this time due to the introduction of a law instead of a (temporary) failure of a data supplier to deliver.

### 2.4. Precision dairy farming: sharing valuable information

This example [6] shows the risks involved in sharing commercially sensitive data.

As dairy farming concerns livestock, it involves many different risks, for example, risks involving food safety and animal health. Assessing risks in detail requires having information on cows, their surroundings, their food, etc. In the past it was difficult to retrieve this information as it was not recorded. But as sensors and ICT have become more affordable, farms are becoming places where a multitude of sensors gather measurement data for interpretation by experts that work on improving milk production, cow welfare, etc. Also, laws and regulations have changed in order to ensure safety and care for the animal and the environment.

The information gathered by these systems however is not easily available for everyone. This is because it provides insight into 'secrets of the trade' and not every animal expert wants to share its findings or data/information. There is the risk of 'teaching your competitors'. Discovering cause and effect relationships requires long-term observation of cows in their contexts. This can be difficult as cows can 'switch farms'. There is the risk of data on a cow being unavailable.

Recently, the InfoBroker concept [7] was developed: 'a platform to make real-time sensor data from different farms available, for model developers to support dairy farmers in Precision Livestock Farming. The data has been made available via a standard interface in an open platform in real time at the individual animal level'. The InfoBroker is designed to making data stored in diverse places available in an efficient and controlled manner. Data is not stored centrally, but remains at the source. The InfoBroker is capable of retrieving individual cow data from many sources while at the same time serving a large number of models on demand. 'For each farm it is specified which data may be released by the InfoBroker. This means that the farmer continues to be the owner of the data'.

Newly identified risks are related to the uncertainty of the quality of the sensor data. Not all data sources are created equal. Some sources are more precise or more frequent (and some change over time). Does a sensor measure the weight per 10 kg or per 0.1 kg? Does a sensor measure the activity per day or per min? For some applications, this is irrelevant. For example, for a dashboard application, typically the data is presented as is, without any qualitative indication. For others, especially model-driven applications, it is essential to know if the data from the device is accurate enough to be used. Some scales can be used as weight input; others are not precise enough. Some activity sensors produce frequent enough measurements; others accumulate over too long time periods. It depends on the farm which sensors are used and therefore available to the model. Therefore, there is a risk of drawing the wrong conclusion for some farmers, because they happen to have a sensor with not enough quality.

## 3. Propagating trust in risk assessment

In the previous section, we have provided examples of risks involved in risk management (related) systems. In this section we focus on the propagation of trust between the different parties that design, construct, manage and use these systems. Without the propagation of trust, it is impossible to assess risk using these systems, during construction as well as during its usage. The approach for propagation we describe is largely based on separation of concerns and will be discussed at the logical/functional level of abstraction, from an architectural point of view.

### 3.1. Separate concerns in risk assessment

A basic underlying problem for teams of multidisciplinary experts to vouch for a system as a whole is that they cannot do a proper review of the work of experts from another domain. They simply miss the expertise. For example, how can a mathematical analyst determine if measurements made by another expert can be trusted if he/she has no expertise in the field of making measurements? How can the analyst assess how much risk is involved in using these measurements? If they want to vouch for the system as a whole, they have to trust the other experts involved.

We state that it is important for continuous risk assessment of complex risk management (related) systems to separate the concerns for experts. This means that instead of assessing risk for a large monolithic system with lots of intertwined functional components, a conglomerate of components should be designed with risk management built in the components. Each expert can focus on their component which he or she understands. They connect their own components to others using well-defined interfaces. These interfaces take into account risk assessment issues, which we will be discussed later on. The idea of separation of concerns in distributed systems is sometimes also referred as 'unbundling'. Finding out where to 'draw the lines of separation' is beyond the scope of this chapter, as it is a topic of its own in distributed systems design. However, we can state that the following categorisation of types of expertise provides an indication of where to separate:

1. **Measurement.** For example, creating an accurate reflection of the physical reality often requires very specific know-how of sensors.

2. **Analysis and modelling.** For example, analysis and modelling require very specific mathematical expertise.

3. **Computation.** For example, parallelisation of computation requires specific knowledge of applied computational science.

4. **Presentation.** For example, letting people draw the right conclusion from information inside a computer requires specific knowledge of human-machine interfaces.

Once a design is separated into components that are understood by experts in their domain, the next step is to make information on risks that involved explicit. Experts should provide risk assessment-related information about the (un)certainty of the (delivery) of data/information of their component to other components explicitly. In this way it is possible to assess the risks involved of the system as a whole. Note that we do not consider the decomposition of a system into parts as something new: we use this as a stepping stone for the next sections.

### 3.2. Propagation of trust

Experts can provide risk assessment-related information about the output of their component, based on their knowledge of how the component transforms input into output. In order to do so, they also need risk assessment-related information with regard to the input of their component. If this is provided, it is possible to have (a certain level of) trust to propagate throughout the system, if components are designed, constructed and used according to the following rules:

1. Components determine if input is provided as promised. They establish a level of trustworthiness for each supplying component.

2. Components include risk assessment (related) information in their output. They provide information for establishing their trustworthiness to other components.

3. Components are auditable by third parties. Noninvolved experts can assess the risk involved in using a component.

In the next sections, we will describe the first two rules in more detail. If these rules are followed, the likelihood increases of being able to trace back potential root causes in case of incorrect behaviour. A more detailed description of the third rule is beyond the scope of this chapter as it involves auditing and certificate practices (**Figure 1**).

#### *3.2.1. Receiving as promised*

Delivery of output as input to another component (e.g. measurement data, analysis results, model-based estimations of probability, etc.) has different aspects. For each of these aspects, risks can be identified. We provide a non-exhaustive list of aspects:

1. **Completeness.** Whether or not all input was received, having everything that is needed.

2. **Timeliness.** Whether or not all input example was received in time. Sometimes data becomes useless if it arrives too late. For example, receiving a warning about potential

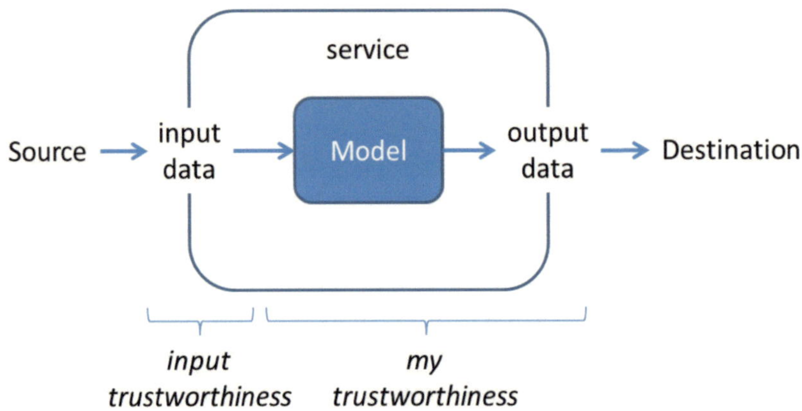


**Figure 1.** Basic trustworthiness model.

flooding 3 days after the flood is a problem for a system that provides information for people who have to decide on a possible evacuation. With respect to behaviour in time, the following types can be identified in terms of reliability (**Figure 2**):

i. **Perfect world**: input is always delivered on time.

ii. **Realistic OK**: sometimes, input arrives somewhat later than promised, but still in time to be useful.

iii. **Offline period**: sometimes, certain input is not delivered at all for a certain amount of time. After the period of silence, the input that should have been delivered during that time is not sent. This affects completeness.

iv. **Delay burst**: sometimes, bursts of input are delivered after the time it was useful to process. This does not affect completeness.

3. **Correctness**. The way in which input conforms to fact or truth. All the aspects of the correctness apply to both the arrival time and contents of the input. Aspects which play a role are:

4. **Accuracy and precision**: both aspects relate to the difference between the content of the input and the actual (physical) state of what the input pretends to provide information about (**Figure 3**). We define accuracy to be the size of the difference, whereas precision relates to the average difference between input and truth (e.g. a measurement and actual physical state).

5. **Consistency**: input (e.g. a set of measurements) is consistent if it is free of conflicts. For example, measurement data that shows that a normal household freezer would jump 20° C in 1 sec is not consistent. Note that a consistent set might be consistent and not accurate at the same time.

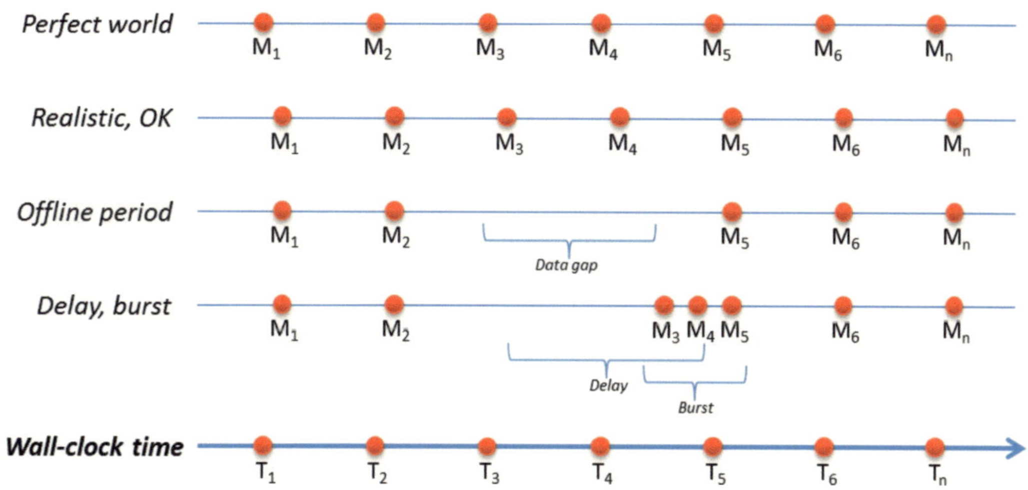


**Figure 2.** Aspects of timeliness. The display dots (M1-Mn) represent input data created at wall clock time T1-Tn.

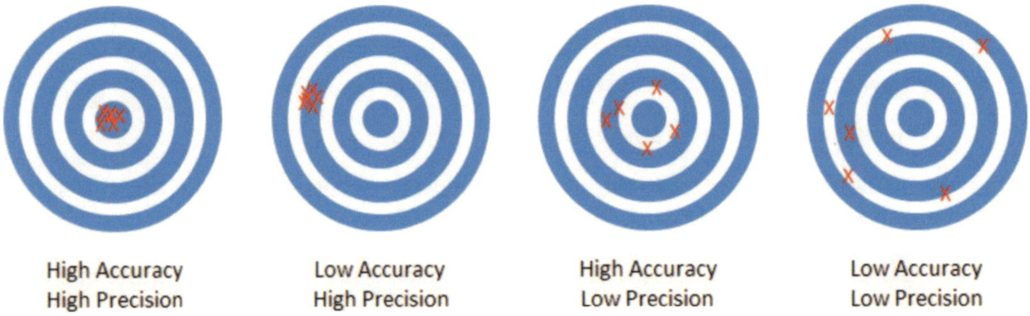


**Figure 3.** Relationship between precision and accuracy.

6. **Validity**: the degree to which input conforms to agreements on syntax and semantics. For example, if the temperature of a freezer is noted in degree Fahrenheit, it might be accurate, precise and consistent. However, if the agreement was to report in degree Celsius, the data is not valid.

Using the concepts of completeness, timeliness and correctness, we can define a tree of trustworthiness for a component as seen in **Figure 4**.

The trustworthiness of a component is determined by its availability to other components and the quality of the output it provides (in time). That quality can be described in terms of 'quality of transport' (QoT) and 'quality of data' (QoD), each from the viewpoint of completeness, timeliness and correctness. From a QoT perspective, the received input is considered as black box, and the focus is on the arrival in time. From a QoD perspective, the content, the data, is considered.

This tree of trustworthiness could also be used to design components that explicitly filter input based on the different aspects: timeliness, completeness and correctness (see **Figure 5)**. Depending on the impact of using input that, for example, did not arrive on time or was partially incomplete, a component might decide not to produce any output. This could in turn result into a cascade of components that stop producing output, thereby signalling the end-user that the system as a whole can no longer be trusted at the same level as before. Whether or not the system as a whole should show this kind of behaviour depends on the specific purpose

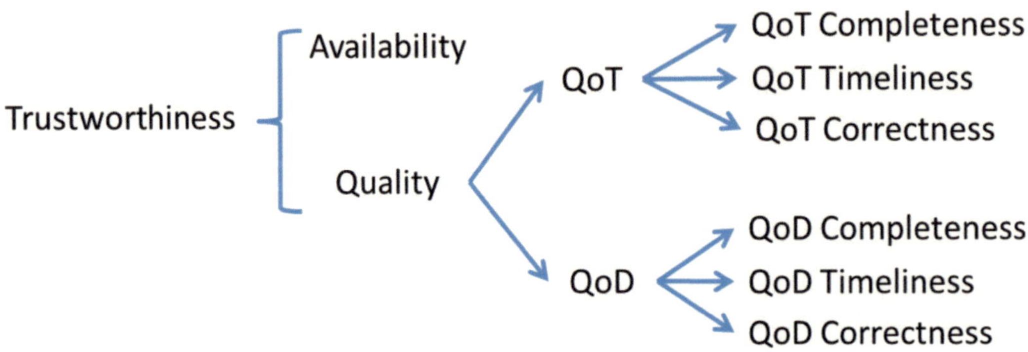


**Figure 4.** Tree of trustworthiness.

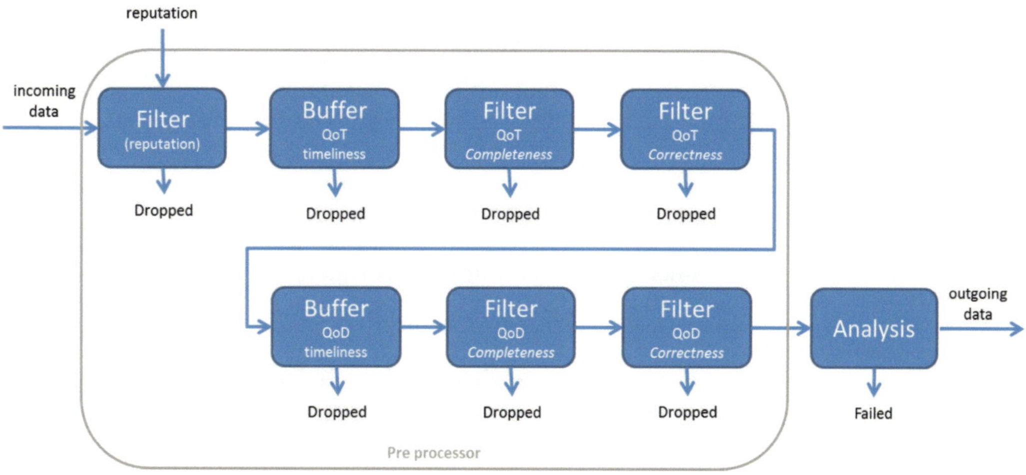


**Figure 5.** Filtering of input based on risk assessment-related aspects.

of the risk management system. It might also be possible to come up with 'best effort' output and explicitly communicate this with the output. This is covered in the next subsection.

### 3.2.2. *Including risk assessment information for trust*

In the previous subsection, we focussed on assessing risk by components that receive input. As described before experts from one domain of expertise cannot review the work of experts in another domain (they do not master). Therefore, components need to include information about the inputs that were used into their output. From a viewpoint of intercomponent communication, this can be done in two ways:

1. **In band**. When data/information is delivered to or retrieved by another component, information on the source and method for creating that data/information is included. This can result in a massive overhead of communication.

2. **Out of band**. Receiving components can ask a component for information about the source and method for specific data/information. This is less overhead in communication but poses more of a risk as this meta-risk assessment information is separated from the data/information that is communicated between components.

Wherever risk assessment information about produced data/information is made available (in band or out of band), there need to be agreements on the syntax and semantics of accuracy, probability, etc.

## 4. Conclusion

In this chapter we have discussed the concept of trust propagation in information and communication technology-based systems that are used for risk management, from a risk assessment point of view. We have provided examples of such risk management systems and shown the possible types of risks involved. Furthermore, we have provided suggestions on how to enhance assessment of risk of these systems, by applying the concept of 'separation of concerns' and making risk assessment information explicitly available.

## Author details

Kristian Helmholt*, Matthijs Vonder, Bram Van Der Waaij, Elena Lazovik and Niels Neumann

*Address all correspondence to: kristian.helmholt@tno.nl

Netherlands Organisation for Applied Scientific Research (TNO), The Hague, Netherlands

## References

[1] Helmholt K, Courage W. Risk management in large scale underground infrastructures. In: 2013 IEEE International Systems Conference (SysCon); 2013. Orlando, Fl: IEEE; 2013. pp. 902-908. DOI: 10.1109/SysCon.2013.6549991

[2] van den Heuvel F, Schouten M, Abspoel L, Courage W, Kruse H, Langius E. InSAR for risk-based asset management of pipeline networks (poster). In: European Space Agency Living Planet Symposium; 9–13 May 2016. Prague; 2016

[3] de Bruijn R. et al. Differential subsidence and its effect on subsurface infrastructure: Predicting probability of pipeline failure (STOOP project). In: 19th EGU General Assembly, EGU2017; 23–28 April, 2017. Vienna, Austria; 2017. p. 15924

[4] Helmholt KA et al. A structured approach to increase situational awareness in low voltage distribution grids. In: 2015 IEEE Eindhoven PowerTech, Eindhoven; 2015. Eindhoven: IEEE; 2015. pp. 1-6. DOI: 10.1109/PTC.2015.7232779

[5] Helmholt KA, Broenink EG. Degrees of freedom in information sharing on a greener and smarter grid. In: ENERGY 2011: The First International Conference on Smart Grids, Green Communications and IT Energy-Aware Technologies; May 22–27, 2011. Venice, Italy; 2011

[6] van der Weerdt CA, Kort J, de Boer J, Paradies GL. Smart dairy farming in practice: Design requirements for user-friendly data based services. In: Conference Proceedings for the International Conference on Precision Dairy Farming. Leeuwarden, Netherlands; 21–23 June, 2016. 427-432

[7] Vonder MR, van der Waaij BD, Harmsma EJ, Donker G. Near real-time large scale (sensor) data provisioning for PLF. In: Guarino M, Berckmans D, editors. European Conference on Precision Livestock; 15 September 2015 through 18 September; 2015. pp. 290-297

Chapter 9

# Risk Assessment for Collaborative Operation: A Case Study on Hand-Guided Industrial Robots

Varun Gopinath, Kerstin Johansen and
Johan Ölvander

Additional information is available at the end of the chapter

http://dx.doi.org/10.5772/intechopen.70607

**Abstract**

Risk assessment is a systematic and iterative process, which involves risk analysis, where probable hazards are identified, and then corresponding risks are evaluated along with solutions to mitigate the effect of these risks. In this article, the outcome of a risk assessment process will be detailed, where a large industrial robot is used as an intelligent and flexible lifting tool that can aid operators in assembly tasks. The realization of a collaborative assembly station has several benefits, such as increased productivity and improved ergonomic work environment. The article will detail the design of the layout of a collaborative assembly workstation, which takes into account the safety and productivity concerns of automotive assembly plants. The hazards associated with hand-guided collaborative operations will also be presented.

**Keywords:** hand-guided robots, industrial system safety, collaborative operations, human-robot collaboration, risk assessment, hazards

## 1. Introduction

In a manufacturing context, collaborative operations refer to specific applications where operators and robots share a common workspace [1, 2]. This allows operators and industrial robots to share assembly tasks within the pre-defined workspace—referred to as collaborative workspace—and this ability to work collaboratively is expected to improve productivity as well as the working environment of the operator [3].

As pointed out by Marvel et al. [1], collaborative operation implies that there is a higher probability for occurrence of hazardous situations due to close proximity of humans and industrial robots. The hazardous situations can lead to serious injury and, therefore, safety needs to be guaranteed while developing collaborative applications [4].

ISO 10218-1 [5] and ISO 10218-2 [6] are international standards aimed at specifying requirements for safety on the design of industrial robots and robotic systems, respectively. They recognize collaborative applications and list four specific types of collaborative operations, namely (1) safety-rated monitored stop, (2) hand-guiding, (3) speed and separation monitoring, and (4) power and force limiting that can be implemented either individually or as a combination of one or more types.

As industrial robots and robotic systems are designed and integrated into specific manufacturing applications, the safety standards state that a risk assessment needs to be conducted is to ensure safe and reliable operations. Risk assessment, as standardized in ISO 12100 [7], is a detailed and iterative process of (1) risk analysis followed by (2) risk evaluation. The safety standards also state that the effect of residual risks needs to be eliminated or mitigated through appropriate risk reduction measures. The goal of a risk assessment program is to ensure that operators, equipment as well as the environment are protected.

As pointed out by Clifton and Ericson [8], hazard identification is a critical step, where the aim is the cognitive process of hazard recognition, whereas the solutions to mitigate the risks are relatively straightforward. Etherton et al. noted that designers lack a database of known hazards during innovation and design stages [9]. The robot safety standards (ISO 10218-1 [5] and ISO 10218-2 [6]) also have tabulated a list of significant hazards whose purpose is to inform risk assessors of probable inherent dangers associated with robot and robotic systems. Therefore, a case study [10] is used to investigate the characteristics of hazards and the associated risks that are relevant for collaborative operation. The study is focused on a collaborative assembly station, where large industrial robots and operators are to share a common workspace enabled through the application of a systematic and standardized risk assessment process followed by risk reduction measures.

This article is structured as follows: in Section 2, an overall description of the methodology used to conduct the research will be presented along with limitations; Section 3 will detail theoretical background; and Section 4 will present the results of the article followed by discussion of the result and conclude with remarks on future work.

### 1.1. Background

Recently, there have been many technological advances within the areas of robot control which aims to solve perceived issues associated with robot safety [11]. A safe collaborative assembly cell, where operators and industrial robots collaborate to complete assembly tasks, is seen as an important technological solution for several reasons including (1) ability to adapt to market fluctuations and trends [12]. (2) Have the possibility to decrease takt time [13, 14]. (3) Improving working environment by decreasing the ergonomic load of the operator [15].

An automotive assembly plant is typically separated into three units: (1) the highly automated body-in-white unit where industrial robots are used to weld sheet metal parts that form the chassis; (2) the body painting unit and (3) the final assembly unit where various components of an automotive are assembled sequentially. The final assembly plants within the automotive industry can be characterized as:

1. having a high production rate, where the capacity of the plant can vary significantly depending on several factors, such as variant, plant location, etc.
2. being dependent on manual labor as the nature of assembly tasks require highly dexterous motion with good hand-eye coordination along with general decision-making skills.

Though, operators are often aided by powered tools to carry out assembly tasks such as pneumatic nut-runners as well as lifting tools, there is a need to improve the ergonomics of their work environment. As pointed by Ore et al. [15], there is demonstrable potential for collaborative operations to aid operators in various tasks including assembly and quality control.

Earlier attempts at introducing automation devices, such as cobots [13, 16], have resulted in custom machinery that functions as ergonomic support. Recently, industrial robots specifically designed for collaboration such as UR10 [17] and KUKA iiwa [18] are available that can be characterized as: (1) having the ability to detect collisions with any part of the robot structure; and (2) having the ability to carry smaller load and shorter reach compared to traditional industrial robots. This feature coupled with the ability to detect collisions fulfills the condition for power and force limiting.

Industrial robots that does not have power and force limiting feature, such as KUKA KR210 [18] or the ABB IRB 6600 [19], have traditionally been used within fenced workstations. In order to enter a robot workspace, the operator was required to deliberately open a gate, which is monitored by a safety device that stops all robot and manufacturing operations within the workstation. As mentioned before, the purpose of the research project was to explore collaborative operations where traditional industry robots are employed for assembly tasks. These robots have the capacity to carry heavy loads with long reach that can be effective for various assembly tasks. However, these advantages correspond to an inherent source of hazard that needs to be understood and managed with appropriate safety focused solutions.

## 2. Working methodology

To take advantage of the physical performance characteristics of large industrial robots along with the advances in sensor and control technologies, a research project ToMM [20] comprising of members representing the automotive industry, research, and academic institutions were tasked with understanding and specifying industry-relevant safety requirements for collaborative operations.

### 2.1. Industrial relevance

The requirements for safety that are relevant for the manufacturing industry are detailed in various standards such as ISO EN 12100 and ISO EN 10218 (parts 1 and 2) which are maintained by various organizations such as International Organization for Standardization (ISO [21]) and International Electrotechnical Commission (IEC [22]). Though these organizations do not have the authority to enforce the standards, a legislatory body such as the European Union, through the EU Machinery directive mandates compliance with normative standards [23] which are prefixed with an EN before their reference number.

### 2.2. Problem study and data collection

Objective of the research was to understand the safety requirements for high-volume assembly stations when industrial robots are to be used in a collaborative manner. A case-based approach [10] was followed, where the initial study was focused on an assembly station where a heavy engine component is assembled on an engine block. To gain a better understanding and knowledge of the case study, the following methods were employed:

1. Regular meeting in order to have detailed discussion with engineers and line managers at the assembly plant [24].
2. Visits to the plant allowed the researchers to directly observe the functioning of the station. This also enabled the researchers to have informal interviews with line workers regarding the assembly tasks as well as the working environment.
3. The researchers participated in the assembly process, guided by the operators, allowed the researchers to gain intuitive understanding of the nature of the task.
4. Literature sourced from academia, books as well as documentation from various industrial equipment manufactures were reviewed.

### 2.3. Integrating safety in early design phase

Introduction of a robot into a manual assembly cell might lead to unforeseen hazards whose potential to cause harm needs to be eliminated or minimized. The machinery safety standard [7] suggests the practice of conducting risk assessment followed by risk reduction measures to ensure the safety of the operator as well as other manufacturing processes. The risk assessment process is iterative that concludes when all probable hazards have been identified along with solutions to mitigate the effects of these hazards have been implemented. This process is usually carried out through a safety program and can be documented according to [25].

**Figure 1** depicts an overview of the safety-focused design strategy employed during the research and development phase. The case study was analyzed to understand the benefits of collaborative operations done through a conceptual study, where the overall robot, operator, and collaborative tasks were specified. Employing the results of the conceptual study, the risk assessment methodology followed by risk reduction was carried out where each phase was supported by the use of demonstrators. Björnsson [26] and Jonsson [27] have elaborated the principles of demonstrator-based design along with their perceived benefits and this methodology has been employed in this research work within the context of safety for collaborative operations.

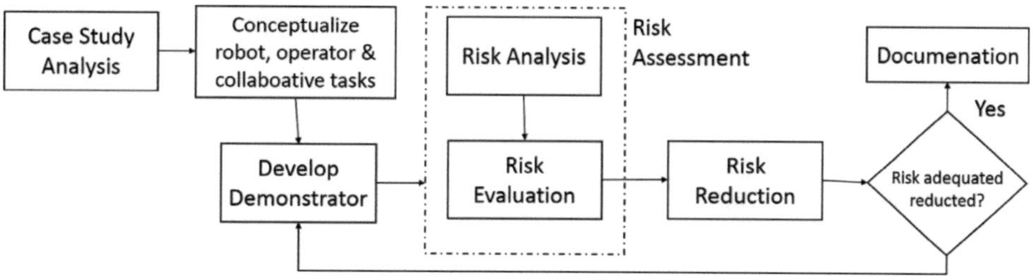


**Figure 1.** Overview of the demonstrator-based design methodology employed to ensure a safe collaborative workstation.

## 3. Theoretical background

In this section, beginning with an overview of industrial robots, concepts from hazard theory, industrial system safety and reliability, and task-based risk assessment methodology will be detailed.

### 3.1. Industrial robotic system and collaborative operations

An industrial robot is defined as an automatically controlled, reprogrammable, multipurpose manipulator, programmable in three or more axes, which can be either fixed in place or mobile for use in industrial automation applications [28]. **Figure 2(A)** shows an illustration of an articulated six-axis manipulator along with the control cabinet and a teach pendant. The control cabinet houses various control equipment such as motor controller, input/output modules, network interfaces, etc.

The teach pendant is used to program the robot, where each line of code establish the robot pose—in terms of coordinates in x, y, z and angles A, B, C—which when executed allow the robot to complete a task. This method of programming is referred to as position control, where individual robot poses are explicitly hard coded. In contrast to position control, sensor-based control allows motion control to be regulated by sensor values. Examples of sensors include vision, force and torque, etc.

On a manufacturing line, robots can be programmed to move at high speed undertaking repetitive tasks. This mode of operation is referred to as automatic mode, and allows the robot controller to execute the program in a loop, provided all safety functions are active. Additionally, ISO 10218-1 [5] has defined manual reduced-speed to allows safe programming and testing of the intended function of the robotic system, where the speed is limited to 250 mm/s at the tool center point. The manual high-speed allows the robot to be moved at high speed, provided all safety functions are activate and this mode is used for verification of the intended function.

The workspace within the robotic station where robots run in automatic mode is termed Robot Workspace (see **Figure 2(B)**). In collaborative operations, where operators and robots can share a workspace, a clearly defined Collaborative Workspace is suggested by [29]. Though the robot can be moved in automatic mode within the collaborative workspace, the speed of the robot is limited [29] and is determined during risk assessment.

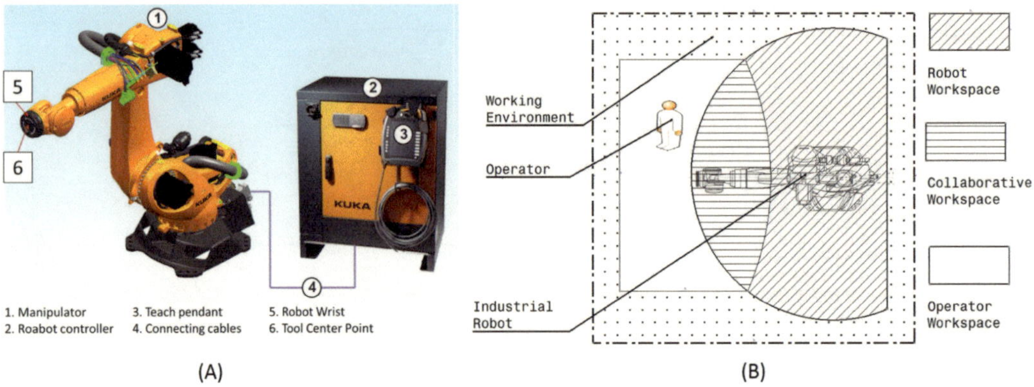


**Figure 2.** (A) An example of a manipulator along with the control box and the teach pendant. Examples include KUKA KR-210 [18] and ABB IR 6620 [19]. (B) Illustrates the interaction between the three participants of a collaborative assembly cell within their corresponding workspaces [3].

Robot safety standards recognize the implementation of one or more of the following four different modes of collaborative operation:

1. Safety-rated monitored stop stipulates that the robot ceases its motion with a category stop 2 when the operator enters the collaborative workspace. In a category stop 2, the robot can decelerate to a stop in a controlled manner.
2. Hand-guiding allows the operator to send position commands to the robot with the help of a hand-guiding tool attached at or close to the end-effector.
3. Speed and separation monitoring allows the operator and the robot to move concurrently in the same workspace provided that there is a safe separation distance between them which is greater than the prescribed protective separation distance determined during risk assessment.
4. Power and force limiting operation refers to robots that are designed to be intrinsically safe and allows contact with the operator provided it does not exert force (either quasi-static or transient contact) larger than a prescribed threshold limit.

### 3.2. Robotic system safety and reliability

An industrial robot normally functions as part of an integrated manufacturing system (IMS) where multiple subsystems that perform different functions operate cohesively. As noted by Levenson (page 14 [30]), safety is a system property (not a component property) and needs to be controlled at the system level. This implies that safety as a property needs to be considered at early design phases, which Ericson (page 34 [8]) refers to as CD-HAT or Conceptual Design Hazard Analysis Type. CD-HAT is the first seven types of hazard analysis types, which needs to be considered during various design phases in order to avoid costly design rework.

http://dx.doi.org/10.5772/intechopen.70607

To realize a functional IMS, a coordinated effort in the form of a system safety program (SSP [8]) which involve participants with various levels of involvement (such as operators, maintenance, line managers, etc.) are carried out. Risk assessment and risk reduction processes are conducted in conjecture with the development of an IMS, in order to promote safety, during development, commissioning, maintenance, upgradation, and finally decommissioning.

#### 3.2.1. *Functional safety and sensitive protective equipment (SPE)*

Functional safety refers to the use of sensors to monitor for hazardous situations and take evasive actions upon detection of an imminent hazard. These sensors are referred to as sensitive protective equipment (SPE) and the selection, positioning, configuration, and commissioning of equipment have been standardized and detailed in IEC 62046 [31]. IEC 62046 defines the performance requirements for this equipment and as stated by Marvel and Norcross [32], when triggered, these sensors use electrical safety signals to trigger safety function of the system. They include provisions for two specific types: (1) electro-sensitive protective equipment (ESPE) and (2) pressure-sensitive protective equipment (PSPE). These are to be used for the detection of the presence of human beings and can be used as part of the safety-related system [31].

Electro-sensitive protective equipment (ESPE) uses optical, microwaves, and passive infrared techniques to detect operators entering a hazard zone. That is, unlike physical fence, where the operators and the machinery are physically separated, ESPE relies on the operators to enter a specific zone for the sensor to be triggered. Examples include laser curtains [33], laser scanners [34], and vision-based safety systems such as the SafetyEye [35].

Pressure-sensitive protective equipment (PSPE) has been standardized in parts 1–3 of ISO13856, and works on the principle of an operator physically engaging a specific part of the workstation. These include: (1) ISO 13856-1—pressure sensitive mats and floors [36]; (2) ISO 13856-2—pressure sensitive bars, edges [37]. (3) ISO 13856-3—bumpers, plates, wires, and similar devices [38].

#### 3.2.2. *System reliability*

Successful robotic systems are both safe to use and reliable in operation. In an integrated manufacturing system (IMS), reliability is the probability that a component of the IMS will perform its intended function under pre-specified conditions [39]. One measure of reliability is MTTF (mean time to failure) and ranges of this measure has been standardized into five discrete level levels or performance levels (PL) ranging from a to e. For example, PL = d refers to a $10^{-6} > \text{MTTF} \geq 10^{-7}$, which is the required performance level with a category structure 3 ISO 10218-2 (page 10, Section 5.2.2 [6]). That is, in order to be viable to the industry, the final design of the robotic system should reach or exceed the minimum required performance level.

### 3.3. Hazard theory: hazards, risks, and accidents

Ericson [8] states that a mishap or an accident is an event which occurs when a hazard, or more specifically hazardous element, is actuated upon by an initiating mechanism. That is, a hazard is a pre-requisite for an accident to occur and is defined as a potential source of harm [7] and is composed of three basic components: (1) hazardous element (HE), (2) initiating mechanism (IM), and (3) target/threat (T/T).

A hazardous element is a resource that has the potential to create a hazard. A target/threat is the person or the equipment directly affected when the hazardous element is activated by an initiating mechanism. These three components, when combined together, can be referred to as a hazard (see **Figure 3(A)**) and are essential components for it to exist. Based on these definitions, if any of the three components are removed or eliminated, by any means (see Section 3.4.2), it is possible to eliminate or reduce the effect of the hazard.

To better illustrate these concepts, consider the fatal accident that took place on July 21, 1984, where an experienced operator entered a robotic workstation while the robot was in automatic mode (see **Figure 3(B)**). The robot was programmed to grasp a die-cast part, dip the part in a quenching tank and place it on an automatic trimming machine. According to Lee et al. [40], the operator was found pinned between the robot and a safety-pole by another operator of an adjacent die-cast station who became curious after hearing the hissing noise of the air-hose for 10–15 min. The function of the safety pole was to limit robot motion and together with the robot-arm can be considered to be a hazardous element. The hazard was initiated by the operator who intentionally entered the workstation either by jumping over the rails or through a 19-inch unguarded spacing and caused the accident. The operator was the target of this unfortunate accident and was pronounced dead after 5 days of the accident.

A hazard is designed into a system [8, 30] and for accident to occur depends on two factors: (1) unique set of hazard components and (2) accident risk presented by the hazard components, where risk is defined

$$Risk = Probability \times Severity \qquad (1)$$

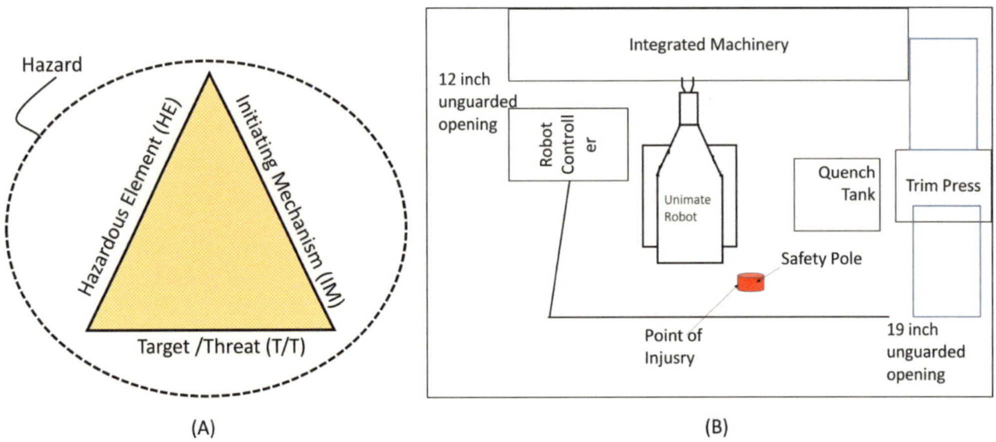


**Figure 3.** (A) The hazard triangle where the three components of hazards—hazardous element, initiating mechanism, and target/threat—are essential and required for the hazard to exist (adapted from page 17 [8]). (B) Shows the layout of the robotic workstation where a fatal accident took place on July 21, 1984 [40].

Ericson notes that a good hazard description can support the risk assessment team to better understand the problem and therefore can enable them to make better judgments (e.g., understanding the severity of the hazard), and therefore suggest that the a good hazard description needs to contain the three hazard components.

### 3.4. Task-based risk assessment and risk reduction

Risk assessment is a general methodology where the scope is to analyze and evaluate risks associated with complex system. Various industries have specific methodologies with the same objective. Etherton has summarized a critical review of various risk assessment methodologies for machine safety in [41]. According to ISO 12100, risk assessment (referred to as MSRA—machine safety risk assessment [41]) is an iterative process which involves two sequential steps: (1) risk analysis and (2) risk evaluation. ISO 12100 suggests that if risks are deemed serious, measures should be taken to either eliminate or mitigate the effects of the risks through risk reduction as depicted in **Figure (4)**.

#### *3.4.1. Risk analysis and risk evaluation*

Within the context of machine safety, risk analysis begins with identifying the limits of machinery, where the limits in terms of space, use, time are identified and specified. Within this boundary, activities focused on identifying hazards are undertaken. The preferred context for identifying hazards for robotics systems is task-based, where he tasks that needs to be undertaken during various phases of operations are first specified. Then the risk assessors specify the hazards associated with each tasks. Hazard identification is a critical step and ISO 10218-1 [5] and ISO 10218-2 [6] tabulates significant hazards associated with robotic systems. However, they do not explicitly state the hazards associated with collaborative operations.

Risk evaluation is based on a systematic metrics where severity of injury, exposure to hazard and avoidance of hazard are used to evaluate the hazard (see page 9, RIA TR R15.306-2014 [25]). The evaluation results in specifying the risk level in terms of negligible, low, medium-high, and very-high, and determine risk reduction measures to be employed. To support the activities associated with risk assessment, ISO TR 15066 [29] details information required to conduct risk assessment specifically for collaborative applications.

#### *3.4.2. Risk reduction*

When risks are deemed serious, the methodology demands measures to eliminate and/or mitigate the risks. The designers have a hierarchical methodology that can be employed to varying degree depending on the risks that have to be managed. The three hierarchical methods allow the designers to optimize the design and can choose either one or a combination of the methods to sufficiently eliminate/mitigate the risks. They are: (1) inherently safe design measures; (2) safeguarding and/or complementary protective measures; and (3) information for use.

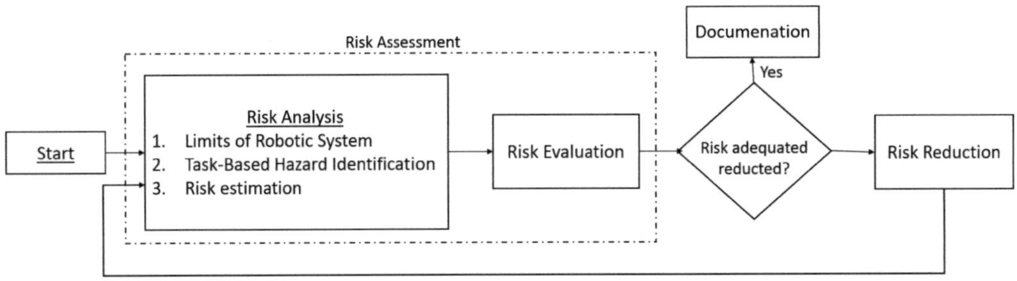


**Figure 4.** An overview of the task-based risk assessment methodology.

# 4. Result: demonstrator for a safe hand-guided collaborative operation

In this section, the development and functioning of a safe assembly station will be detailed, where a large industrial robot is used in a hand-guided collaborative operation. In order to understand potential benefits with hand-guided industrial robots, an automotive assembly station will be presented as a case study in Section 4.1. With the aim to improve the ergonomics of the assembly station and increase the productivity, the assembly tasks are conceptualized as robot, operator, and collaborative task where the collaborative task is the hand-guided operation and is described in Section 4.2. The results of the iterative risk assessment and risk reduction process (see Section 3.4) will be detailed in Section 4.3. The final layout and the task sequence will be detailed in Section 4.4, and **Table 1** will document the hazards that were identified during risk assessment that were used to improve the safety features of the assembly cell.

## 4.1. Case study: manual assembly of a flywheel housing cover

The assembly task is to install a flywheel housing cover (FWC) on the engine block with an intermediate step between the picking of the FWC from the material rack and securing it on the engine block with fasteners. The assembly of FWC, which weighs 20 kg, is a manual operation and these tasks are carried out by one or more operators (see **Figure 5(A)**) and can be described as follows:

1. An operator picks up the flywheel housing cover (FWC) with the aid of a lifting device from position P1. The covers are placed on a material rack and can contain upto three part variants.
2. This operator moves from position P1 to P2 by pushing the FWC and installs it on the machine (integrated machinery) where secondary operations will be performed.
3. After the secondary operation, the operator pushes the FWC to the engine housing (position P3). Here, the operator needs to align the flywheel housing cover with the engine block with the aid of guiding pins. After the two parts are aligned, the operator pushes the flywheel housing cover forward until the two parts are in contact. The operator must exert force to mate these two surfaces.
4. Then the operators begin to fasten the parts with several bolts with the help of two pneumatically powered devices. In order to keep low takt time, these tasks are done in parallel and require the participation of more than one operator.

| No. | Hazard description | Hazardous element (HE) | Initiating mechanism (IM) | Target/threat (T/T) | Risk reduction measure |
|---|---|---|---|---|---|
| 1. | The operator can accidentally enter robot workspace and collide with the robot moving at high speed | Fast moving robot | Operator is unaware of the system state | Operators | 1. A light curtain to monitor the robot workspace. 2. A lamp to signal the system state |
| 2. | In collaborative mode, sensor-guided motion is active. Robot motion can be triggered unintentionally resulting in unpredictable motion | Crushing | Operator accidentally activate the sensor, | Operator(s) and/or equipment(s) | An enabling device, when actuated, will start sensor-guided motion. An ergonomically designed enabling device can act as a hand-guiding tool |
| 3. | The operator places their hands between the FWC and the engine, thereby crushing their hands | Crushing | Operator distracted due to assembly task | Operator | An enabling device can ensure that the operator's hands are at a predefined location. |
| 4. | While aligning the pins with the holes, the operator can break the pins by moving vertically or horizontally | Imprecise hand-guided motion | Operator fails to keep steady motion | Operators | 1. Vertical hand-guided motion needs to be eliminated. 2. Operator training |
| 5. | The robot collides with an operator while being hand-guided by another operator | Collision | Designated operator is not aware of others in the vicinity | Operators | The designated operator has clear view of the station |
| 6. | An operator accidentally engages mode-change button though the collaborative task is incomplete | Error in judgment of the operators | Engaging the mode-change button | Operator/ equipment | A button on the hand-guiding tool that the operator engages before exiting the workspace |

**Table 1.** The table describes the hazards that were identified during the risk assessment process.

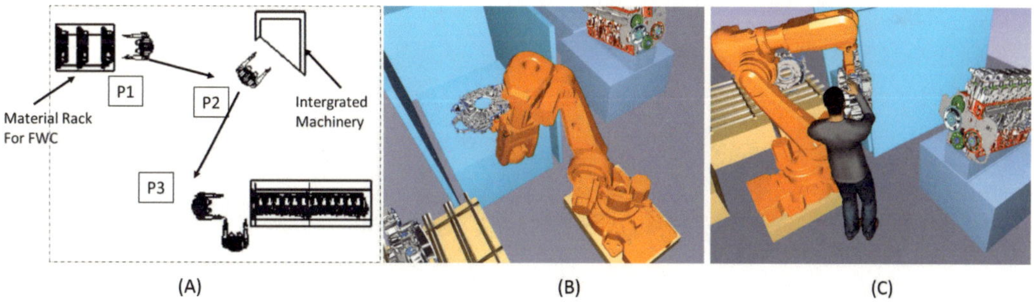


**Figure 5.** (A) Shows the manual workstation where several operators work together to assemble flywheel housing covers (FWC) on the engine block. (B) Shows the robot placing the FWC on the integrated machinery. (C) Shows the robot being hand-guided by an operator thereby reducing the ergonomic effort to position the flywheel housing cover on the engine block.

## 4.2. Task allocation and conceptual design of the hand-guiding tool

**Figure 5(B)** and **(C)**, shows ergonomic simulations reported by Ore et al. [15] and shows the operator being aided by an industrial robot to complete the task. The first two tasks can be automated by the robot, i.e., picking the FWC from Position 1 and moving it to the integrated machine (position P2, **Figure 5(B)**). Then, the robot moves the FWC to the hand over position where the robot will come to a stop and signal to the operator that the collaborative mode is activated. This allows the operator to hand-guide the robot by grasping the FWC and directing the motion towards the engine block.

Once the motion of the robot is under human control, the operator can assemble the FWC onto the engine block and proceeds to secure it with bolts. After the bolts have been fastened, the operator then moves the robot back to the hand-over position and reactivates the automatic mode which starts the next cycle.

### 4.3. Safe hand-guiding in the collaborative workspace

The risk assessment identified several hazardous situations that can affect the safe functioning during the collaborative mode—that is when the operator goes into the workstation and hand-guides the robot to assemble the FWC—and has been tabulated in **Table 1**.

**Figure 6(A)** and **(B)** shows two versions of the end-effector that was developed to support hand-guided robotic assembly. The safety focused design of the hand-guiding tool shown in **Figure 6(A)** has been detailed by Gopinath et al. [42] where the interfaces are part of the end-effector. That is, in an open enclosure (without physical fences—not shown), the location for the interfaces and control devices would optimally be a design feature of the end-effector. However, risk assessment pointed out that an open enclosure might require the following safety measures:

1. The robot needs to be programmed to move at slow speed so that it can stop (in time) according to speed and separation monitoring mode of collaborative operation.
2. To implement speed and separation monitoring, a safety rated vision system might be probable solution. However, this may not be viable solution on the current factory floor.

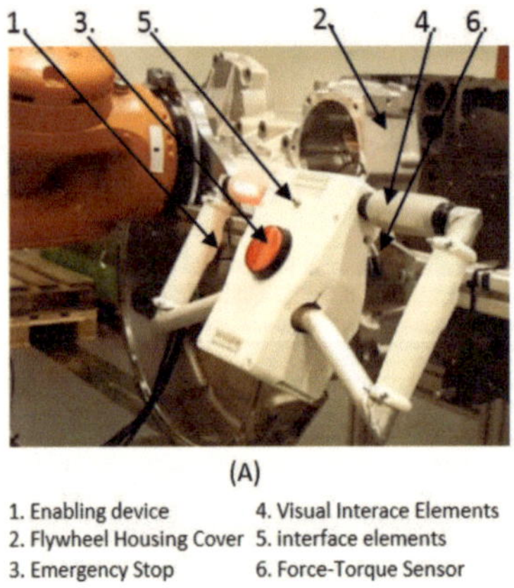


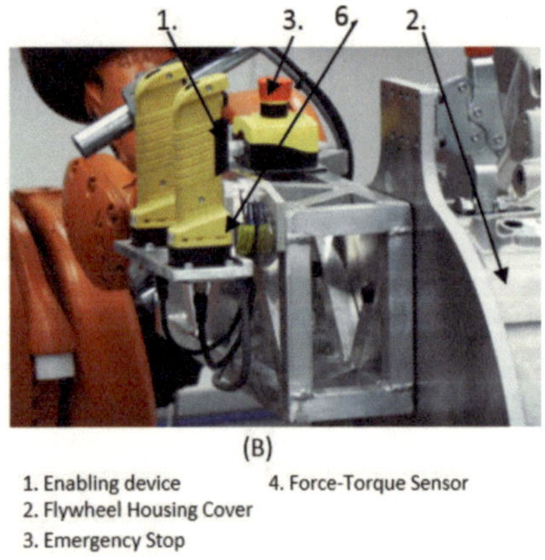


**Figure 6.** (A) and (B) are two versions of the end-effector that was prototyped to verify and validate the design.

The limited space, high volume and the nature of the hazards puts severe restriction on the type of safety solution that can be considered. An enclosed station is shown in **Figure 7**, where physical fences are being used as a safeguarding measure to limit personnel movement, thereby eliminating the possibility of operator accidently entering the robot workspace. The layout of this collaborative station has been detailed by Gopinath et al. [43] and in **Table 2**, a comparison of the design features has been discussed. The change from Design A to Design B was motivated by change in requirements namely:

1. A change in design that would allow the operator to visually align the pins on the engine block with the mating holes on the FWC.
2. A change in design to improve reliability as well as avoid tampering through the use of standardized components. Ensure that the operator feel safer during hand-guiding by ensuring that the robot arms are not close to the operator.

### 4.4. Demonstrator for a safe hand-guided collaborative assembly workstation

**Figure 7** shows a picture of the demonstrator developed in a laboratory environment. Here, a KUKA KR-210 industrial robot is part of the robotic system where the safeguarding solutions include the use of physical fences as well as sensor-based solutions.

**Figure 8** describes the sequence of task necessary to complete the assembly operations. These tasks have been separated into three, i.e., robot, operator, and collaborative tasks and can be described as follows:

1. The robot tasks, which are preprogramed tasks undertaken in automatic mode. When the robot tasks are completed, it is programmed to stop at the hand-over position.
2. The collaborative task which begins when the operators enters the monitored space and takes control of the robot using the hand-guiding device. The collaborative mode is complete when the operator returns the robot to the hand-over position and restarts the automatic mode.
3. The operator task is the fastening of the bolts required to secure the FWC to the engine block. The operators need to fasten several bolts and therefore use pneumatically powered tool (not shown here) to help them with this task.

#### *4.4.1. Safeguarding*

With an understanding that operators are any personnel within the vicinity of hazardous machinery [7], the physical fences can be used to ensure that they do not accidentally enter a hazardous zone. The design requirements stated that the engine block needs to be outside the enclosed zone, meant that the robot needs to move out of the fenced area during collaborative mode (see **Figure 8**). Therefore, the hand over position is located inside the enclosure and the assembly point is located outside of the enclosure and both these points are part of the collaborative workspace. The opening in the fences is monitored during automatic mode using laser curtains.

**Figure 7.** The layout of the physical demonstrator installed in a laboratory environment.

| Design feature | Design A | Design B | Design evaluation |
|---|---|---|---|
| 1. Orientation of the end-effector | End-effector is parallel to the robot wrist | End-effector is perpendicular to the robot wrist. | In design A, the last two links of the robot are close to the operator which might make the operators feel unsafe. Design B might allow for an overall safer design due to use of standardized components |
| 2. Position of Flywheel housing cover (FWC) | The FWC is positioned left to the operator | The FWC is positioned in front of the operator | Design A requires more effort from the operator to align the locating pins (on the engine block) and the mating holes (on the FWC). The operator loses sight of the pins when the two parts are close to each other. In Design B, it is possible to align the two parts by visually aligning the outer edges |
| 3. Location of Emergency stop | Good location and easy to actuate | Good location and easy to actuate | In design A, it was evaluated that the E-stop can be accidentally actuated which might lead to unproductive stops |
| 4. Location of visual interfaces | Good location and visibility | No visual interfaces | Evaluation of design A resulted in the decision that interfaces need to be visible to all working within the vicinity |
| 5. Location of physical interfaces | Good location with easy reach. | Minimal physical interfaces | Evaluation of design A resulted in the decision that interfaces are optimally placed outside the fences area |
| 6. Overall ergonomic design | The handles are angled and is more comfortable | The distance between the handles is short | Designs A and B have good overall design. Design B uses standardized components. Design A employs softer materials and interfaces that are easily visible |

**Table 2.** Feature comparison of two versions of the end-effector shown in **Figure 6(A)** and **(B)**.

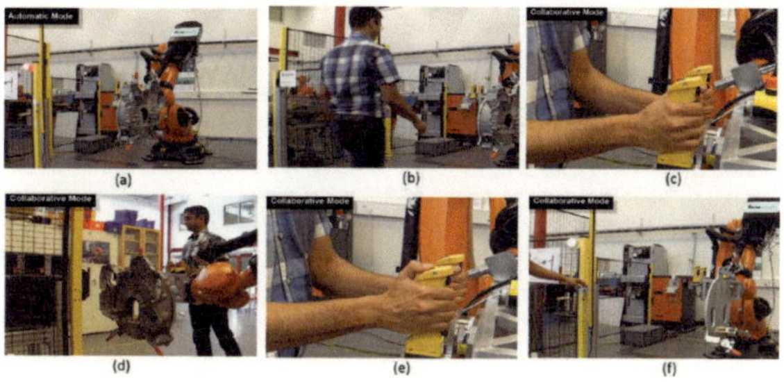

**Figure 8.** The figure describes the task sequence of the collaborative assembly station where an industrial robot is used as an intelligent and flexible lifting tool. The tasks are decomposed into three — Operator task (OT), Collaborative task (CT) and Robot task (RT) — which are detailed in **Table 3**.

### *4.4.2. Interfaces*

During risk evaluation, the decision to have several interfaces was motivated. A single warning LED lamp (see **Figure 8**) can convey that when the robot has finished the preprogrammed task and waiting to be hand-guided. Additionally, the two physical buttons outside the enclosure has separate functions. The Auto-continue button allows the operator to let the robot continue in automatic mode if the laser curtains were accidentally triggered by an operator and this button is located where it is not easily reached. The second button is meant to start the next assembly cycle (see **Table 1**). **Table 1** (Nos. 2 and 3) motivates the use of enabling devices to trigger the sensor guided motion (see **Figure 6(B)**). The two enabling devices provide the following functions: (1) it acts as a hand-guiding tool that the operator can use to precisely maneuver the robot. (2) By specifying that the switches on the enabling device are engaged for hand-guiding motion, the operators hands are at a prespecified and safe location. (3) Additionally, by engaging the switch, the operator is deliberately changing the mode of the robot to collaborative-mode. This ensures that unintended motion of the robot is avoided.

| Tasks | Task description |
|---|---|
| 1. Robot task | The robot tasks are to pick up the flywheel housing cover, place the part on the fixture and when the secondary operators are completed, pick up the part and wait at the hand-over position. During this mode, the warning lamp is red, signaling automatic mode. The hand-over position is located inside the enclosed area and is monitored by laser curtains. The robot will stop if an operator accidentally enters this workspace and can be restarted by the auto-continue button (**Figure 7**) |
| 2. Operator task | Enter collaborative space: When the warning lamp turns to green, the laser curtains are deactivated; the operator enters the collaborative workspace |
| 3. Collaborative task | Engage enabling switch: the operator begins hand-guiding by engaging both the enabling switches simultaneously. This activates the sensor-guided motion and the operator can move the robot by applying force on the enabling device. If the operator releases the enabling switch, the motion is deactivated (see point 2 in **Table 1**). To reactivate motion, the operator engages both the enabling switches |
| 4. Collaborative task | Hand-guide the robot: the operator moves the FWC from the hand-over position to the assembly point. Then removes the clamp and return the robot back to the hand-over position |

| Tasks | Task description |
|---|---|
| 5. Collaborative task | Engage automatic mode: before going out of the assembly station, the operator needs to engage the three-button switch. This deliberate action signals to the robot that the collaborative task is complete |
| 6. Robot task | The operator goes out and engages the mode-change button. Then, the following sequence of events is carried out: (1) laser curtains are activated, (2) warning lamp turns from green to red, and (3) the robot starts the next cycle |

**Table 3.** The table articulates the sequence of tasks that were formulated during the risk assessment process.

# 5. Discussion

In this section, the discussion will be focused on the application of the risk assessment methodology and the hazards that were identified during this process.

## 5.1. Task-based risk assessment methodology

A risk assessment (RA) is done on a system that exists in a form that can function as a context within which hazards can be documented. In the case study, a force/torque sensor was used to hand-guide the robot and this technique was chosen at the conceptual stage. RA based on this technique led to decision of introducing enabling devices (No. 2 in **Table 1**) to ensure that, while the operator is hand guiding the robot, the hands are within a predetermined safe location and is engaged. Another industrially viable solution is the use of joysticks to hand-guide the robot but this option was not explored further during discussion as it might be less intuitive than force/torque based control. Regardless, it is implicit that the choice of technique poses its own hazardous situation and the risk assessors need a good understanding of the system boundary.

Additionally, during risk assessment, the failure of the various components was not considered explicitly. For example, what if laser curtains failed to function as intended? The explanation lies in the choice of components. As stated in Section 3.2.2, a robotic system to be considered reliable, the components must have a performance level PL = d, which implies a very low probability of failure. Most safety-equipment manufactures publish their MTTF values along with their performance levels and the intended use.

## 5.2. Hazards

The critical step in conducting risk assessment (RA) is hazard identification. In Section 3.3, a hazard was decomposed into three: (1) hazardous element (HE), (2) initiating mechanism (IM), and (3) target/threat (T/T). The three sides of the hazard triangle (Section 3.3) have lengths proportional to the degree with which these components can trigger the hazard and cause an accident. That is, if the length of IM side is much larger than the other two, then the most influencing factor to cause an accident is IM. The discussion on risk assessment (Section 3.4) stresses on eliminating/mitigating hazards which implies that the goal of risk assessment can

http://dx.doi.org/10.5772/intechopen.70607

be understood as a method to reduce/remove one or more of the sides of the hazard triangle. Therefore, documenting the hazards in terms of its components might allow for simplified and straightforward downstream RA activities.

The hazards presented in **Table 1** can be summarized as follows: (1) the main source of hazardous element (HE) is slow/fast motion of the robot. (2) The initiating mechanism (IM) can be attributed to unintended actions by an operator. (3) The safety of the operator can be compromised and has the possibility to damage machinery and disrupt production. It can also be motivated, based on the presented case study, that through the use of systematic risk assessment process, hazards associated with collaborative motion can be identified and managed to an acceptable level of risk.

As noted by Eberts and Salvendy [44] and Parsons [45], human factors play a major role in robotic system safety. There are various parameters that can be used to better understand the effect of human behavior in system such as overloaded and/or underloaded working environment, perception of safety, etc. The risk assessors need to be aware of human tendencies and take into consideration while proposing safety solutions. Incidentally, in the fatal accident discussed in Section 3.3, perhaps the operator did not perceive the robot as a serious threat and referred to the robot as Robby [40].

In an automotive assembly plant, as the production volume is relatively high and requires collaborating with other operators, there is a higher probability for an operator to make errors. In **Table 1** (No. 6), a three-button switch was specified to ensure unintentional mode change of the robot. It is probable that an operator can accidentally engage the mode-change button (see **Figure 7**) while the robot is in collaborative mode or the hand-guiding operator did not intend the collaborative mode to be completed. In such a scenario, a robot operating in automatic mode was evaluated to have a high risk level, and therefore the decision was made to have a design change with an additional safety-interface—the three-button switch—that is accessible only to the hand-guiding operator.

Informal interviews suggested that the system should be inherently safe for the operators and that the task sequence—robot, operator, and collaborative tasks—should not demand constant monitoring by the operators as it might lead to increased stress. That is, operators should feel safe and in control and that the tasks should demand minimum attention and time.

## 6. Conclusion and future work

The article presents the results of a risk assessment program, where the objective was the development of an assembly workstation that involves the use of a large industrial robot in a hand-guiding collaborative operation. The collaborative workstation has been realized as a laboratory demonstrator, where the robot functions as an intelligent lifting device. That is, the tasks that can be automated have been tasked to the robot and these sequences of tasks are preprogrammed and run in automatic mode. During collaborative mode, operators are responsible for tasks that are cognitively demanding that require the skills and flexibility inherent to a human being. During this mode, the hand-guided robot carries the weight of the flywheel housing cover, thereby improving the ergonomics of the workstation.

In addition to the laboratory demonstrator, an analysis of the hazards pertinent to hand-guided collaborative operations has been presented. These hazards were identified during the risk assessment phase, where the hazardous element mainly stems from human error. The decisions taken during the risk reduction phase to eliminate or mitigate the risks associated with these hazards have also been presented.

The risk assessment was carried out through different phases, where physical demonstrators supported each phase of the process. The demonstrator-based approach allowed the researchers to have a common understanding of the nature of the system and the associated hazards. That is, it acted as platform for discussion. The laboratory workstation can act as a demonstration platform where operators and engineers can judge for themselves the advantage and disadvantages of collaborative operations. The demonstration activities can be beneficial to researchers as it can function as a feedback mechanism with respect to the decisions that have been made during the risk assessment process.

Therefore, the next step is to invite operators and engineers in trying out the hand-guided assembly workstation. The working hypothesis in inviting operators and engineers is that, personnel whose main responsibility during their working time in an assembly plant is to find the optimal balance between various production related parameters (such as maintenance time, productivity, safety, working environment, etc.) might have deeper insight into the challenges of introducing large industrial robots in the assembly line.

## Acknowledgements

The authors would like to thank Björn Backman of Swerea IVF, Fredrik Ore and Lars Oxelmark of Scania CV for their valuable contributions during the research and development phase of this work. This work has been primarily funded within the FFI program and the authors would like to graciously thank them for their support. In addition, we would like to thank ToMM 2 project members for their valuable input and suggestions.

## Author details

Varun Gopinath*, Kerstin Johansen and Johan Ölvander

*Address all correspondence to: varun.gopinath@liu.se

Division of Machine Design, Department of Management and Engineering, Linköping University, Sweden

## References

[1] Marvel JA, Falco J, Marstio I. Characterizing task-based human-robot collaboration safety in manufacturing. IEEE Transactions on Systems, Man, and Cybernetics: Systems. 2015;**45**(2):260-275

[2] Tsarouchi P, Matthaiakis A-S, Makris S. On a human-robot collaboration in an assembly. International Journal of Computer Integrated Manufacturing. 2016;**30**(6):580-589

[3] Gopinath V, Johansen K. Risk assessment process for collaborative assembly—A job safety analysis approach. Procedia CIRP. 2016;**44**:199-203

[4] Caputo AC, Pelagagge PM, Salini P. AHP-based methodology for selecting safety devices of industrial machinery. Safety Science. 2013;**53**:202-218

[5] Swedish Standards Institute. SS-ISO 10218-1:2011—Robots and Robotic Devices—Safety Requirements for Industrial Robots. Part 1: Robot. Stockholm, Sweden: Swedish Standards Institute; 2011

[6] Swedish Standards Institute. SS-ISO 10218-2:2011—Robots and Robotic Devices—Safety Requirements for Industrial Robots. Part 2: Robot Systems and Integration. Stockholm, Sweden: Swedish Standards Institute; 2011

[7] Swedish Standards Institute (SIS). SS-ISO 12100:2010: Safety of Machinery - General principles of Design - Risk assessment and risk reduction. Stockholm, Sweden: Swedish Standards Institute (SIS); 2010. 96 p

[8] Clifton A, Ericson II. Hazard Analysis Techniques for System Safety. Hoboken, New Jersey, USA: John Wiley & Sons; 2015

[9] Etherton J, Taubitz M, Raafat H, Russell J, Roudebush C. Machinery risk assessment for risk reduction. Human and Ecological Risk Assessment: An International Journal. 2001;**7**(7):1787-1799

[10] Robert K.Yin. Case Study Research - Design and Methods. 5th ed. California, USA: Sage Publications Inc; 2014. 282 p

[11] Brogårdh T. Present and future robot control development – An industrial perspective. Annual Reviews in Control. 2007;**31**(1):69-79

[12] Krüger J, Lien TK, Verl A. Cooperation of human and machines in assembly lines. CIRP Annals - Manufacturing Technology. 2009;**58**(2):628-646

[13] Siciliano B, Khatib O, editors. Springer Handbook of Robotics. Secaucus, NJ, USA: Springer-Verlag New; 2007

[14] Krüger J, Bernhardt R, Surdilovic D. Intelligent assist systems for flexible. CIRP Annals - Manufacturing Technology. 2006;**55**(1):29-32

[15] Ore F, Hanson L, Delfs N, Wiktorsson M. Human industrial robot collaboration—Development and application of simulation software. International Journal of Human Factors Modelling and Simulation. 2015;**5**:164-185

[16] J. Edward Colgate, Michael Peshkin, Witaya Wannasuphoprasit. Cobots: Robots for collaboration with human operators. Proceedings of the ASME Dynamic Systems and Control Division. Atlanta, GA. 1996;**58**:433-440

[17] Universal Robots. Universal Robots [Internet]. Available from: https://www.universal-robots.com/ [Accessed: March 2017]

[18] KUKA AG. Available from: http://www.kuka.com/ [Accessed: March 2017]

[19] ABB AB. Available from: http://www.abb.com/ [Accessed: January 2017]

[20] ToMM2—Framtida-samarbete-mellan-manniska-och-robot/. Available from: https://www.vinnova.se/ [Accessed: June 2017]

[21] The International Organization for Standardization (ISO). Available from: https://www.iso.org/home.html [Accessed: June 2017]

[22] International Electrotechnical Commission (IEC). Available from: http://www.iec.ch/ [Accessed: June 2017]

[23] David Macdonald. Practical machinery safety. 1st ed. Jordan Hill, Oxford: Newnes; 2004. 304 p

[24] Leedy PD, Ormrod JE. Practical Research: Planning and Design. Upper Saddle River, New Jersey: Pearson; 2013

[25] Robotic Industrial Association. RIA TR R15.406-2014: Safeguarding. 1st ed. Ann Arbour, Michigan, USA: Robotic Industrial Association; 2014. 60 p

[26] Andreas Björnsson. Automated layup and forming of prepreg laminates [dissertation]. Linköping: Linköping University; 2017

[27] Marie Jonsson. On Manufacturing Technology as an Enabler of Flexibility: Affordable Reconfigurable Tooling and Force-Controlled Robotics [dissertation]. Linköping, Sweden: Linköping Studies in Science and Technology. Dissertations: 1501; 2013

[28] Swedish Standards Institute. SS-ISO 8373:2012—Industrial Robot Terminology. Stockholm, Sweden: Swedish Standards Institute; 2012

[29] The International Organization for Standardization. ISO/TS 15066: Robots and robotic devices—Collaborative robots. Switzerland: The International Organization for Standardization; 2016

[30] Leveson NG. Engineering a Safer World: Systems Thinking Applied to Safety. Engineering Systems ed. USA: MIT Press; 2011

[31] The International Electrotechnical Commision. IEC TS 62046:2008 – Safety of machiner-Application of protective equipment to detect the presence of persons. Switzerland: The International Electrotechnical Commision; 2008

[32] Marvel JA, Norcross R. Implementing speed and separation monitoring in collaborative robot workcells. Robotics and Computer-Integrated Manufacturing. 2017;**44**:144-155

[33] SICK AG. Available from: http://www.sick.com [Accessed: December 2016]

[34] REER Automation. Available from: http://www.reer.it/ [Accessed: December 2016]

[35] Pilz International. Safety EYE. Available from: http://www.pilz.com/ [Accessed: May 2014]

[36] The International Organization for Standardization. ISO 13856-1:2013 – Safety of machinery – Pressure-sensitive protective devices – Part 1: General principles for design and testing of pressure-sensitive mats and pressure-sensitive floors. Switzerland: The International Organization for Standardization; 2013

[37] The International Organization for Standardization. ISO 13856-2:2013 – Safety of machinery – Pressure-sensitive protective devices – Part 2: General principles for design and testing of pressure-sensitive edges and pressure-sensitive bars. Switzerland: The International Organization for Standardization; 2013

[38] The International Organization for Standardization. ISO 13856-3:2013 – Safety of machinery – Pressure-sensitive protective devices – Part 3: General principles for design and testing of pressure-sensitive bumpers, plates, wires and similar devices. Switzerland: The International Organization for Standardization; 2013

[39] Dhillon BS. Robot reliability and Safety. New York: Springer-Verlag; 1991

[40] Sanderson LM, Collins JW, McGlothlin JD. Robot-related fatality involving a U.S. manufacturing plant employee: Case report and recommendations. Journal of Occupational Accidents. 1986;**8**:13-23

[41] Etherton JR. Industrial machine systems risk assessment: A critical review of concepts and methods. Risk Analysis. 2007;**27**(1):17-82

[42] Varun Gopinath, Kerstin Johansen, and Åke Gustafsson. Design Criteria for Conceptual End Effector for Physical Human Robot Production Cell. In: Swedish Production Symposium; Göteborg, Sweden;2014

[43] Gopinath V, Ore F, Johansen K. Safe assembly cell layout through risk assessment—An application with hand guided industrial robot. Procedia CIRP. 2017;**63**:430-435

[44] Eberts R, Salvendy G. The contribution of cognitive engineering to the safe. Journal of Occupational Accidents. 1986;**8**:49-67

[45] McIlvaine Parsons H. Human factors in industrial robot safety. Journal of Occupational Accidents. 1986;**8**(1-2):25-47

Chapter 10

# Managing Technogenic Risks with Stakeholder Cooperation

Riitta Molarius

Additional information is available at the end of the chapter

http://dx.doi.org/10.5772/intechopen.70903

**Abstract**

Risks involved in new technologies or arising from novel configurations of old technologies recurrently result in major accidents. For example, the new bioleaching technology to extract nickel from ore was taken into use in Finland in 2008. Later, one of the personnel died as a victim of hydrogen sulfide exposure and there were unplanned releases of process waters that contaminated lakes and rivers. Several risk analyses were performed but none of them considered the local climate and surrounding environmental circumstances. A comprehensive risk assessment process combining the knowledge of different stakeholders, authorities, and citizens would have helped to avoid the sad outcome. A single enterprise has a very clear picture of the risk figure on its own, but is reluctant to reveal commercially sensitive information to others, and even incapable of understanding all the expectations and constraints that the natural and built environments may impose. Only governmental authorities are in a position to form a comprehensive picture of all the risks. This paper presents a new approach for a proactive risk identification method based on collaborative integrated assessment. It states that by implementing this method society is able to utilize the science-based information in an efficient way for managing the emerging technogenic risks.

**Keywords:** risk identification, risk assessment, technogenic risk, stakeholder cooperation, hydrogen, fuel cell

## 1. Introduction

New technologies have been involved in several accidents. For example, in Germany, the novel magnetic levitation train collided with a maintenance vehicle and killed 23 people in 2006 [1]. Ten years later, in 2016, two trains collided again in Germany, killing 11 people and injuring 85 people despite the automatic braking system that ought to have been at work [2].

In 2016, the autonomous car claimed its first victim when the radar system of the autopilot failed to recognize a truck that was crossing the road in Florida [3]. In June 2017, at least 80 people died in a huge fire that took place in a London tower block that was covered with a new type of cladding, which included polyethylene foam [4]. In 2008, the new bioleaching technology used to extract nickel from ore was taken into commercial use in Finland, the first to adopt it in Europe [5]. Four years later, in 2012, one of the personnel died (as a victim of hydrogen sulfide exposure) due to a lack of safety equipment [6]; additionally, there were significant challenges associated with the management of the process waters, which consequently resulted in the company filing for bankruptcy [7]. Today drones hit people all over the world and collisions with helicopters and planes are just a matter of time [8].

Sociologist Ulrich Beck warned us over 20 years ago about the "risk society" where society is gradually exposed to the risks it creates and finally the negative effects of the progress become greater than the positive impacts [9]. The fast change and development of new kinds of technologies have increased likelihood and probability of technogenic risks even if there is a strong attempt to identify and anticipate risks. The term "technogenic risk" stands here for risks whose origin is in man-made technology, also including newer technologies, such as nanotechnology, biotechnology and information technology. In this article, technogenic risks not only mark accidental risks but also the creeping effects of risks to society, such as gradual land pollution or effects to human welfare.

Today's risk landscape consists of many interlinked elements, including interdependency, complexity, uncertainty, ambiguity, and cascading effects, which are all amplified by an increased dynamic of globalization [10, 11]. Advances in information and communication technology as well as in other kinds of technologies have increased these linkages and connections between states, institutions, corporations, civil society, and individuals. As a result of this process, the amount of interdependencies between persons, nations, markets, and societies is bigger than ever before [10]. Due to the complexity of the systems, it is difficult or even impossible to identify or quantify causal links between causes and the adverse effects of the unwanted phenomenon. One initial event may inflict different consequences in different parts of the state or world. Uncertainty is an inevitable part of different and distinct components of risks, such as statistical variation, measurement errors, ignorance, and indeterminacy [12]. Uncertainty reduces confidence in the estimated cause and effect chains. Ambiguity implies different interpretations based on human observations or data assessments. It strengthens the effect of cultural differences on risk assessment. Cascading effects describe the second, third, etc. step consequences of the initial risks to society as a whole, which, unfortunately, due to the previous elements are difficult to assess. It has been stated that in the near future different technologies, such as nanotechnology, biotechnology, information technology, and cognitive science (NBIC), will converge [13–15]. Due to this, the technogenic risks of new technologies grow even harder to understand and manage.

Complete risk management is challenged by a lack of knowledge. Although our knowledge of the world around us grows day by day, we are not aware of what pieces of information are still missing [16]. It has been pointed out that we do not have enough knowledge of, or we do not have enough understanding of climate change, technological innovations, wars, human behavior in different circumstances or changes in markets [12]. Moreover, even if the required information exists somewhere, we do not want to accept it due to our personal biases. Nicolas Taleb presented, in 2008, the concept of "Black Swans," which are highly consequential but unlikely events [16]. They are easily explainable but, unfortunately, only after the event. The existence of Black Swans highlights human nature: even if we have all the required information, we do not see—or we do not want to see—what is coming. Risk is a very adaptable and flexible unit, and for some, it appears to be a threat and at the same time for others an opportunity; risk is relative and individually defined. Renn has stated that the more ambiguous risk is, the more need there is for interpretation, and it also creates more cognitive, evaluative or normative conflicts [17]. These conflicts cannot be solved only by pure scientific knowledge, as even the scientific opinions of complex issues differ. There is a need for multidimensional discussion and collaborative multidisciplinary risk assessment.

In a standard risk management procedure, risk identification is the first step toward holistic risk management. It creates the basis for reasonable, effective, and comprehensive risk assessment. Risk identification ensures the quality of risk assessment and finally the effectiveness of the whole risk management process. Thus, the risk identification stage should be done with considerable care.

In spite of all the scientific knowledge about risk identification, assessment, and management, the main responsibility for managing technogenic risks in the European Union is put on operators, corporations, manufacturers, and suppliers. The latest SEVESO III directive (Directive 2012/18/EU) insists that operators and corporations should cooperate to identify the domino or cascading effects of initial risks. This is a challenging task to perform, as often in the same industrial area there are many competing companies that are not willing to share commercially sensitive information. In addition, at least in Finland, the latest technogenic risks ensued from the following initial faulty or poor solutions [18]:

- The new technology (train steering system) was not compatible with the existing one (old tracks and steering methods).
- The built environment gradually accumulated risks, while new technology was combined with the old one. Due to this, liquid ammonia was released inside the factory leading to evacuation.
- There was not enough communication between the authorities and the company and thus the identified risks were not known. This resulted in damages to natural gas pipelines.
- There were too many subcontractors, and no one understood the entirety of the new building construction. Due to this, several roofs of sports halls and commercial buildings collapsed during the winter.
- There was not enough knowledge of what happens in elderly peoples' homes during long electricity blackouts. When it happened during the winter, society was not prepared to evacuate people as soon as it should be done.
- The authorities did not have enough knowledge of new technologies, and therefore, they made faulty decisions with environmental licenses.

None of these technogenic risks could have been prevented by the cooperation of single companies but may have been by broader stakeholder collaboration.

Due to the rapid change in new technologies and a shrinking and convergent world, it is clear that no person or organization alone is capable of identifying all the emerging technogenic risks. There is a need for effective collaboration between all the different stakeholders, authorities, scientists, politicians, and civilians. The commonly admitted and approved solutions should be retrieved with cooperation and a common valuation of differing values. The future is not defined in advance, but we are all able to change and reconstruct it a little toward the plausible future. Using the methods of future studies, we can create new methods to manage future risks, and in this way, we can define what kind of future we want to have.

Despite all the requirements, it is not a very conventional task to collect all the requisite stakeholders together or combine their knowledge to focus on identifying the technogenic risks of future technologies. The next chapters present not only the method developed for this risk identification task but also the basis for the solution.

## 2. Research process and methods

To ensure that societies are prepared against technogenic risks due to new technologies, the aim of this research process was to develop a risk identification process that is able to combine information of new technologies from different disciplines as well as from different stakeholders to anticipate future risks.

The research process included two main steps:

1. Development of the risk identification tool. This stage of the process started from interviews with authorities and literature research to find out the most suitable methods and to select the best one for stakeholder cooperation. The tool was then tested in a real-life situation to find out the risks to society initiating from hydrogen and fuel cell technology.

2. Development of the risk identification procedure for risk identification workshops. To arrange effective collaboration in workshops, a broad literature study was done to tackle the worst mistakes preventing fruitful cooperation and to create the guidelines for authorities for workshops.

As a result of the research, a new risk identification method called anticipation of future technogenic risks identification (FuTecRI) was developed. This risk identification method was developed in close cooperation with Finnish authorities, such as the Finnish Chemical and Safety agency (TUKES), the Pirkanmaa and Uusimaa Centres for Economic Development, Transport and the Environment, the Rescue Services of Helsinki, the City of Virrat and the Council of Tampere Region, for example. These authorities were interviewed, they took part in workshops and they evaluated the method after the workshop. They also did self-assessment of their accrued knowledge of hydrogen and fuel cell technology before and after the workshop.

## 3. FuTecRI: a method for risk identification

The developed future technogenic risks identification (FuTecRI) tool is based on the future studies method called the Futures Wheel that was developed in 1971 by Jerome Glenn. It was developed to organize future events in a reasonable order. The method is a visual one, and it helps one to receive a comprehensive picture of the issue discussed. The Futures Wheel is mainly used to present thoughts about future development or trends [19].

In this study, the FuTecRI method was developed mainly for the authorities' needs. Their opinion was taken into account when selecting a suitable method for risk identification. The Futures Wheel was selected as a base ground for the tool development among 22 different risk assessment and future studies methods for six main reasons:

1. It can be used for studying technical systems and their connections with the natural and built environment.
2. It is suitable for studying future aspects of risks.
3. It is suitable for collaborative brainstorming.
4. It is easy to learn and take into use.
5. It does not require much time resources from professionals, authorities, or scientists.
6. It does not necessarily require any facilitator services.

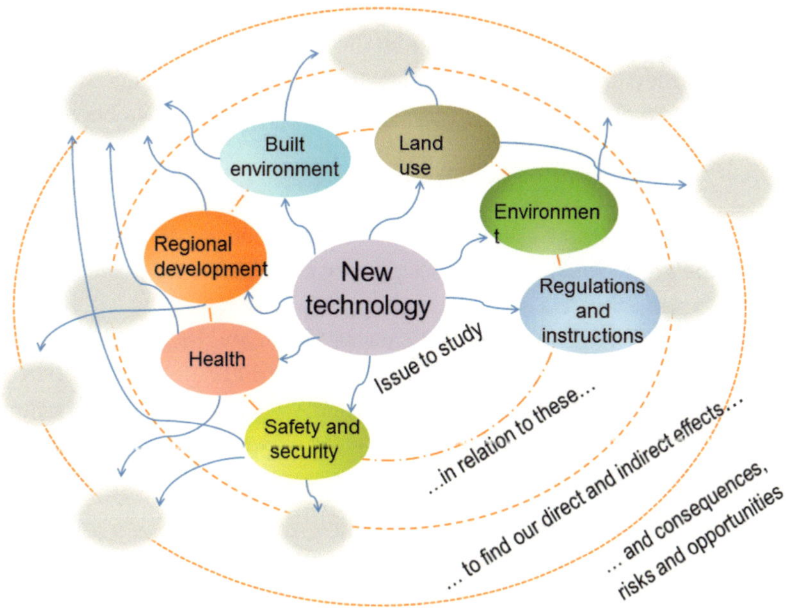


**Figure 1.** FuTecRI tool for identification of the future risks introduced by new technologies [18].

The Futures Wheel approach was further developed to identify the future negative effects of new technologies In the first stage, four different prior new technologies (chimney, match-stick, steamboat and train, electricity, and mobile phone) were analyzed to find out what tor tens of years ago. These changes were categorized to select key words for the FuTecRI tool against which the risk of new technologies should be identified and evaluated. The break-through times of these old technologies were different (chimney 400 years, mobile phone 34 years), which also indicates the amount of changes needed in the regional culture to accept new technologies. The selected key words were health, safety and security, environment, built environment, regulations and instructions, land use, and regional development of the area.

The developed FuTecRI tool is a fill-in-the-blanks diagram presented in **Figure 1**. The central term in the figure describes the discussed new technology that should be evaluated in relation to the key aspects (surrounding it) to find out the potential risks.

## 4. Challenges of collaborative brainstorming

The risk identification tool alone does not ensure effective risk identification, but great attention must be paid to the participants of the workshops. To find out the rules for selection of these participants and for facilitation of the workshops, another broad literature research was done.

Cognitive sciences highlight that multiprofessional and multidisciplinary cooperation is a key for collaborative grading through which a group can achieve better results in problem solving than by working individually [20] and in group work people are able to solve problems that are unsolved by working alone [21]. This is because in difficult decision-making situations people use other actors and people's knowledge to widen their own knowledge and understanding, and thus, they can effectively solve challenging problems [22, 23]. When people are working in groups, they are also forced to recognize the shortcomings of their own knowledge and they can even change their opinions accordingly [24]. It has been pointed out that when all the knowledge dealing with a common target from different disciplines is combined, it is possible to find solutions that cannot be found by any single discipline alone [25].

However, group work does not automatically ensure better results than working alone. To work effectively, the group must be multidisciplinary or at least multiprofessional. The participants need to understand the idea of the workshop, they have to engage with the work and they should deliver their knowledge to other participants. Bohm and Beat have presented a method for creative dialogue insisting that all the participants should be flexible and they should have the ability to negotiate about their opinions with others [26]. The result of this negotiation is not a compromise but rather a creative solution that will be acceptable to all the participants. The creative dialogue can be received if the participants present broadly different disciplines.

One of the main challenges for the multiprofessional group work comes from the competition between professions. Profession is defined here as an authorized status or post in relation to other professionals with specialized knowledge, and the ability to use one's own discretion at work [27]. It seems that professional competition is very common and it becomes apparent when discussing who has the power to make decisions in a certain context [28]. Very often, the professions feel cooperation with other professions is a threat to their own knowledge and authority. People are also unwilling to receive new knowledge if it contradicts their existing knowledge [29]. One of the best situations in which to exceed professional competition and to share and receive new knowledge is in multidisciplinary workshops that are targeted at sharing knowledge with other professionals [30].

Even when the participants are ready to take part in the workshops and share their knowledge, there are still barriers to overcome. Different work cultures may prevent or hinder cooperation; for example, in hierarchically arranged organizations, such as the police and in hospitals, the valuable knowledge of subordinates might stay invisible. Also, the different paradigms, beliefs, terminologies, and methods will prevent common understanding [26, 31].

Finally, there are some other difficulties that may prevent cooperation. Especially dominant persons may hinder discussion and prevent the group from sharing all information and knowledge they have [32]. It is important that all the participants are interested in the subject they are discussing and they have booked enough time for the work. To get through all these impediments, it is important to create a safe and trustful environment for the work groups. The participants must feel that they can trust not only the expertise of other participants but also their behavior, which should be appreciative, friendly, and predictable [33].

## 5. Risk identification procedure

As a result of the analyses of the advantages and disadvantages of the collaborative group work, a range of conclusions were made and the working procedure for the authorities was produced. One remarkable note was that all the participants taking part in the workshops should have a personal interest in the discussed topic. However, the authorities are overworked in these days, and therefore, an interest is not enough; they also need to have enough time to take part in the workshops. This led to the conclusion that all the workshops arranged to identify technogenic risks should be strongly justified. This means that for one new technology there should be only 1–2 workshops in the whole country, and they should be planned to combine broadly all scientific knowledge, professional data, and stakeholder opinions in an efficient way. **Figure 2** presents the decision frame to start a new risk identification workshop as well as the main steps of the risk identification workshop. All stakeholders have the possibility to start a new risk identification workshop if they feel that it is needed for ensuring safety and security of society in the future. In the beginning, they have to answer three questions, and if the answer is always "yes," they should start the process. If this condition is not met, the FuTecRI method should not be used, but other methods may be more useful. For example, if the impacts of the new technology are only local, traditional risk assessment methods are more suitable.

The participants of the workshops should be selected in a reasonable way taking into account the following viewpoints [18]:

- Multiprofessionality—the participants should represent all the different authorities in charge of preventing the risks of new or novel technologies, or of preserving or maintaining a safe and secure society.
- Multidisciplinary—the participants should represent the latest academic and scientific knowledge of the technology in question.
- Personal features—the participants need to be personally interested in the technology discussed, they should be open-minded and responsible persons, and they should want to find good solutions for everyone.

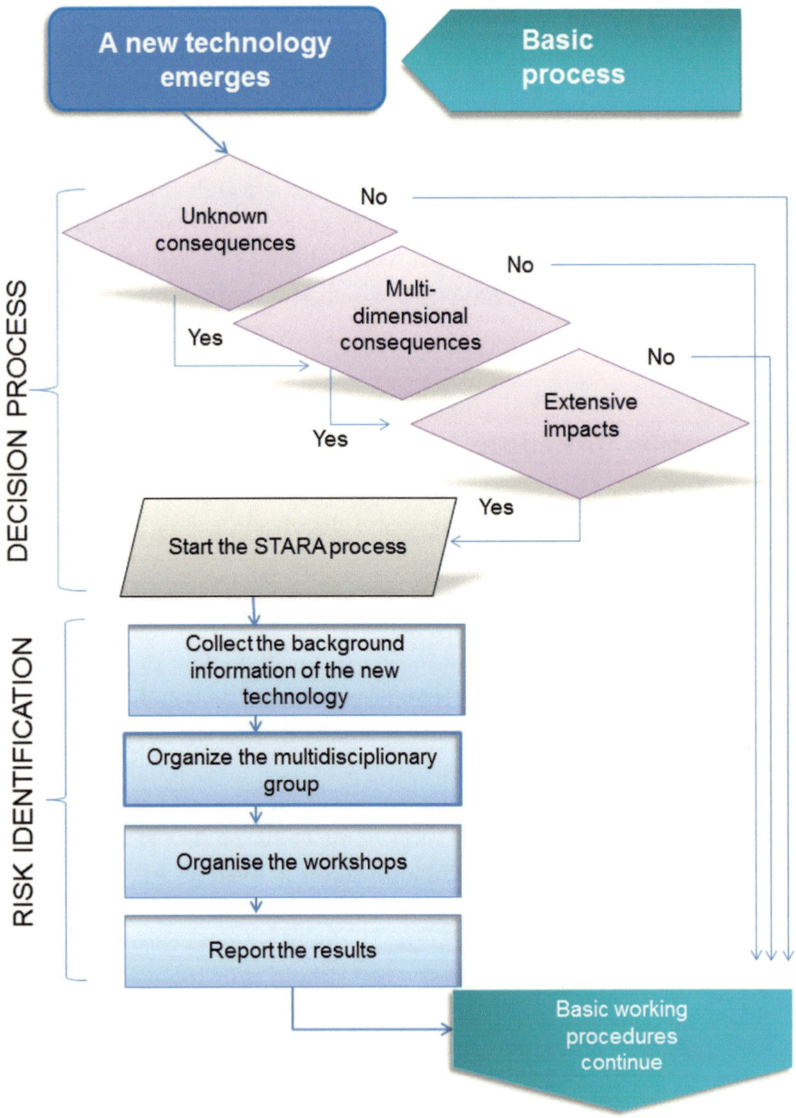


**Figure 2.** Decision process to start the FuTecRI workshop and the main steps for arranging it [18].

The main question for a successful FuTecRI workshop is how to ensure that the authorities especially, but also other stakeholders, put aside their professional and official position during the workshop because it may prevent them from sharing their personal knowledge and receiving new information. This is an issue that should be clearly discussed at the beginning of the workshop. The participants must understand that they are not in the workshop because of their official status but because of the knowledge they have, and their role is to deliver this knowledge to other participants.

To focus the group on the same target, the flow of the FuTecRI workshop must be well organized and the frame and content should be clear (**Figure 3**). At the beginning of the workshop, it is essential to highlight the importance of risk identification as well as to justify the working method. The importance of the identification of the risks of technogenic risks can be stated by Geel's theory of sociotechnical change [34, 35]. According to this theory, new technologies may break through and spread all over if external circumstances are favorable to them. In these cases, they may affect huge changes in culture, infrastructure, regulation, and markets, for example. A good example of this kind of change took place when mobile phones came into the market. It is clear that some technologies have the potential to change the whole society, and society should be able to handle and remove risks before they are actualized.

To ensure that the topic of the workshop is clear, there should be a state-of-the-art presentation of the discussed technology before the collaborative work. In this presentation, all the known aspects of the technology in question, pros and cons, should be given to the participants, and therefore, this presentation should be given by academic or research institutes. The challenge is to present the technology with terminology and concepts that are understandable to all stakeholders and authorities, professionals, and nonprofessionals.

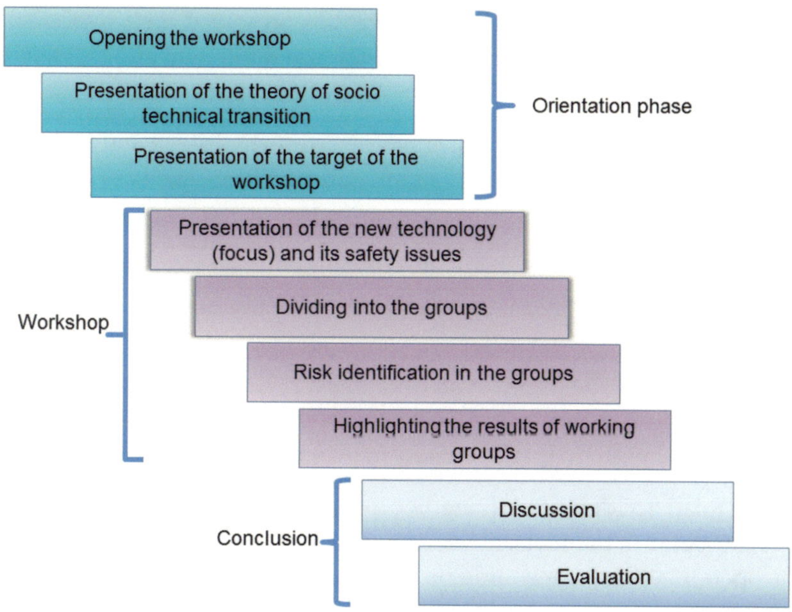


**Figure 3.** The flow of the FuTecRI workshop [18].

Finally, when risk identification starts, the participants should be arranged into smaller working groups. The group should include only five to seven participants, as this has been proved to be the most effective group size for cooperation [32]. In larger groups, nonparticipation increases because people easily forget their role in the workshop as a knowledge deliverer. It is also important that no notes of who-said-what are taken but all the notes are done as a group. This kind of working removes official roles and gives room for expertise. Each group may have a recorder of their own, but it is also possible for all the participants to make notes on the working tool.

## 6. Piloting the method: risks of hydrogen and fuel cell technology

The developed FuTecRI tool and working procedure were tested in a workshop arranged by hydrogen and fuel cell researchers from the VTT Technical Research Centre of Finland. Hydrogen and fuel cell technology is undergoing strong development work, and the researchers wanted to know if there is an understanding of what kind of requirements the new technology imposes on society. The participants of the workshop consist of representatives from the Finnish Chemical and Safety Agency (TUKES), the Pirkanmaa and Uusimaa Centres for Economic Development, Transport and the Environment, the Rescue Services of Helsinki and VTT; all together 11 participants were arranged into two working groups. The workshop followed the procedure presented in **Figure 3**. The new technology was presented by VTT researchers to the participants.

The workshop took 5 h, the first 2 h of which were discussions of dealing with the theory of sociotechnical change and the hydrogen and fuel cell technology and its current state. After the lunch break, people worked in groups for 2 h to explore what kind of risk hydrogen and fuel cell technology might create. Finally, both groups presented their results to each other. The results of the workshop were combined into the one mind map and delivered to the participants for their later use.

The workshop results brought out several issues regarding hydrogen and fuel cell technology that need to be discussed and managed at a governmental level, such as [18]:

- Environmental issues. The production of hydrogen and fuel cells requires platinum as a raw material. This will improve material recycling but also increase mining actions. The positive effects include emission-free fuel and quiet traffic.

- Built environment. The fuel distribution stations are not covered by any legislation. Thus, the hydrogen fuel can be delivered even from delivery trucks, which can cause dangerous situations. There is a need for new legislation. In addition, underground parking places need to be equipped with hydrogen sensors to avoid explosions.

- Safety issues. Road accidents involving hydrogen and fuel cell cars may cause danger to rescue services because the fuel cell vehicles do not visibly differ from other vehicles but the rescue operations vary depending on the fuel type of the vehicle. In addition, fuel cell vehicles move quietly, which may increase road accidents, as people may not hear the arriving cars especially in the winter.

- Land use. For safe land use, there is a need to plan regional hydrogen pipelines that are later suitable for different kinds of hydrogen use.

The workshop participants also took part into two different surveys. The first one was done in two parts, at the beginning of the workshop and immediately after it. This survey focused on evaluating the change in participants' knowledge of hydrogen and fuel cell technology. The idea was that people made a self-assessment at the beginning of the workshop evaluating their own knowledge on a scale of 5–10 (5: weak knowledge; 10: excellent knowledge). This scale was selected as it was used for a long period in Finnish schools, and therefore, it was easy for the participants to understand. After the workshop, they were asked to evaluate themselves a second time. They then had to answer two questions: What did they now think their knowledge was at the beginning of the workshop? and What did they think their knowledge level is after the workshop?

The results were very interesting (**Table 1**). At the beginning of the day, participants thought that their knowledge was at a considerably low level, and only one participant (perhaps a researcher) thought that he had excellent knowledge. During the workshop, they understood that their knowledge level was not even at that level, and a comparison between the morning and afternoon estimations indicates that all of the participants lowered their estimations. The very interesting thing is that the workshop brought a lot of new knowledge to all participants. Even the hydrogen and cell fuel researchers received a lot of information regarding the impacts of new technology on society and the built environment. It seems that the FuTecRI workshop worked as it was planned to, it stimulated the participants to share their knowledge and accumulated new information on top of the old information, and in that way, it made it possible to also identify the risks of new technology.

Evaluation criteria: 5 = poor knowledge, 10 = excellent knowledge [18].

The other survey handled the content of the workshop. The survey was sent to the participants about 1 month after the workshop. The participants were asked how they later thought of the position of the workshop in their minds. The first questions dealt with the reliability and validity of the distributed information dealing with hydrogen and fuel cell technology. The participants were convinced that it was the latest and most up-to-date information they received. However, one researcher pointed out that private companies very often have the newest knowledge, but they are not willing to share it even with research organizations.

| **Self-evaluation of the level of knowledge, hydrogen and fuel cell technology** | **5** | **6** | **7** | **8** | **9** | **10** | **Av.** |
|---|---|---|---|---|---|---|---|
| Estimated knowledge level before the FuTecRI workshop evaluated before the workshop | 2 | 2 | 3 | 1 | | 1 | 6.8 |
| Estimated knowledge level before the FuTecRI workshop evaluated after the workshop | 3 | 3 | 2 | 1 | | | 6.1 |
| Estimated knowledge level after the FuTecRI workshop evaluated after the workshop | | | 3 | 5 | 1 | | 7.8 |

**Table 1.** The results of the survey dealing with the increase of knowledge in FuTecRI workshop.

The next questions concerned the functioning of the workshop. All participants were satisfied with the workshop proceedings. They felt that because of the small groups they were consulted and it was easy for them to bring their own knowledge into the process. The results of the brainstorming work were written directly on to a wide paper sheet where the main words were ready-written in the middle of the paper. This helped people to immediately start the brainstorming process, and the fear of the empty paper was tackled. The participants were very active in discussing the hydrogen and fuel cell technology, which was surprising to all.

## 7. Discussion

The FuTecRI method was developed to help authorities to be prepared for future technogenic risks introduced by new technologies. The use of the method requires effective stakeholder cooperation at least from authorities and scientists. The results may be even better if the companies developing new technology could also take part in the process. According to the results of the FuTecRI workshops, it is possible to steer the development of society toward a safe and secure future through the use of, for example, new regulations or improved land use planning.

To work well, the method should involve not only the authorities and other stakeholders but also researchers from academia and research institutes. It is especially important that the focus of the workshop, the new technology, is presented by special researchers who are specialist in the technology in question. Otherwise, the result of the workshop might be just guesswork and no future solutions can be built on it.

Because the FuTecRI method involves a large group of professionals and scientists, it is important that no workshops are performed in vain, because it will reduce the motivation to take part into the FuTecRI method. Therefore, the results of each workshop should be delivered to all essential authorities through their own information networks.

However, this kind of workshop works only as a starting point to manage the risks of new technologies. The method should be further developed to also produce guidelines on how to analyze the highlighted risks or take them into account in different kinds of processes, such as environmental or chemical licenses, or land use planning.

## Acknowledgements

This research and article presents the main findings of the thesis of the author [18]. I am greatly thankful to the representatives of the Finnish authorities who helped me through this work and gave me valuable information. The work was carried out during the years 2012–2015, and it was not the main duty either for these authorities or for me. Because both parties also had other duties to perform, the collaboration was important to the success of this work.

## Author details

Riitta Molarius

Address all correspondence to: riitta.molarius@vtt.fi

Technical Research Centre of Finland, Ltd., Finland

## References

[1] The Guardian. Mag-lev Crash Kills 23 [Internet]. Sep 22, 2006. Available from: https://www.theguardian.com/technology/2006/sep/22/news.germany [Accessed: Jul 14, 2017]

[2] BBC News. Germany Train Crash: Controller 'Distracted by Computer Game' [Internet]. Apr 12, 2016. Available from: http://www.bbc.com/news/world-europe-36025951 [Accessed: Jul 14, 2017]

[3] New Scientist. Tesla Driver Dies in First Fatal Autonomous Car Crash in US [Internet] June 1, 2018. Available from: https://www.newscientist.com/article/2095740-tesla-driver-dies-in-first-fatal-autonomous-car-crash-in-us/ [Accessed: Jul 14, 2017]

[4] BBC News. London Fire: What Happened at Grenfell Tower? [Internet]. June 15, 2017. Available from: http://www.bbc.com/news/uk-england-london-40272168 [Accessed: Jul 14, 2017]

[5] Riekkola-Vanhanen M. Talvivaara mining company – From a project to a mine. Minerals Engineering. 2013;**48**:2-9

[6] Yle Uutiset. Työtapaturma Talvivaaran kuolemantapauksen takana [Internet]. 2012. [Internet]. Mar 15, 2013. Available from: https://yle.fi/uutiset/3-5072844 [Accessed: Jul 14, 2017]

[7] Yle Uutiset. Talvivaaran Kaivososakeyhtiötä haetaan konkurssiin [Internet]. Mar 23, 2017. Available from: https://yle.fi/uutiset/3-9527477 [Accessed: Jul 14, 2017]

[8] Clothier R, Greer DA, Greer DG, Mehta A. Risk perception and the public acceptance of drones. Risk Analysis. 2015;**35**(6):1167-1183

[9] Beck U. Risk Society: Towards a New Modernity. 1st ed. London: SAGE Publications Ltd; 1992 272 p

[10] Habegger B. Risk analysis and management in a dynamic risk landscape. In: Habegger B, editor. International Handbook on Risk Analysis and Management. Professional Experiences. Switzerland: Center for Security Studies; 2008. pp. 13-32

[11] Klinke A, Renn O. Systemic risks as challenge for policy making in risk governance. Forum qualitative sozialforschung (Forum: Qualitative Social Research). 2006;**7**(1): Art 33, p13

[12] van Asselt M. Perspectives on Uncertainty and Risk. The PRIMA Approach to Decision Support. 1st ed Amsterdam: Kluwer Academic Publishers; 2000. 434 p

[13] Nordmann A. Converging Technologies – Shaping the Future of European Societies. Interim Report of the Scenarios Group, High Level Expert Group, 3. 1st ed. Luxembourg: Office for Official Publications of the European Communities; 2004 64 p

[14] Roco M. Possibilities for global governance of converging technologies. Journal of Nanoparticle Research. 2008;**10**(1):11-29

[15] Wolbring G. Why NBIC? Why human performance enhancement? Innovation: The European Journal of Social Science Research. 2008;**21**(1):25-40. DOI: 10.1080/13511610802002189

[16] Taleb N. The Black Swan: The Impact of the Highly Improbable. 1st ed. Random House: New York; 2007 400 p

[17] Renn O. The challenge of integrating deliberation and expertise. Participation and discourse in risk management. In: McDaniels T, Small M, editors. Risk Analysis and Society. 1st ed. New York: Cambridge University Press; 2003. p. 289-366

[18] Molarius R. Foreseeing risks associated with new technologies (in Finnish) [dissertation]. VTT Technical Research Centre of Finland: Tampere; 2016. 208 p Available from: http://www.vtt.fi/inf/pdf/science/2016/S120.pdf

[19] Glenn J. The futures wheel. In: Glenn J, Gordon T, editors. Futures Research Methodology. 3rd ed. Washington: The United Nations University - The Millennium Project; 2009

[20] Leathard A. Introduction. In: Leathard A, editor. Interprofessional Collaboration. From Policy to Practice in Health and Social Care. Hove and New York: Brunner & Routledge; 2003. p. 3-11

[21] Roschelle J. Learning by collaborating: Convergent conceptual change. Journal of Learning Sciences. 1992;**2**(3):235-276

[22] Hutchins E. Distributed cognition. In: Smelser N, Baltes P, editors. International Encyclopedia of Social & Behavioral Sciences. Elsevier, New York, Amsterdam; 2001. p. 2068-2072

[23] Giere R, Moffat B. Distributed cognition: Where the cognitive and the social merge. Social Studies of Science. 2003;**33**(2):301-310. DOI: 10.1177/03063127030332017

[24] Vygotsky L. Mind in Society: The Development of Higher Psychological Processes. Cambridge: Harvard University Press; 1978 159 p

[25] Kline SJ. Conceptual Foundations for Multidisciplinary Thinking. Stanford: Stanford University Press; 1995 337 p

[26] Bohm D, Peat F. Science, Order and Creativity. London and New York: Routledge; 1992 330 p

[27] Puustinen S. Suunnittelijaprofessio refleksiivisyyden puristuksessa. Yhteiskuntasuunnittelu. 2001;**39**(1):26-45

[28] Abbott A. Linked ecologies: States and universities as environment of professions. Sociological Theory. 2005;**23**(3):245-275

[29] Von Krogh G, Ichijo K, Nonaka I. Enabling Knowledge Creation. How to Unlock the Mystery of Tacit Knowledge and Release the Power of Innovation. New York: Oxford University Press; 2000 292 p

[30] Pärnä K. Developmental interprofessional collaboration as a process – Possibilities for early intervention to support families. (in Finnish) [dissertation]. Turku: University of Turku; 2012 245 p. Available from: http://www.doria.fi/bitstream/handle/10024/77506/AnnalesC341Parna.pdf

[31] Edmondson A, Nembhard I. Product development and learning in project teams: The challenges are the benefits. Journal of Production Innovation Management. 2009;**26**:123-138

[32] Kuivalahti M. Yksilön oppiminen ryhmässä. Tapaustutkimus systeemisuunnittelun ryhmätöistä [dissertation]. Tampere: University of Tampere; 1999

[33] Lin C-P. To share or not to share: Modeling tacit knowledge sharing, its mediators and antecedents. Journal of Business Ethics. 2007;**70**:411-428. DOI: 10.1007/s10551-006-9119-0

[34] Bijker W. Of Bicycles, Bakelites, and Bulbs: Towards a Theory of Sociotechnical Change. Cambridge MA: MIT Press; 1997 390 p

[35] Geels F. From sectoral systems of innovation to socio-technical systems: Insights about dynamics and change from sociology and institutional theory. Research Policy. 2004;**33**(6):897-920

Chapter 11

# Risks, Safety and Security in the Ecosystem of Smart Cities

Stig O. Johnsen

Additional information is available at the end of the chapter

http://dx.doi.org/10.5772/intechopen.70740

**Abstract**

We have performed a review of systemic risks in smart cities dependent on intelligent and partly autonomous transport systems. Smart cities include concepts such as smart transportation/use of autonomous transportation systems (i.e., autonomous cars, subways, shipping, drones) and improved management of infrastructure (power and water supply). At the same time, this requires safe and resilient infrastructures and need for global collaboration. One challenge is some sort of risk based regulation of emergent vulnerabilities. In this paper we focus on emergent vulnerabilities and discussion of how mitigation can be organized and structured based on emergent and known scenarios cross boundaries. We regard a smart city as a software ecosystem (SEC), defined as a dynamic evolution of systems on top of a common technological platform offering a set of software solutions and services. Software ecosystems are increasingly being used to support critical tasks and operations. As a part of our work we have performed a systematic literature review of safety, security and resilience software ecosystems, in the period 2007–2016. The perspective of software ecosystems has helped to identify and specify patterns of safety, security and resilience on a relevant abstraction level. Significant vulnerabilities and poor awareness of safety, security and resilience has been identified. Key actors that should increase their attention are vendors, regulators, insurance companies and the research community. There is a need to improve private-public partnership and to improve the learning loops between computer emergency teams, security information providers (SIP), regulators and vendors. There is a need to focus more on safety, security and resilience and to establish regulations of responsibilities on the vendors for liabilities.

**Keywords:** safety, security, resilience, smart cities, software ecosystems

# 1. Introduction

This paper contains a discussion and review of safety, security and resilience of smart cities, considered as software ecosystem (SEC). The purpose is to provide an overview of research in the field, identify emergent risks in a systemic perspective and identify possible issues that existing literature is not addressing adequately. The article is initiated by a discussion of the concept of software ecosystems and the need for safety, security and resilience in smart cities.

## 1.1. Smart cities and software ecosystems

In Ref. [1] there is a fairly general definition of a smart city, described as a place when investments in human and social capital and traditional (transport) and modern (ICT) communication infrastructure, fuel sustainable economic growth and a high quality of life, with a wise management of natural resources, through participatory governance. From [2], looking at specific systems, smart cites are described as a city that monitors and integrates conditions of its critical infrastructures, traffic (including roads, bridges, tunnels, rails, subways, airports, seaports), communications, water, power and major buildings. All this is done in order to optimize resources, plan its preventive maintenance and monitor security aspects while maximizing services to its citizens.

A literature review of software ecosystems in general was performed in [3] identifying 90 papers in the period from 2007 to 2012. The review identified the software ecosystem (SEC) as a fruitful systemic perspective. The review inspired us to find papers discussing safety, security and resilience of SEC published in the period from 2007 to 2016.

A software ecosystem (SEC) describes the complex environment of a smart city. A SEC will consist of components developed by actors both internally and externally, and solutions will spread outside the traditional borders of software companies to a group of companies, private persons and entities. In [3] they defined a software ecosystem as: "*the interaction of a set of actors on top of a common technological platform that results in a number of software solutions or services. Each actor is motivated by a set of interests or business models and connected to the rest of the actors and the ecosystem as a whole with symbiotic relationships, while, the technological platform is structured in a way that allows the involvement and contribution of the different actors….*"

When discussing software ecosystems, we include the legal and organizational framework in addition to applications and supporting infrastructure as described in **Figure 1**. Scope of digital ecosystems.

<table>
<tr><td colspan="2">Legal and organizational framework</td></tr>
<tr><td colspan="2">Applications and Architecture</td></tr>
<tr><td>Components</td><td>Data/ Digital Content</td></tr>
<tr><td colspan="2">Infrastructure</td></tr>
</table>

**Figure 1.** Scope of software ecosystems.

http://dx.doi.org/10.5772/intechopen.70740

Software ecosystems are often based on the internet as infrastructure. The internet economy makes up a significant part of the GDP in 2016 since it is 5.3% of GDP [4]. SEC has gained more importance due to mobile platforms such as iPhone and android. Examples of SEC are:

- Digital learning environments;
- Mobile systems (phone applications);
- Shopping and payments systems;
- Social networking systems;
- Personal wellness and healthcare (training, food, medical equipment and surveillance…);
- Smart cities with transportation, infrastructure control (water, power). A part of Smart cities is intelligent transport systems (ITS) controlling vehicles, traffic management systems, electronic payments…;
- "Smart farming" with systems to track livestock and harvest/yield; outside cities.

Arguments for discussing software ecosystems has been the speed of development; increased competition and reduction of development costs due to the opening up of development outside of organizational silos. Some of the software ecosystems are critical, in that a malfunction can severely affect the functioning of society or personal well-being. Examples are systems used in transportation, car control systems and health systems (such as pacemakers).

### 1.2. Safety, security and risks

In this paper we have used the definition of safety as a state, as described by Department of Defense - [5], *"freedom from those conditions that can cause death, injury, occupational illness, damage to or loss of equipment or property, or damage to the environment."* This definition describes safety as being free from conditions causing a mishap or accident, i.e., safety is in some sense a "non-event."

Security is used to describe conditions of intentional harm. The relationships to safety are discussed in [6]. Security is defined as *"the degree to which malicious harm is prevented, reduced and properly reacted to"* and safety is defined as *"the degree to which accidental harm is prevented, reduced and properly reacted to"* from [7]. In information systems, there has often been a focus on *"information security."* Information security is defined as *"protecting information and information systems from unauthorized access, use, disclosure, disruption, modification, or destruction in order to provide: integrity, confidentiality and availability"*, from [8]. Since software ecosystems not only handle information, but also actual critical processes in smart cities, in automobiles and other applications the software ecosystems must both be secure (i.e., protected from malicious harm) and safe (i.e., protected from accidental harm). The systems must be able to handle unanticipated risks, and the ecosystem must be able to handle breakdowns and ensure that the systems has a safe state and/or a secure state.

In [9], the following definition of risk is given, *"Risk: two dimensional combination of the consequences (of an activity) and associated uncertainties (what will the outcome). Probabilities are used to express the uncertainties. When risk is quantified in a risk analysis, this definition is in line with the ISO/IEC Guide 73 (2002) standard definition* [10]: *combination of the probability of an event and its consequence.*" Related to new emerging risks and complexities of interactions, there may be challenging to establish the probability of an event since they may be unanticipated.

Systemic risk is defined as "*Probability of loss or failure common to all members of a class or group or to an entire system.*" When discussing systemic risks related to smart cities we are exploring failures common to members of a smart city.

A key element when assessing systemic risks are the scope and prioritization of systems to be evaluated. We have focused on critical systems of common interest in a city. In the following we have discussed risks and protection of what is defined as part of critical infrastructure. Definition and protection of critical infrastructure has been a key concern in the US and EU. In the US the establishment of the national infrastructure protection plan (NIPP) from 2009 has been updated systematically. The latest, [11], has the title "Partnering for Critical Infrastructure – Security and Resilience." The successive NIPP has identified specific areas of concern such as interdependencies, cyber security and the international nature of threats. The risk management framework of NIPP is interesting since it is broad and systemic including physical, cyber and human elements. In the EU, the directive 114/08 on the identification and designation of European critical infrastructure and the assessment of the need to improve their protection was established as a council directive in 2008 [12]. In **Table 1**, we have described critical infrastructure sectors.

The main elements and areas of this list of critical infrastructure, is highly relevant when discussing smart cities, of special interest are smart city applications related to: Transportation systems, energy systems (power supply), bank and finance; communication systems; Technologies of information (including navigation systems); Water supply and health systems. The following areas are critical when impacted by loss or failures (**Table 2**).

The criticality or potential loss due to failures, breakdowns or attacks increases the need to be able to support critical operations even when the system is under stress or may fail, thus the ability to handle unanticipated incidents (or ability to go to a safe and secure state) are gaining importance. The concept of resilience engineering is an important strategy to handle these unanticipated incidents. In [13] resilience is defined as "*the intrinsic ability of a system to adjust*

| US-NIPP sectors | EU 114/08 sectors |
|---|---|
| Energy | Energy—electricity, oil, natural gas |
| Transportation systems | Transport—roads and highways, railroads, aviation, inland waterways, shipping and ports |
| Agriculture and food; bank and finance; communications; military installation and defense; technologies of information; national monuments and icons; drinking water treatments plans | (NA—not applicable yet) |

**Table 1.** Critical infrastructure sectors in US-NIPP and EU 114/08.

| Areas impacted by smart cities | Examples of critical loss or failures |
|---|---|
| Transportation systems—Intelligent transportation, management and control of transport being impacted by surveillance, Google Map used to recommend routes, autonomous transport, traffic control centers managing transport and road signals | Breakdown of transport; impacting emergency services (police, fire, ambulance) or flow of material. (trough exploiting vulnerabilities in autonomous car systems; or influencing traffic control systems/halting control systems) |
| Energy systems—smart metering and improved management through centralized or more technology-based power grid systems | Breakdown of power supply – breakdown of power grids through exploiting vulnerabilities in new technology, on a control level and in smart metering systems |
| Bank and finance—many areas—one example payments systems integrated in toll roads and parking to manage traffic (reduce traffic in selected areas) and to make parking more user friendly and efficient | Breakdown of payment systems, halting/delaying key services |
| Communication systems and technologies of information (including navigation systems)—autonomous cars, traffic control, route planning and signaling systems dependent on information technology and navigation systems. | Breakdown (accidents) in transportation due to nonfunctioning supporting technology |
| Water supply—managed and optimized by control systems | Breakdown, contamination of water supply due breakdown (hacking) of control systems |
| Health systems—dependent on sensors (such as in pacemakers) and communication systems (sending and receiving data from health professionals), "just in time" management of health, | Breakdown/errors in health systems used to manage critical operations, such as in peacemakers |

**Table 2.** Systemic risks in smart cities.

*its functioning prior to or following changes and disturbances, so that it can sustain operations even after a major mishap or in the presence of continuous stress."* In [14], Woods focuses on unanticipated disturbances and adaptations, and describes resilience as: *"How well can a system handle disruptions and variations that fall outside of the base mechanisms/model for being adaptive (adaptive defined as the ability to absorb or adapt to disturbance, disruption and change)."* The handling of the unanticipated and continued functioning has been a key property of resilient systems.

In the European Union, safety, security and resilience are prioritized in the cybersecurity strategy [15]. Three of the top five strategic issues mentioned are: Develop the industrial and technological resources for cybersecurity; Achieving cyber resilience; and establish a coherent international cyberspace policy for the EU. Thus, safety, security and resilience of smart cities are important issues that should be explored further. In addition, it is important to understand how risk governance of smart cities is addressed and established in order to support a coherent cyberspace policy of the key issues.

## 2. Problem definition and methods

Based on the preceding introduction, and the summary above, the three research questions we wanted to explore are:

- RQ1: How is safety, security and risks of smart cities (software ecosystems of cities) framed and defined?
- RQ2: How is risk governance of smart cities (software ecosystems of cities) addressed?
- RQ3: What are key issues in Governance of the ecosystem?

In the following we have described some of the challenges and problems of these research questions and our methodology (i.e., approach).

### 2.1. Challenges and problems

There is often poor focus on emerging risks, safety and security. These issues have been identified late when vulnerabilities have been exploited and unwanted incidents have been published. The suppliers and vendors (software vendors) seldom has to pay for unwanted incidents even if they are due to poor quality issues in the systems such as safety, security or resilience. The bill has been given to the users, the organizations and/or society.

Critical infrastructure is in most cases regulated by the authorities. Safety and security regulation is often reactive, and lags technological innovation. New software is implemented and societal consequences are discussed later. Internet of things (IoT) is an example of new technology that are introduced in software ecosystems that may affect operations of critical infrastructure. IoT has introduced a broad set of vulnerabilities and can challenge safety, security and resilience of software ecosystems. As an example the Mirai botnet was used in a Denial of service (DoS) attack on the internet firm Dyn, Ref. [16], using unsecured devices on a large scale. The attack affected Dyn's clients such as Twitter, Reddift, Spotify, and SoundCloud. The cyber-attacks caused outages across the whole East Coast in the US in October 2016. When discussing vulnerabilities in a software ecosystem such as in smart cities, one challenge is that there is not one single supplier, but a set of suppliers that must be involved. Incident handling moves to a broader area where it can be difficult to identify responsibilities and manage competencies. This is relevant, in [17] the author points out that there are serious vulnerabilities (poor quality control) in systems used in smart cities (i.e., traffic control systems), which could be used to cause traffic jams or collisions.

### 2.2. Methodology

The literature review started by a keyword search based on combination of "software ecosystems" "smart cities" and "safety, security, resilience." Using Google Scholar and then searching the ACM Digital Library, IEEE Explore, Springer Link and Science Direct. The literature body was selected based on that software ecosystems/smart cities and safety (security and resilience) was the main theme. In addition, papers were selected based on a set of criteria i.e., have been peer reviewed and published in a scientific context (journal, conference), available in English, and more than one-page long. Since software ecosystems involves governmental rules, relevant white papers were also identified. The identified literature body is gathered in Section 5, numbered from [18–29] [LIT BODY:13] and [LIT BODY:14]. In addition, we have listed other general references that could not be included in the literature body, in Section 6.

The concept of Risk and Risk Governance has been an issue in the review, and we have structured papers based on risk governance, see [30], starting with problem framing; then risk appraisal (hazards and vulnerabilities); risk judgment; risk communication and risk management.

## 3. Findings and reflections

We found 14 papers in total, 13 papers published in the interval 2007–2016; and we included a paper from 2003 that had an illuminating discussion of resilience of systems. The following three sections are based on our research questions (RQ1 to RQ3 as described in Section 2) and have been used as title of the chapter:

- Framing of safety, security and risks in smart cities (RQ1)
- Risk Governance of smart cities (software ecosystems) (RQ2)
- What are key issues in Governance of the ecosystem? (RQ3)

### 3.1. Framing of safety, security and risks

In [31] there is a discussion of the convergence of safety and security, pointing out that a successful integration of both requirements needs the collaboration of both safety and security disciplines, aided by a common understanding. In [28, 32], it is pointed out that both safety and security issues must be assessed to build trustworthy software ecosystems. Issues identified through security analysis (i.e., threats) must be combined with issues from safety analysis (i.e., hazards). In [33] there is a focus on of the development of industrial control systems and how safety and security must be integrated in the development methodologies. These control systems are similar to control systems employed in smart cities. An overview and comparison of methodologies is given.

In [34], a broad overview of security and safety challenges of digital systems are given based on an ecological perspective. Ecology is used both as a metaphor to learn from the development in the nature, but also to have a more holistic perspective of systems involving human actors in a society. The ecosystem perspective is as an important viewpoint when discussing safety and security in a changing word, and especially when exploring risks and risk governance of smart cities.

In [35] there is a discussion of infrastructure resilience from an organizational context. Adaptive capacity, resource robustness is discussed related to infrastructure and a conceptual framework for assessing resilience is outlined. The conceptual framework seems to be useful when discussing resilience in software ecosystems, especially of critical (infrastructure) ecosystems. In [36] different elements of resilience are discussed. The paper presents a framework for system resilience, consisting of five aspects: time periods, system types, events, resilience actions and properties to preserve. It is followed by principles for emergence, and factors affecting resilience, including improving resilience, trade-offs, and loss of resilience.

Not many textbooks (that can be used in teaching) have been found related to implementation of security and resilience in control systems of smart cities. However, in [37] guidelines for secure and resilient software development are discussed. The development guidelines are targeted toward software ecosystems; and the goal is to improve developer skills related to security and resilience. It is pointed out that security and resilience must be integrated from concept/early design, it reviews security design methodologies and suggests how to measure the development process. The discussion of security in Industrial Automation setting, is discussed in [38], including the challenges of adapting general software security principles to industrial automation and control systems.

In [29] resilience and cyber security of the ecosystems is seen as a part of the maturity of governance and collaboration between industry and government. Thus, cyber resilience is seen as the next step of cyber security.

In [24] there is a discussion of the security dynamics of software ecosystem (SEC), pointing out that SEC reduces cost and are increasing efficiencies for the software producers while society get the costs of software failures (i.e., issues related to security, safety and poor resilience). The paper has a quantitative examination of 27,000 vulnerabilities disclosed over the past decade (1996–2008). The paper identifies the interest of several stakeholders in the market of software vulnerabilities such as the vendors, safety experts/consultants, security information providers (SIP), and criminals. The paper explores several policies such as security trough obscurity, responsible disclosure of vulnerabilities (as a suggested policy) or security trough transparency. One of the key insights is that secrecy prevents people from assessing their own risks, which contributes to a false sense of security. The process of responsible disclosure is that the researcher discloses full information to the vendor, expecting that a patch is developed within a reasonable timeframe. An increasing number of vendors and security organizations have adopted some form of responsible disclosure. The role of security information providers (SIP) as risk-communicators is discussed in the vulnerabilities market.

In summary, there has been a positive development in identifying the need to explore both safety and security in development and to use resilience as a mitigating strategy. The concept of software ecosystems benefits the developers and industries, but it seems that at present that society gets the costs of software failures. Responsible disclosure of vulnerabilities to the vendors, expecting a patch, seems to be a beneficial policy. The role of actors in the software ecology, such as security information providers should be explored further.

### 3.2. Risk governance of smart cities (software ecosystems) –vulnerabilities and risks

In [26] a set of vulnerabilities in cars are pointed out such as the possibility to control a wide range of automotive functions and completely ignore driver input from dashboard, including disabling the brakes, selectively braking individual wheels on demand, stopping the engine, and so on. Attacks were easy to perform and the effects were significant. It is possible to bypass rudimentary network security protections within the car, and perform attack that embeds malicious code in the car that will completely erase any evidence of its presence (after a crash). There is a discussion of the challenges in addressing these vulnerabilities while considering the existing automotive ecosystem.

http://dx.doi.org/10.5772/intechopen.70740

In [27] Semi-autonomous and fully autonomous cars are described as coming from the development stage to actual operations. The autonomous systems are creating safety and security challenges. These challenges require a holistic analysis, under the perspective of ecosystems of autonomous vehicles. The perspective of ecosystems is seen as useful to understand and mitigate security and safety challenges. These systems will become important critical information infrastructures, simultaneously featuring connectivity, autonomy and cooperation. Threat analyses and safety cases should include both (random) faults and (purposeful) attacks.

In [39], there is a discussion of Cyber-Physical infrastructure risks in the future smart cities. Several examples of unwanted incidents are described in transportation systems (autonomous vehicles; Trains; …) in electricity distribution and management and in the water and wastewater systems sector. It is suggested to the regulator to work with standards and regulations in addition to communication and increased engagement by giving direct assistance. Challenges mentioned are the need to establish goal based standards and regulations as new technology is implemented and to focus on dissemination of best practices in combination with systematic education.

In [17] there is an empirical evaluation of "smart cities" looking at a broad set of technologies of traffic control, management of energy/water/waste and security. Known vulnerabilities are in traffic control systems, mobile applications used by citizens, smart grids/smart meters and video cameras. The issues are in line with peer-reviewed papers, i.e., lack of cyber security testing and approval, lack of encryption, lack of City Computer Emergency Response Teams (CERT), and lack of cyber attack emergency plans. There are reasons to anticipate that we establish potential for serious incidents, if these issues are not addressed and mitigated.

In [20] there is a discussion of the expanded use of federated embedded systems (FES) in automotive and process automation. Expected benefits include the possibility of third-party actors developing add-on functionality; a shorter time to market for new functions; and the ability to upgrade existing products in the field. This is a substantial area for innovation and change, the responsibilities of the manufacturer will change, and a key challenge will be ecosystem management. However, it is suggested that the liabilities and responsibilities of the total product must rest with the manufacturer. The regulator has a key role to define responsibilities. These issues highlight the need for Risk Governance of systems to be used in smart cities.

In [21] open software ecosystem is proposed as an approach to develop software for embedded systems in the automotive industry. The focus is on the need to deliver functionality to customers faster. The paper describes quality attributes and defines a reference architecture. Both safety, security and dependability are explored.

In [22] they model the architecture of a cloud-based ecosystem, showing security patterns for its main components; and discuss the value of such an approach. The ecosystem approach provides a holistic view and is valuable in security, by indicating places where security mechanisms can be attached. Holistic views are seen as important to combine quality factors such as safety and reliability with security. By using this abstraction level, it is argued in the paper that this unified approach reduces complexity, one of the important weaknesses used by attackers and can enable analysis of the propagation of threats and data leaks.

In [28] they cover research on Enterprise Architectures of ecosystems (i.e., software ecosystems) discussing resilience and adaptability as a key area and suggest reference architectures mentioning security. However, safety is not mentioned.

In [25] there is a discussion on how to build robust and evolvable resilient software systems, discussing redundant data structures, transformer middleware and service-oriented communities. The use of transformer middleware may lead to more complex systems and higher costs or latency. Exploration of service-oriented communities may support adaptation and spontaneous emergence of resilience, but may lead to higher costs due to high degree of redundancy and challenges with deterministic behavior.

In summary, there has been documented several vulnerabilities in smart cities, intelligent transport systems and autonomous cars. However, software ecosystems have beneficial elements since more actors are developing functionality and enabling a shorter time to market. Liabilities must rest on the manufacturer and the regulator must define responsibilities. The ecosystem provides a holistic view that is seen as important to combine safety and reliability with security. It is argued in several papers that this approach reduces complexity; one of the weaknesses used by attackers and can enable improved analysis of propagation of threats and data leaks.

### 3.3. Key issues in governance—responsibilities, management and communication

International governance of security of the infrastructure of software ecosystems is addressed through several channels such as standards (ISO, IEC) or international bodies such as OECD, EU, NATO and UN. Software Ecosystems are international—involving many actors with different agendas. In [40] there is a discussion of governance of emerging technology (such as IoT) as it is integrated into critical infrastructure. It is suggested that manufacturers should follow the principle of privacy and security by design, when developing new products, and must be prepared to accept legal liability for the quality of the technology they produce. Buyers should collectively demand that manufacturers respond effectively to concerns about privacy and security. Governments can play a positive role by incorporating minimum security standards in their procurement. It is suggested that government regulations should require routine, transparent reporting of technological problems to provide the data required for a transparent market-based cyber-insurance industry. It is suggested to establish an agreement (a compact) based on collaboration between government, industry and private society supporting evidence based decision making.

In [19] the focus is on software assurance of safety-critical and security-critical software (i.e., conceptualized as SEC). The perception is that the use of current methods has not achieved the wished-for level of protection, and that there are missing security principles and standards. The industry continues to see an expansion of major breaches occurring in both the public and private sectors. There need to be incentives or regulations for implementing protective and immunizing measures. Such measures could be a mandatory part of the security architecture of all applications. A formal requirement could be that implementation of protective and immunizing measures is included in any certification process. On governance it is suggested to establish software assurance standards at the UN level; to have a risk based approach; to share best of breed methods; and the need to discuss liabilities for damages occurring as a result of an attack or security-related errors.

In [18] the issue of Information security is highlighted in national governance. They propose a comprehensive conceptual framework for building a robust, resilient and dependable Information Security Infrastructure, based on the perspective of software ecosystems.

Development of security and resilience is seen as a maturity process in [41], referencing the CERT Resilience Management Model (CERT RMM) from the Carnegie Mellon Software Engineering Institute. Resilience as a strategy is not simple to implement, in [42], the analysis of resilience strategies in the US agencies revealed that most of the plans only focus on a few of the stages of resilience. Plans do not focus on resilience in the information and social domain, and do not consider long-term adaptation.

In [23] there is a discussion of resilience as a high level design principle. There is an argument for resilience in systems, i.e., distributed systems composed of independent yet interactive elements may deliver equivalent or better functionality with greater resilience. Guidelines for resilience are given such as robustness through resilience rather than resistance, and intervention rather than control. It is argued for the perspective of resilience and to use an ecological perspective in system design and deployment, thus this article describes a design methodology on the ecology level based on resilience principles.

In [43] there is a discussion of development of software-systems, and ignoring some perspectives of software ecosystems. If we want systems that are secure and reliable, both security and reliability must be built together. Applications, middleware and operating systems must be built in the same way, to get systems that are inherently secure and that can withstand attacks from malicious applications and resist errors. The suggested approach is based on security patterns that are mapped through the architectural levels.

### 3.4. Key issues related to methods of risk assessments

The papers identified that the risk assessment was complex, thus there is a need to use methods that integrates the following issues:

- Technology, ensuring that scope of methodology includes safety issues (such as described by IEC 61508 - Functional Safety of Electrical/Electronic/Programmable Electronic Safety-related Systems) security issues (ISA/IEC-62443 to ensure secure Industrial Automation and Control Systems) in addition to a Certification Framework such as the European IACS components Cybersecurity Certification Framework (ICCF).
- Human Elements: Risk understanding and perceptions of the involved actors, including understanding of both safety issues and security issues
- Organizational issues, ensuring that responsibilities are in place, and that necessary organizational structures are in place both to specify, implement, operate and handle deviations critical components of the system.

In summary, if we want systems that are secure and reliable, both security and reliability must be built together. The suggested approach is based on security patterns that are mapped through the architectural levels of the system. However, there is missing international regulation or compacts based on private public partnerships to ensure privacy, safety, security and resilience. Vendors must ensure this quality by design, and must be prepared to accept legal liability for the quality of the technology they produce. Regulations should require routine, transparent reporting of technological problems to provide data for a transparent market-based cyber-insurance industry. There is an argument for resilience in systems, composed of independent yet interactive elements that may deliver equivalent or better functionality with greater resilience. However, the maturity of resilience in use is varying.

## 4. Summary

In this review, we have used the concept of software ecosystems on systems used in smart cities. Our review indicates that the "smart cities" concept are vulnerable and subject to increased emerging risks due to introduction of new technology (such as autonomous transport systems) unsecured components and new connections that has not been foreseen or thought of. Threats, new vulnerabilities and new unwanted incidents are emerging and can be observed through media attention and exploration.

Software assurance of safety-critical and security-critical software (conceptualized as the software system ecosystem) is strongly needed. Current methods have not achieved the necessary level of protection, and are missing security principles and standards. The industry continues to see an expansion of major breaches occurring in both the public and private sectors. Incentives or regulations are needed to implement protective and immunizing measures.

The ecosystem approach seems a promising approach since it provides a holistic view of security needs, by indicating places where security mechanisms can be attached. This approach reduces complexity; one of the important weaknesses used by attackers and can enable analysis of the propagation of threats and data leaks.

Due to the increased proliferation of the IoT and the vulnerability of the Internet, there is a strong need to establish a social compact (agreement) ensuring that the Internet continues to be accessible, inclusive, secure and trustworthy. To ensure that all actors in the value-chain understand the vulnerabilities and the risks, a silo-based "need to know" principle must be replaced by transparent and open reporting. This may support a market based cyber-insurance industry.

In the literature body and in [32] there is an increased understanding of the need for collaboration between the safety and security disciplines to understand and mitigate risks and vulnerabilities. The differences in perspectives between security and safety are due to different adversity models. The security community addresses threats (directed, deliberate, hostile acts) and the safety community addresses hazards (undirected events). Software ecosystems are so pervasive across all sectors of economic activity that this silo approach can no longer be regarded as acceptable.

There is a need for international rulemaking and regulation. This may be difficult to achieve. Vendors must ensure safety, security and resilience by design, and must be prepared to accept legal liability for the quality of the technology they produce. Prescriptive and detailed rule-making on a national level is missing and is difficult to achieve. This is an international challenge. The Mira denial of service (DoS) attack was due to components produced in China but used in the US. No penetration testing, acceptance or testing of robustness was performed prior to release of the product.

In general, there is a need to establish functional standards, responsibilities of liabilities and practices cross-countries. There must be a specific responsibility of the producer to ensure safety, security and resilience, and ideally, a formal process of product acceptance or certification or safety case exploration before a product can be sold or offered. Thus, there is a need for regulatory action from government to set minimum standards, establish responsibility, and follow up of incidents/accidents. The suppliers should establish a proactive focus on (best practice) safety/security standards.

In **Table 3**, we have exemplified critical ecosystems such as smart cities/intelligent transport systems. Based on our review so far, these critical systems have no mandated test criteria (neither safety cases nor security cases, thus it is described as "Poor") and there are no organizations such as CERTS to handle and systematize unwanted incidents.

| Ecosystem | Vulnerability | Test | CERT |
|---|---|---|---|
| Smart cities and intelligent transport systems | Disruption of services (transport, power, water) | Poor | No |

**Table 3.** Critical digital ecosystems and learning.

Development of safety has often been dependent on exploration of publicized accidents and incidents, and a systematic learning loop between users, the regulator and industry. An important component in the learning loop of software systems has been structured reporting and analysis of incidents trough computer incident response teams, i.e., CERTS. There is a need to regulate and ensure that new technology is approved/tested (has some sort of quality control/safety case examination) and that there is some sort of a structured learning process when incidents happen.

Software ecosystems will be exposed to new strains as new unsecured technology are introduced—thus there must be an increased focus on how to handle surprises i.e., resilience and adaptability in software ecosystems to ensure that new demands/stress/failures are not impacting the infrastructure in a catastrophic way. In the review of resilience, [44], there is an increased use of resilience in papers from 2006 on. The resilience concepts are in development, and there is a need to be careful not to place the responsibility of resilience on the individual (i.e., expecting resilience from an individual only). Resilience is the integrated ability of the ecosystem as a whole consisting of an interplay between technology abilities, organizational abilities and human abilities. During the review process, several issues have not been addressed adequately, and are in need of further research, such as:

- There has not been an exploration of the different actors that can affect safety, security and resilience in smart cities (i.e., software ecosystems). Such an exploration should give insight into how to improve safety, security and resilience of systems, and how liabilities should be placed
- There has been no systematic discussion of the maturation of resilience in smart cities (specifics in software ecosystems) discussing technology, organizations and human awareness/ human actions together
- There have been few definitions of patterns of resilience in smart cities and related software ecosystems and how these can be used at an architectural level. There is a missing discussion on how smart cities/critical ecosystems can become resilient, based on patterns
- How to perform ecosystem management of development of federated embedded systems (FES) used in smart cities (i.e., transportation, automotive systems…)

## Author details

Stig O. Johnsen[1,2]*

*Address all correspondence to: stig.ole.johnsen@ntnu.no

1 NTNU, Faculty of Information Technology, Trondheim, Norway

2 SINTEF Technology and Society, Safety Research, Trondheim, Norway

## References

[1] Caragliu A, Del C, Nijkamp P. Smart cities in Europe. Journal of Urban Technology. 2011;**18**(2):65-82. DOI: 10.1080/10630732.2011.601117

[2] Chourabi H, Nam T, Walker S, Gil-Garcia JR, Mellouli S, Nahon K, et al. Understanding smart cities: An integrative framework. In: IEEE Computer Society, Proceedings of the 45th Hawaii International Conference; 2012. pp. 2289-2297

[3] Manikas K, Hansen KM. Software ecosystems–a systematic literature review. Journal of Systems and Software. 2013;**86**(5):1294-1306

[4] Boston Consulting Group. The Internet Economy in the G-20. 2012 Retrieved from: www.bcgperspectives.com

[5] DoD. U.S. Department of Defense, "Standard Practice for System Safety" MIL-STD-882D 2000

[6] Pietre-Cambacedes L, Chaudet C. The SEMA referential framework: Avoiding ambiguities in the terms "security" and "safety". International Journal of Critical Infrastructure Protection. 2010;**3**:55-66

[7] Firesmith DG. "Common Concepts Underlying Safety, Security, and Survivability Engineering", Technical Note CMU/SEI-2003-TN-033. Carnegie Mellon University; Pittsburg, USA, 2003

[8] U.S. Code Title 44, Chapter 35, Subchapter III, § 3542 retrieved at 2006-12-31. Uscode.House.Gov/Download/Pls/44c35.Txt

[9] Aven T. Risk Analysis: Assessing Uncertainties beyond Expected Values and Probabilities. Wiley; Chichester, UKT, 2008

[10] ISO Guide 73, I. E. C. (2002): Risk Management. Vocabulary. Guidelines for Use in Standards. Geneva, Switzerland

[11] NIPP (2013) Department of Homeland Security: Partnering for critical infrastructure security and resilience. Washington D.C.: Department of Homeland Security. Retreived from www.dhs.gov

[12] EU (114/08) On the identification and designation of European critical infrastructure and the assessment of the need to improve their protection. Council directive (2008)

[13] Hollnagel E, Woods D, Leveson N. Resilience Engineering. Ashgate; Hampshire, England, 2006

[14] Woods D. In: Hollnagel E, Woods D, Leveson N, editors. Essential Characteristics of Resilience. "Resilience Engineering" Ashgate; Hampshire, England, 2006

[15] EU (2013) European Commission, Cybersecurity Strategy of the European Union: An Open, Safe and Secure Cyber Space (7 February 2013)

[16] CSM. 2016. Retrieved from: www.csmonitor.com/World/Passcode/2016/1026/What-you-need-to-know-about-the-botnet-that-broke-the-internet

[17] Cerrudo C. (2015). An Emerging US (and World) Threat: Cities Wide Open to Cyber Attacks. Securing Smart Cities – White paper – IOActive. www.ioactive.com/pdfs/IOActive_HackingCitiesPaper_cyber-security_CesarCerrudo.pdf

[18] Ranjan S, Maurya MK, Malviya AK, Yadav R, Gupta R, Mishra M, Rai S. Building an information security infrastructure-a comprehensive framework towards a robust, resilient and dependable infrastructure. International Journal of Computer Science Issues. 2012;**9**:414-419

[19] Axelrod CW. Reducing software assurance risks for security-critical and safety-critical systems. In: Systems, Applications and Technology Conference (LISAT), 2014 IEEE Long Island, IEEE; 2014, May. pp. 1-6

[20] Axelsson J, Papatheocharous E, Andersson J. Characteristics of software ecosystems for federated embedded systems: A case study. Information and Software Technology. 2014;**56**(11):1457-1475

[21] Eklund U, Bosch J. Architecture for embedded open software ecosystems. Journal of Systems and Software. 2014;**92**:128-142

[22] Fernandez EB, Yoshioka N, Washizaki H. Patterns for security and privacy in cloud ecosystems. In: Proceedings of the 23rd IEEE International Requirements Engineering Conference, Ottawa, ON, Canada; 2015. pp. 24-28

[23] Fiksel J. Designing resilient, sustainable systems. Environmental Science & Technology. 2003;**37**(23):5330-5339

[24] Frei S, Schatzmann D, Plattner B, Trammell B. Modeling the security ecosystem-the dynamics of (in) security. In: Economics of Information Security and Privacy. Springer Boston, MA; 2010. p. 79-106

[25] De Florio V. Robust-and-evolvable resilient software systems: Open problems and lessons learned. In: Proceedings of the 8th workshop on Assurances for self-adaptive systems, ACM; 2011, September. pp. 10-17

[26] Koscher K, Czeskis A, Roesner F, Patel S, Kohno T, Checkoway S, Savage S. Experimental security analysis of a modern automobile. In: 2010 IEEE Symposium on Security and Privacy; 2010, May. pp. 447-462

[27] Lima A, Rocha F, Völp M, Esteves-Veríssimo P. Towards safe and secure autonomous and cooperative vehicle ecosystems. In: Proceedings of the 2nd ACM Workshop on Cyber-Physical Systems Security and Privacy, ACM; 2016. pp. 59-70

[28] Zimmermann A, Schmidt R, Jugel D, Möhring M. Evolving enterprise architectures for digital transformations. DEC. 2015;**15**:25-26

[29] Sharkov G. From cybersecurity to collaborative resiliency. In: Proceedings of the 2016 ACM Workshop on Automated Decision Making for Active Cyber Defense, ACM; 2016, October. pp. 3-9

[30] Renn O. (2005). Risk governance—Towards an integrative approach, white paper no.1—International risk governance council

[31] Piggin R. S. H. (2013). Process safety and cyber security convergence: Lessons identified, but not learnt? In: IET Conference Proceedings. The Institution of Engineering & Technology

[32] Bryant IRC. Towards a trustworthy software ecosystem. International Software Quality Management Conference (SQM); 2012. pp. 1-7

[33] Kriaa S, Pietre-Cambacedes L, Bouissou M, Halgand Y. A survey of approaches combining safety and security for industrial control systems. Reliability Engineering & System Safety. 2015;**139**:156-178

[34] Carlsson B, Jacobsson A. (2012). Om säkerhet i digitala ekosystem. Studentlitteratur AB, Lund, Sverige

[35] Longstaff PH, Armstrong NJ, Perrin K, Parker WM, Hidek MA. Building resilient communities: A preliminary framework for assessment. Homeland Security Affairs. 2010;**6**(3):1-22

[36] Sheard S, Mostashari A. A framework for system resilience discussions. In: Proc Eighteenth Annual Int Symp INCOSE; 2008

[37] Merkow MS, Raghavan L. Secure and Resilient Software Development. CRC Press; London, UK, 2010

[38] Krutz RL. Industrial Automation and Control System Security Principles. ISA; NC, USA, 2013

[39] DHS. Department of Homeland Security, Office of Cyber and Infrastructure Analysis: The Future of Smart Cities: Cyber-Physical Infrastructure Risk. 2015

[40] GCIG. 2016 Global Commission on Internet Governance, "One Internet" www.ourinternet.org

[41] Payette J, Anegbe E, Caceres E, Muegge S. Secure by design: Cybersecurity extensions to project management maturity models for critical infrastructure projects. Technology Innovation Management Review. 2015;**5**(6):26

[42] Larkin S, Fox-Lent C, Eisenberg DA, Trump BD, Wallace S, Chadderton C, Linkov I. Benchmarking agency and organizational practices in resilience decision making. Environment Systems and Decisions. 2015;**35**(2):185-195

[43] Fernández EB, Washizaki H, Yoshioka N, Van Hilst M. An approach to model-based development of secure and reliable systems. In: Sixth international conference on availability, reliability and security, ARES, Vienna. 2011. pp. 260-265. DOI: 10.1109/ARES.2011.45

[44] Bergström J, Winsen R, Henriqson E. On the rationale of resilience in the domain of safety: A literature review. Reliability Engineering & System Safety. 2015;**141**:131-141

Section 3

# Financial Risk

Chapter 12

# Implementation of Basel and Solvency Risk Assessment Standards in Banks and Insurance Companies of Southeastern Europe Countries

Safet Kozarevic, Emira Kozarevic,
Pasqualina Porretta and Fabrizio Santoboni

Additional information is available at the end of the chapter

http://dx.doi.org/10.5772/intechopen.70605

**Abstract**

Since banks and insurance companies are financial institutions dealing with risks in various aspects of their business, it is important to make a general approach to the issue of risk and risk management from the perspective of these institutions. In that regard, we made a short overview of the risks these institutions are faced to as well as the methods they use for risk assessments. In order to limit our research to specific financial markets, we provided the basic characteristics of bank and insurance markets of Southeastern Europe (SEE), presenting the core indicators of development level of these markets and their legal and institutional environment related to new regulation proposed in Europe. We analyzed the level of implementation of Basel and Solvency risk assessment standards in these countries and tried to anticipate direction(s) where these markets are going to move in the coming period.

**Keywords:** risk, risk assessment, banks, insurance companies, Basel, solvency, Southeastern Europe (SEE)

## 1. Introduction

The global financial crisis (2007–2009) severely tested the ability of intermediaries to effectively manage and control the various types of risk, revealing the incoherence between the risks taken and perceived as well as profound weaknesses in risk governance. Since as early as 2009, the OECD, the Senior Supervisors Group, the G-30, and the European Commission itself have denounced the close interdependence between the serious deficiencies in banks' organizational and control systems, and the equally serious inadequacy of risk governance. In 2009, the Basel Committee on Banking Supervision (BCBS) issued a policy document "Enhancements to the Basel II framework - Supplemental Pillar 2 practices" (part of the package is better known as B2.5). The document highlighted the need to "overcome organizational silos between business lines" and the urgent need to enhance "firm-wide oversight, risk management and controls." The interventions it indicated as necessary in order to improve the overall risk governance system included: strengthening of the board's and senior management's role of guidance and control, including through definition of the risk appetite framework (RAF), risk management and internal control systems integrated at entire group level, and an appropriate management information system (MIS) at business unit and group level.

This undoubtedly provided the seedbed for the ongoing proliferation of standards on risks and capital we have witnessed in recent years, intended, desirably, to lay the foundations for solidity and stability of individual banks and, consequently, ensure resilience of the financial system in its entirety. A proliferation which might, at times, appear excessive considering the great amount of sound practices, principles, regulatory frameworks regarding risks and capital produced by the regulator in recent years. The new regulatory framework, known as Basel III, introduced an in-depth reform of banking regulation and constitutes a significant step forward, resulting in interaction between supervision of individual banks (known as "microprudential supervision") and supervision of the banking and financial system as a whole ("macroprudential supervision") [1].

With regard to capital, through the new Basel III regulations, the supervisory authorities saw fit to bring the concept of supervisory capital closer to that of capital from an accounting perspective, since the set had previously included tools that did not constitute capital in the strict sense and which were therefore unusable for covering operating losses recorded during the bank's operation. Furthermore, Basel II's concept of supervisory capital included financial resources treated differently in the various European countries, and this caused a problem in comparing the capital of the various banking institutions.

The new rules were intended to render capital structure simpler and more transparent as well as improve the quality and quantity of supervisory capital. In particular, it was established that the new concept of capital would include just two levels: Tier 1 (comprising common equity Tier 1 capital and additional Tier 1 capital) and Tier 2. The first was intended to ensure corporate continuity (going concern), while the second would absorb losses in the event of serious difficulties and liquidation (gone concern). Tier 3 capital was therefore eliminated in order to ensure that market risk was covered by the same capital resources used to cover the other risks. To improve the quality of supervisory capital, it was established that, in order to perform its function, Tier 1 would have to be composed predominantly of common equity (ordinary shares and retained earnings, therefore excluding preference shares and savings shares) and constitute the most subordinate component in redemption and remuneration (i.e., the most stable) as well as the one most available to absorb management losses. Other capital resources may be included in Tier 1, to a limited extent and on condition that they are subordinate, do not have obligatory remuneration, do not have a

http://dx.doi.org/10.5772/intechopen.70605

maturity or incentives to redeem, and are adequate to absorb management losses.[1] Besides, to improve the quantity of supervisory capital, Basel II provided that the overall required capital provision had to be equal to at least 8% of risk-weighted assets, while, in accordance with the new Basel II regulations, it must be equal to at least 10.5% of risk-weighted assets. In particular, it was established that common equity must be equal to at least 7% of risk-weighted assets and supplemented by two buffers which would be introduced gradually: a conservation buffer and a countercyclical buffer. The conservation buffer must be equal to 2.5% of the bank's overall risk exposure and be composed of common equity. It must be intended for conservation of corporate continuity, in other words absorption of losses during crisis periods. The countercyclical buffer may be equal to a maximum of 2.5% of risk-weighted assets and must be built up during favorable periods of the economic cycle, when risk is accumulated, for possible use when conditions deteriorate and risk materializes. This must be observed by setting aside common equity or other capital resources capable of fully absorbing losses, and its purpose is to achieve the broader macroprudential objective of protecting the banking sector from phases of excessive total credit expansion in relation to GDP (contain the procyclicality phenomenon). It is not obligatory for this buffer to always be built up like the first one, and it may even be less than 2.5%, at the discretion of the national authorities.

For implementation of the Basel III Accord, the European Commission transposed its content into the following two legislative acts:

- CRD IV (Capital Requirement Directive IV)[2], which concerns, among other things, the conditions for taking-up and pursuit of the business of banks, right of establishment and freedom to provide services, the prudential control process and additional capital reserves. In order to be applicable, this directive must be transposed into the legislation of the various nations;
- CRR (Capital Requirement Regulation) [2], which governs first-tier prudential supervisors and the rules on public disclosure (capital, leverage ratio (LR), counterparty credit risk, and liquidity risk). The inclusion of prudential rules in a regulation responds to the objective of creating a set of rules valid throughout Europe which are as harmonized as possible for intermediaries operating in the single market. This regulation does not need to be transposed into the respective national legislations and is therefore directly applicable in the various Member States, thus rendering the regulation and supervision process faster and simpler.

In this new regulatory context, the risk management function (risk assessment, measurement, monitoring, and reporting) of undertakings and their various businesses assumes strategic importance.

[1] A list has been prepared of 14 criteria which must be met, individually and jointly, in order for a resource to be included under common equity.

[2] In the case of smaller banks, in particular, it is observed that the types of risk classified as key risks often comprise only credit risk, market risk, and operational risk and that supervisory risk measurement methods are also deployed internally to measure these risks.

## 2. Role of risk assessment in banks and insurance companies

The Concise Oxford English Dictionary defines risk as the hazard, the chance of a bad consequence, loss or exposure to an accident. This is actually the definition of downside risk ("bad" or "real" risk, the risk of loss), which we often encounter in everyday life. The upside risk is opposite to the downside risk, which rarely happens and relates to the potential for gain.

Financial risk is considered *sui generis*, and it can be defined as any event or activity that may adversely affect the ability of an organization, commonly a bank or an insurance company, to achieve its objectives or implement its strategies. Alternatively, risk is also defined as measurable probability of loss or reduced expected return [3]. In fact, contemporary understanding of the financial risk refers to the downside risks associated with financial institutions, that is, banks and insurance companies, and the volatility related to their returns.

Since the banks and insurance companies are professional risk takers and they operate on the basis of transferred powers and trusts (by depositors and insureds), their risk management (including assessment) is subject to regulatory constraints, especially in the sense of applying the Basel and Solvency standards, in order to reduce the systemic risk (the primary objective of regulation in banking) and to protect the insureds (the primary objective of regulation in insurance sector). The most recognized type of risk in banking is probably the market risk, or the risk of a change in the value of a financial position due to variation of the value of the underlying factors, such as stock and bond prices, exchange rates, commodity prices, and so on. The next important category is credit risk (or default risk), which is the risk of not receiving promised payments on executed investments (such as loans and bonds) due to borrower's default. Another risk category, which in recent times has been receiving increasing attention, is operational risk, which is the risk of loss resulting from inadequate management of information telecommunications and other systems as well as from inappropriate internal procedures, processes and controls, weaknesses, omissions, employee irregularities and errors, external illegal actions or unforeseen external events that have a potential to cause physical damage to assets of the bank, and so on.

Special risk-type insurers are faced to is underwriting risk, that is, the risk inherent to insurance policies. Risk factors which are important here are as follows: changes of the ways of national catastrophes, demographic changes, changes in customer behavior (e.g., prepayment), and so on.

It is very difficult if not impossible to identify all types of risks that could possibly occur in the operations of today's financial institutions. This is even a bigger challenge when we consider the complexity due to interactions between various types of risks and the ever-changing environment in which the financial institutions operate. However, the taxonomy of the risks associated with financial services identifies the risk types presented in **Table 1**.

The only effective way to successfully manage financial risks is in a holistic, integrated approach that takes into account all types of risks and their interactions. Traditional "silo" approach has been less effective as it assumes isolated treatment of certain types of financial risks.

http://dx.doi.org/10.5772/intechopen.70605

| | |
|---|---|
| Market risk | • Interest rate risk<br>• Exchange rate risk |
| Credit risk | • Active vs. passive credit risk<br>• Credit risk specific to the company vs. systemic credit risk |
| Counterparty risk | • The risk of an ineffective client or a trading partner |
| Liquidity risk | • Normal liquidity risk<br>• Crisis liquidity risk |
| Operational risk | • Technical risks<br>• Organizational risks |
| Risk inherent with insurance policies | • Negative (wrong) selection risk<br>• Moral hazard |
| Legal risk | • Risk related to changes in law, court judgements, or regulations<br>• Fraud and violations of regulatory violations by employees of a financial institution |
| Model-related risk | • Risks associated with bad models when determining the value of financial instruments<br>• Inaccurate risk assessment of trading activities due to bad models for decision-making |
| Event-related risk | • Stock market collapse<br>• Sudden changes in the tax system<br>• Sudden changes in regulatory policy<br>• Currency exchange rate extreme volatility<br>• Terrorist attacks<br>• Various disasters (wars, revolutions, earthquakes, extreme weather) |
| Strategic risk | • Risks associated with the business plans and strategies of financial institutions |
| Business risk | • Business cycle risks<br>• Revenue fluctuations (commission based, for example) |
| Reputational risk | • Risk associated with bad reputation and negative publicity that leads to loss of income |
| Systemic risk | • Interest rates variations<br>• Base risk<br>• Inflation |
| Solvency risk | • Definitive and aggregate risk of financial institutions |

Source: [4].

**Table 1.** Financial risks taxonomy.

## 3. New regulatory risk assessment framework in Europe

The regulation associated with the financial risk management stems from early finance history, such as Venetian banks and the early insurance companies that emerged from London coffee shops in the eighteenth century. However, the key developments that have led to the current international regulatory framework and financial risk management are a typical twentieth-century story.

### 3.1. Path to regulation of financial risks

Most regulations associated with the financial (especially banking) risks resulted from the BCBS. While the BCBS does not have any formal supranational supervisory authority, it formulates some general supervisory standards and guidelines. It also defines and recommends the best practice, with the expectation that relevant authorities will take the necessary steps to implement it through detailed arrangements (prescribed by law or otherwise), best suited to their respective national systems. It is important to note that the original guidelines referred to only "internationally active" banks, which has since changed.

On 11 July 1988, the BCBS issued "International Convergence of Capital Measurement and Capital Standards," or "Capital Accord," and then on 19 January 1996, adopted a corresponding amendment ("Amendment to the Capital Acord to Incorporate Market Risks"). The Capital Accord, including the Amendment, is primarily focused on the protection of banks against credit and market (interest and currency) risks. On one hand, banks require linking between the credit risks and prescribed capital reserves and capital adequacy ratios, while, on the other hand, they are free to determine their capital reserves for market risks by using internal models, primarily so-called Value-et-Risk (VaR) model.

Over the recent years, we have witnessed a growing regulatory pressure on the insurance sector, linked with the combination of the two regulatory frameworks: institutional or methodological. For example, the Joint Forum on Financial Conglomerates was founded in early 1996, under the patronage of the BCBS, the International Organization of Securities Commissions (IOSCO), and the International Insurance Supervision Association International Association of Insurance Supervisors (IAIS). This body was formed with the aim to take over the so-called Tripartite Group, which handled the supervisory issues related to financial conglomerates since 1993 and until July 1995. The Joint Forum includes the same number of supervisors of older banks, insurance companies, and financial services companies (investment banks, brokers, and dealers), representing each supervisory unit, or three parent organizations.

A complete set of baseline rules for prudential supervision, also well known as Basel I, has significantly stepped up with regard to the effective determination of the minimum international capital standards. This historic agreement was signed in July 1988 by the member countries of Group 10, Luxembourg and Spain. It came into full force in January 1993. To date, these rules have been accepted by over 100 countries. There is an understanding and expectation that banking credibility is not satisfactory if Basel rules about business operations are not implemented. Basel I accepted a standard division of total banking capital into two basic components (**Table 2**).

| |
|---|
| I Primary capital (core capital) |
| • Share capital |
| • Retained profits |
| II Secondary capital (supplementary capital) |
| • Subordinated bonds with a maturity of over 5 years |
| • General reserves for loss coverage |
| • Latent reserves |
| • Valorization reserves |
| III (I + II) Total capital |
| Source: [5]. |

**Table 2.** Basic structure of bank capital.

The current rate of the required regulatory capital for a bank to qualify as adequately capitalized implies the following [6]:

- The percentage of the primary capital ratio (CR) and the total weighted risk must be at least 4%.
- The percentage of total capital to total risk-weighted assets must be at least 8%, with secondary capital limited to 100% of the primary capital.

Despite having historically assisted in the development of an international framework for the regulation of financial risks, Basel I has also been subjected to criticism. Two main areas come to focus as limiting: a fairly general approach to credit risk and the failure to appreciate the impact of market risk. In the context of strengthening connections between banks and financial markets (the so-called market banking) and the increased banking risks in the field of financial instruments, foreign exchange instruments, and derivative positions, the BCBS published some new proposals in 1993. They focused on the requirement for banks that faced higher degree of exposure to market risk that they hold a higher quantum of capital with regard to the size of their assets. The Basel I amendments in 1996 prescribed a standardized market risk model, while at the same time allowing larger and more sophisticated banks to choose an internal model based on VaR (an in-house developed model). Legal implementation of these amendments was achieved in 1998.

Instead of a standard methodology, banks may alternatively use internal models when evaluating market risks if they have adequate internal market risk management systems. However, these systems must be explicitly accepted by the appropriate supervisory agency. This agency prescribes general qualitative and quantitative standards, which must be integrated into internal models for the assessment of market risks and the formation of an adequate level of capital. For example, a 10-day VaR of EUR 20 million at a confidence level of 99% means that the market portfolio will cause a loss of EUR 20 million or more with a probability of 1% until the end of the 10-day holding period, if the composition of the portfolio remains unchanged during the same period. The choice of the holding period (10 days) and the level of reliability

(99%) lie in the hands of the regulator only when VaR is used to calculate regulatory capital[3]; otherwise, it is arbitrated. The main qualitative standards are as follows [7]:

- Internal models must be integrated into daily bank risk management processes.
- The board of directors and senior management of the bank should be actively involved in the risk control processes.
- There must be an independent risk control service within the bank.
- Internal models must contain appropriate limits that restrict the exposure to market risks at operational level.
- The corrective multiplier must be at least three and can be increased to four, which depends on the results of the backtesting.
- The bank must implement the regulatory program of backtesting of its internal market risk model. Namely, we are talking about ex post checks of the efficiency of the internal model by comparing the VaR risk assessment with the actual profits and losses of the bank. Backtesting should show whether or not the specific internal model used by the bank is a reliable measure of potential losses. If backtesting shows satisfactory results, the corrective multiplier holds the value of three; otherwise, it can be increased to four.
- The bank should have a readily available rigorous stress-testing program.

The main quantitative standards relate to the following rules:

- The VaR determination is performed on a daily basis with a 99% confidence level.
- The holding period for the risk calculation due to the holding of financial instruments is at least 10 working days.
- The statistical basis for VaR accounts must be at least 1 year, but the audit is done at least at quarterly intervals.
- The supervisory authorities do not prescribe a certain internal model that banks have to adhere to, so it is possible to use various models, ensuring that qualitative and quantitative standards are being followed. An internal model determined by a bank must obtain the explicit consent of the supervisory authority.

On 26 June 2004, the central bank governors and supervisory directors of Group 10 published the changes to capital framework, called Basel II. The key conceptual change within Basel II is the introduction of the concept of three tiers.

### 3.2. The three-tier concept

Tier 1 (risk assessment) requires that the banks calculate the minimum capital charge, which relates to regulatory capital, in order to quantify the minimum capital in relation to the

[3]Bank is required to provide equity for market risk equal to the average VaR estimate for the last 60 days, which is further multiplied by the safety coefficient (usually equal to three).

http://dx.doi.org/10.5772/intechopen.70605

potential economic loss. According to Basel II, there is a cost of capital for credit risk, market risk, and (for the first time!) operational risk. While the treatment of market risk is relatively unchanged from the Basel II Amendment of 1996, the cost of capital for credit risk has been significantly changed. In calculating the cost of capital for credit and operational risk, banks can make a choice between three different approaches in increasing the sensitivity and complexity of the risk: the basic indicator approach, standardized approach, and advanced measurement approach (internal methodology using quantitative and qualitative criteria).

Any quantitative approach to risk management should be embedded in the functional governing structure of an institution. Therefore, the best practice in risk management imposes clear restrictions on the organization of the institution (board of directors, management, employees, internal, and external audit processes). Specifically, the Board of Directors assumes ultimate responsibility for the failure of the "landscape" risk and the risk appetite formulation. This is where the Tier 2 (risk management) comes in. With this important tier, which is designated as a supervisory audit process, local regulators perform various checks, bringing the order of balance. This tier starts from the need for an effective overview of the bank's internal overall risk assessment and ensures that the management trains the "sound" of the assessment and set aside the adequate capital for various risks.

Finally, in order to fulfill the promise that the growing regulation will also reduce the systemic risk, clear guidelines on risk reporting by financial institutions are also required. Tier 3 (risk transparency) requires the establishment of a market discipline through more effective communication of risk measures and other information relevant to risk management.

In analogy to Basel II, the structure of the Solvency 2, regulatory regime in the insurance sector, is also set up as a three-tier structure.

### 3.3. From Solvency 1 to Solvency 2

Until the latter half of the 1980s, the insurance sector was notable for its substantial "staticity," having, for a long time, been sheltered from environmental and regulatory disturbance and not being marked by any particular vivacity in terms of product and process innovation. The shifting environmental conditions of the last 25 years, affecting the structure of the insurance market and the financial system in general (consequent to the process of deregulation and internationalization of the industry that colored the 1990s), have, however, triggered profound changes in ownership structures, forms of competition, and distribution patterns, and, overall, in the management techniques of insurance undertakings [8].

The supervisory system has not been unaffected by this process of profound change, also undergoing significant alteration. The specific nature of this evolution is evident in the move from supervision based essentially on "structural" monitoring methods to other, "prudential" methods as well as in the strengthening of supervision on consolidated foundations and in the increase in cooperation agreements between the Supervisory Authorities at national and international levels.

The insurance (and reinsurance) sector—albeit on a different timescale to the banking one but in fundamentally similar ways—has not been left "untouched" by this evolutionary process either and, following a long "gestation" period, on 1 January 2016, the new regulatory regime known as "Solvency 2," pursuant to Directive 2009/138/EC (containing the general principles), together with the additional levels of the legislation (containing greater detail)[4], was introduced throughout the European Union (EU). While Solvency 1 proved itself to function without causing serious "trauma," given that the financial crisis did not highlight any significant cases of bankruptcy of insurance undertakings, its limitations were evident right from the early 2000s (in other words, from the start-up phase of the draft revision of the current regulatory regime). Moreover, the undisputed merits of its extremely simple and robust regulatory infrastructure were countered by a series of critical issues which called its future applicability into question. In this perspective, Solvency 1—characterized essentially by its solvency margin requirements, its requirements for risk quantification of commitments undertaken, and its quantitative and qualitative limits on assets to cover liabilities—induced scholars, practitioners, and the supervisory system itself to highlight the relative limitations[5] and identify the possible solutions [9].

One of the main criticisms leveled at Solvency 1 is that, being a model based on flat rates, it neither took due account of the specific riskiness of each undertaking nor considered the portfolio of risks to which an undertaking is exposed, on the assets and liabilities sides [10]. Indeed, Solvency 1 not only provided (on the liabilities side) for application of essentially flat rates to technical parameters which often revealed insufficiencies in the rounding of the risks of an undertaking's insurance portfolio, but also did not take account of the quality of the said portfolio or (on the assets side) of the investments made in order to provide adequate cover. This resulted in the clear paradox that two insurance undertakings showing (on the liabilities side) the same riskiness are required to hold the same capital for supervision purposes despite clearly differing (on the assets side) in terms of the riskiness of their investments.

The need to remedy the said limitations spurred European legislators to initiate an in-depth review of the system for supervision of company solvency in insurance undertakings, with a view to comprehensive innovation of the supervisory prudential rules in order to ensure a level playing field within the insurance sector and create a new, risk-oriented regulatory framework (Solvency 2) [11]. In this regard, the end goal aimed for was the definition of a prudential supervisory system involving appropriate capital requirements, in other words "tailored" to the risks to which insurance undertakings are actually exposed and capable of incentivizing more careful measurement and management of these by the intermediaries themselves.

[4]We refer, in particular, to: the delegated acts adopted by the Commission through Regulation 2015/35 of 10 October 2014: amended by Commission Delegated Regulation (EU) 2016/467 of 30 September 2015:, 16 implementing regulations (ITS: Implementing Technical Standards) proposed by the EIOPA (European Insurance and Occupational Pension Authority) and adopted by the Commission and, finally, the EIOPA guidelines which, unlike the previous (directly applicable) regulations, must be transposed into the individual national legal systems.

[5]These limitations may be identified, in summary, as: (a) the lack of transparency of the insolvency margin criterion and its inadequacy in expressing the risks taken by undertakings, (b) the absence of incentives for undertakings to perform effective risk management, and (c) the lack of recognition of the economic situation of transnational groups of undertakings.

### 3.4. Risk assessment in the light of Basel III

The BCBS reached a consensus regarding the Basel III framework on 26 July 2010. On 30 November 2010, the G-20 leaders indorsed the framework at the summit in Seoul, South Korea. Basically, Basel III relies on the Basel II framework. Basel III has three objectives that are articulated by the Committee: to address the issues which led to the global financial crisis and the lessons learned from the crisis, improve risk management and governance, and strengthen bank's transparency and disclosures. The Basel III Accord is characterized as a successive process in which banks are to increase the primary capital ratio from 2 to 7% in several years to come. The total implementation of Basel III is expected on 1 January 2019. Regulators believe that under crisis banks may temporarily reduce the ratio to 4.5%, but are not allowed to pay bonuses and dividends until they return the ratio to 7%.

In addition to the Basel II capital ratio, Basel III requires banks to respect additional ratios, such as leverage ratio, liquidity coverage ratio (LCR), and net stable funding ratio (NSFR). The risk-based capital ratio (CR) requires that 8% of risk-weighted assets are covered by tier 1-capital plus tier 2-capital. The volume-based leverage ratio (LR) requires banks to hold 3% T1-capital against total assets. The net stable funding ratio (NSFR) stipulates that stable-funding weighted assets are to be 100% covered by stable-funding weighted liabilities. The liquidity coverage ratio (LCR) requires that outflows are 100% covered by inflows plus the (haircut-weighted) liquidity reserve [12].

The new Basel III regulatory framework increases the importance of the risk management process, which includes risk assessment. Indeed, risk assessment is a strategic phase of the risk management process. However, new developments in financial markets, financial management practices, operational complexity, and supervisory approaches and rules (Basel II) have, over the last decade, created new risks, cross-type risks, interrelation between traditional risks (such as counterparty risk, credit risk, market risk, operational risk, and liquidity risk) and increased the importance of the risk management process, infrastructure, and governance. All this has presented the risk management function with a great challenge.

Article 86 of Directive 2013/36/EU [13] requires undertakings to have robust strategies, policies, processes, and systems for identification, measurement, management, and monitoring of liquidity risk over an appropriate set of time horizons and management and monitoring of funding positions, in order to ensure that they maintain adequate liquidity buffers and funding (commonly referred to as the "Internal Liquidity Adequacy Assessment Process": ILAAP). Risk assessment is an extremely critical phase of this process and should cover the major sources of financial intermediaries. As we know, risk assessment is the starting point for the "Internal Capital Adequacy Assessment Process" (ICAAP) and, also, for strategic, capital, and liquidity planning, the "Risk Appetite Framework" (RAF)[6] and the "Recovery Plan."

[6]The RAF is the reference framework that defines an undertaking's risk appetite, tolerance thresholds, risk limits, risk management policies, the relevant processes necessary to define, and implement these in line with the maximum risk that may be taken, business model and strategic plan.

The EBA guidelines [14] aim to achieve convergence of supervisory practices in the assessment of ICAAP and ILAAP as required by the SREP Guidelines, by introducing a common set of information that competent authorities will be using in their assessments across the EU. In particular, the guidelines aim to specify what general and ICAAP- and ILAAP-specific information competent authorities should collect from institutions following their minimum engagement model as specified in the SREP Guidelines. In this perspective, the EBA guidelines recommend that financial intermediaries prepare a list of risk categories and subcategories covered by ICAAP, including their definitions and parameters of individual risk categories, a description of their approach to the identification of risks (including risk concentrations) and the inclusions of identified risks within risk categories and subcategories to be covered. On evidencing the implementation of the scope, the general objectives, and the main assumptions underlying ICAAP, competent authorities should ensure that they receive from institutions the following: (a) list of risk categories and subcategories covered by ICAAP, including their definitions and perimeters of individual risk categories; (b) explanations of the differences between the risks covered by ICAAP and the risk appetite framework, where the scope of risks covered is different; and (c) description of any deviations in the ICAAP process and in the key assumptions within the group and the entities of the group, where appropriate. The effectiveness of the risk management process as a whole depends, in fact, on the quality of the risk assessment (thoroughness, coherence with the bank's business model, etc.) and is consolidated over time if there exists, within the financial intermediary, a continual process of reconciliation between capital and liquidity planning, the ICAAP, the RAF and the Recovery Plan, governance, structure and organization. However, the goal of alignment and coherence constraint remains an aspiration for many undertakings because it requires a holistic approach to bank governance.

In this perspective, it is important to understand the way in which undertakings perform their risk assessments. There is no single approach for all banks because they usually have different business models, strategies, capital and liquidity buffers, operational complexity structures and governance, and, therefore, different risks.

In general terms, the risk map should be: exhaustive, transparent, consistent, and well-integrated into the business model (holistic approach). Firstly, the assessment should include all risks to which undertakings are or could be exposed (principle of exhaustiveness). This should include difficult-to-measure risks such as compliance risk and reputational risk. Secondly, the methodology used to assess and measure exposure should be duly documented together with the assumptions used during the risk assessment process (principle of transparency). Nevertheless, the risk assessment process should be consistent and aligned with other institutions' processes and with the nature, size, complexity, and scale of the bank's business activities.

Faced with these procedural principles, we might ask ourselves: which risks do undertakings consider in their assessments? In the case of smaller banks, in general, the types of risk classified as key risks often comprise only credit risk, market risk, and operational risk, and supervisory risk measurement methods are also deployed internally to measure these risks. However, undertakings should, as a minimum, take into account in their assessment the risks recommended by the regulator or supervisory authority. It is important to underline that: (a) the risk mapping recommended by the regulator and supervisory authority are not fully aligned; (b) the regulator/supervisory authority's risk mapping is not to be considered mandatory. It is an undertaking's choice whether and how to combine risk types and risk subcategories. Indeed, the undertaking might classify certain risks differently to the regulator/supervisory authority. In the opinion of the European Central Bank (ECB) [15], it is important to map at least the following risks: (a) credit risk, (b) market risk, (c) operational risk, (d) interest rate risk in the banking book, (e) participation or equity risk, (f) sovereign risk, (g) pension risk, (h) funding risk, (i) risk concentrations, and (j) business and strategic risk. In the case of conglomerates and for material participations (e.g., in insurance undertakings), institutions are also expected to take inherent risks, such as insurance risk, into account in their risk mapping; in the case of other immaterial risks, institutions should explain why these are considered immaterial.

It is desirable, also for regulatory purposes, for institutions to define each of the key risks listed above as well as any other risks identified as key based on the institution's risk profile, how the institution defines the materiality of each key risk, and a description of how each material risk is then quantified for capital allocation purposes, including detailed methodology to specify data, assumptions, and calculations. The bank's assessment of risk indicators should also be reflected in its risk policies (measurement, management, and control). This means, for example, that a bank which considers country risks to be immaterial will subsequently avoid taking on material country risks, that is, the bank will keep activities such as proprietary transactions in foreign securities, interbank trading with international counterparties, or loans to foreign borrowers to a minimum.

It is important to understand that risk assessment must be based as closely as possible on an approach integrated and coherent with the bank's business model and with risk governance as a whole, as previously underlined (principle of holism). This is in order to reduce overlap between businesses and the associated operational inefficiencies, to develop forms of collaboration orientated toward the pursuit operational synergies, and to achieve greater impact during the risk measurement, management, and control phases. This is also a cultural change whose driving force is the risk management function, called upon to perform a role of proactive consultancy in all areas of decision-making and of fruitful dialog with the operational functions. In this perspective, the risk management function must transition from the role of simple "controller" of risk levels and "producer" of information for the benefit of the top management and the board to that of influential partner actively involved in strategic decision-making processes.

In this regulatory and market context, there is a significant risk, for banking intermediaries, of being subjected to the numerous rules and chasing regulatory compliance without succeeding in germinating the seed of change planted in the new vision for the regulatory and supervisory system. In this perspective, it is therefore useful to adopt an integrated, holistic approach to risk management ("integrated risk management"). This is not merely a philosophical stance but, rather, a mindset that marks a moment of significant discontinuity with the past permeating entire banking undertakings capable of setting aside cultural resistance to change and waving of ideological flags. It is an approach that integrates and synergizes:

- The different regulatory frameworks, often managed and implemented in silos by banking intermediaries;
- Risk management under normal and stressed conditions, in line with a backward- and forward-looking vision;
- Internal management for risk supervision purposes, therefore incorporating the new SREP, strategic planning, ICAAP, ILAAP, RAF, Recovery Plan, capital allocation, liquidity allocation, market disclosure, and reporting to the regulator.

### 3.5. The impact and implementation challenges of new regulatory frameworks

According to the Bank for International Settlements (BIS), implementation statistics regarding Basel I as a voluntary standard are impressive. From 2001, over 100 countries worldwide had implemented Basel I. Furthermore, from 2004, these countries intended to implement Basel II—13 were Basel Committee Members and 88 were non-Basel Committee Members. As of 2008, 105 countries had either implemented Basel II (57) or intended to implement Basel II (48), a total of 13 Basel Committee Members and 92 non-Basel Committee Members.

From 2008, the Basel Committee country membership expanded to include 27 Members, emerging countries (such as Argentina and India) as well. What is important, this allows emerging countries to join the negotiations about what the "global" supervisory framework would look like. Although the Basel accords are intended for implementation by internationally active banks and in developed economies, other economies (and, therefore, emerging) are also forced to implement the accords due to international regulatory and competitiveness matters. For example, in Asia, at least nine countries had implemented or expected to implement Basel II by 2010 (e.g., India, Hong Kong, Japan, Philippines, Singapore, Taiwan, and Thailand had implemented several aspects of Basel II as early as by 2008). Moreover, the argument for prevention of crises and maintenance bank soundness is supported by the fact that banking crises in emerging economies have generally exceeded 25% of GDP and are, proportionately, much larger in scale than in developed economies [16].

As far as Solvency 2 is concerned, one of the first absolute innovations is the fact that, while the directives issued prior to establishment of the new regulatory regime focused predominantly on "capital leverage," with a view to defining the requirements necessary to guarantee solvency, the Solvency 2 regulatory framework places risk at the very center of management of an insurance undertaking, imposing assessment of the risks peculiar to each of the businesses they conduct and suggesting risk management criteria.

*"… Regulators and rating agencies expect that companies have a good understanding of their risk profiles and have implemented the appropriate governance structure to mitigate their risks. The insurance industry is everchanging, and it can be challenging for an organization to have a complete understanding of the risks that can pose potential pitfalls to its operations…"* [17].

This new approach also affects the *modus operandi* of the Supervisory Authorities, which are consequently called upon to induce insurance undertakings to graduate their recourse to and use of capital leverage according to the riskiness of their business and to "mold" their organizational structure according to their ability to cope with the risks taken [18]. In this perspective, the new regulatory regime places particular emphasis on an undertaking's risk management system, which consists of the set of strategies, processes, and reporting procedures designed to identify, measure, continually monitor, and thus manage on an ongoing basis the risks, at individual and joint levels, to which undertakings and groups are or could be exposed, as well as the interdependence of these. The main protection provided by the risk management system lies in the identification and measurement (in short, "risk assessment") of the risks to which the undertaking is exposed. Risk assessment is normally formalized by the risk management function which, based on the evidence provided by the other, second-level control functions (internal auditing, compliance, and actuarial function), and on the findings of the control activities carried out by the business functions responsible for management of the above-mentioned risks, highlights their presence and provides a quali-quantitative assessment of them.

As we have already stated, adoption of the regulatory requirements of Solvency 2 is not limited simply to application of a new capital requirement calculation methodology and new accounting standards but, rather, requires significant adjustments in terms of corporate governance and organization. It is therefore necessary for boards of directors to increase their interest in and understanding of the risk profiles inherent in the business of their undertakings. In other words, they must develop a risk-oriented governance framework permitting the necessary diffusion, at all corporate levels, of an appropriate risk culture and ensuring that the board has adequate knowledge and control of decision-making processes, risk management, and operations in a "Solvency 2-compliant" context. In this regard, the introduction of a risk-based capital requirement system is intended to underline the importance of the risk management function and internal control systems [19].

Continuing with our analysis of the innovations introduced by the new regulatory framework, it is essential that we underline a further and important innovation, namely consideration of a broader and more complete range of risks for the purpose of calculating capital requirements as well as quantifying these more accurately. Indeed, whereas, with Solvency 1, regulatory capital measured only the so-called "technical risks," excluding such things as market risks from calculation, the new regime also considers these. Solvency 2, moreover, sets out to overcome a further limitation connected to the fact that Solvency 1 quantified the said technical risks in a more simplistic manner, that is, through application of flat rates, thus preventing due consideration of the size and riskiness of the various portfolios [20].

Assuming that each business, in tackling the transition from the old regulatory regime to the new, will take into account its own specific characteristics in terms of investment position, business mix, and strategic objectives, it may be hypothesized that the above-mentioned innovations constitute, in a general sense, incentives for insurance undertakings to base their practical operations on certain "rules of conduct." In this perspective, it is desirable, first and foremost, for insurance undertakings, as far as the assets side of the balance sheet is concerned, to proceed with careful selection of asset classes, favoring those which—particularly with regard to life business—permit better matching with liabilities (asset-liability management) and are less correlated with market trends over more volatile and/or risky investments, which are more capital-intensive. At the same time, insurance undertakings are obliged to deal with the dynamics of the current economic situation, featuring, in particular, low interest rates, and even negative ones in some markets and for certain maturities. In this regard, identification and adoption of more adequate investment policies, based on which it is possible to select (new?) asset classes capable of guaranteeing an adequate return on the commitments entered into with policyholders, is undoubtedly desirable [21].

Turning our gaze to the liabilities side of the balance sheet, insurance undertakings are then required to implement more careful and selective insurance risk-taking, giving due consideration to aspects relating to its capital-intensiveness, the profitability associated with different businesses, and, more generally, the principles of correct product governance [22]. In other words, the objective of insurance undertakings must be to "build" products and services capable of meeting the needs arising from the relevant market, as well as those of the distributions channels used for the purpose, taking due account of their risk/return constraints. Joint consideration of the factors we have just mentioned therefore requires insurance undertakings to conduct a major review of their product design, pricing, and distribution [23]. For this reason, particularly in life business, we are witnessing—and will increasingly witness in the future—the offering of a different product mix and a restructuring of the sureties offered to policyholders. In truth, we are facing a trend, already in progress in Europe, leading to the replacement of life products involving the provision of certain guarantees by insurance undertaking (such as so-called "with-profit" products) with contracts in which the risk lies with the policy holder ("unit-linked" products); something which also imposes a transition from "traditional" life policies to "multiline" life products, which combine the guarantee of traditional products with the higher returns of unit-linked. It is clear, therefore, that the opportunity to benefit from the positive effects associated with risk diversification provided by Solvency 2 may only be fully seized by those insurance undertakings capable of operating in multiple business lines, since it is, in effect, closed to monoline undertakings.

A further "rule of conduct" upon which insurance undertakings should base their operations concerns greater use of risk mitigation techniques. In practice, we are referring, on one hand, to the use of financial derivatives,[7] which permits adequate coverage of financial risks and reduction of asset/liability mismatch, notwithstanding the new requirements introduced by the European Market Infrastructure Regulation (EMIR)[8] and, on the other, with regard to insurance risks, to the use of both "traditional" reinsurance (benefitting from current soft market conditions) and so-called "Alternative Risk Transfer" (ART) techniques such as, for example, cat bonds. However, implementing Solvency 2 is not simply a question of capital requirements and accounting standards. It requires considerable adjustments to company governance and organization. In the first place, boards of directors must increase their interest in and their understanding of the risk profiles inherent in company activities. This does not mean that boards should only comprise technical experts or that market strategies need not be

[7]It should be remembered that, similarly to Solvency I, the new regulatory framework, too, permits the use of financial derivatives insofar as they contribute to reducing risks or facilitating effective portfolio management (see Art. 132 of the Directive).

[8]EU Regulation No. 648/2012 of 4 July 2012.

http://dx.doi.org/10.5772/intechopen.70605

discussed—for insurance companies, understanding risk is the basis for any informed strategic decision and, to a large extent, it is their business. The entire way that businesses are organized must change in order to introduce the new approach to risk assessment, which should neither be reduced to a cosmetic exercise for supervisory purposes, nor be relegated to the office of the chief risk officer, but it must play a key role in the management of the company. There are two basic logics—bottom-up approach, which is more invasive and starts from the analysis of all business processes to find the risks involved, and top-down approach, which starts from the risk analysis and tries to identify the processes that generated these risks. In any case, the involvement of risk owners is crucial. In the while, the top-down approach cannot ignore the observation of past risk events, while the bottom-up approach is conducted independently from these past risk events. In other words, the top-down approach follows a backward-looking perspective, while the bottom-up approach follows a forward-looking perspective. For this reason, the best approach is the approach that can complement the two methodologies.

The evidence gathered so far from inspections and the initial ORSA findings are not entirely comforting, especially as regards smallest firms. We know that the transition will not be easy, but we are ready to work together with the companies, each assuming our own responsibility, to facilitate change [24].

## 4. Southeastern Europe market review[9]

### 4.1. Banking sector

The region of the Southeastern Europe (SEE) has experienced some significant changes in recent years. These changes can be observed from the political, economical, and demographic aspects, making an indirect influence on the financial system of the region. Croatia became a member of the European Union (EU), and the other countries of the Western Balkans (Albania, Bosnia and Herzegovina, Montenegro, and Serbia) continue with the slow process of integration. Greece is still burdened with a huge public debt, Turkey suffers from problems related to the wars on its eastern border, while Slovenia, Romania, and Bulgaria continue adaptation to the unique EU market.

The financial system of the SEE countries is basically bank-centric. The system is dominated by commercial banks. The size of the banking sector differs between countries of the region, with the largest share of banks' assets of GDP in Greece (220% in 2015) and the lowest in Romania (53% in 2015).

In all observed countries, except Turkey, there has been a high share of foreign ownership in total banking assets (from 68% in Greece to 90.4% in Romania). In Turkey, the banking sector is dominated by domestic banks (74% of total assets are held by domestic banks).

---

[9] *The review is based on the survey fulfilled by national regulators of the SEE countries as well as the information published on their official websites.*

In all observed countries, net loans have the highest share in banks' assets. Cash funds are the second largest item in Bosnia and Herzegovina, Montenegro, Serbia, and Turkey, while government bonds are the second largest in Albania, Croatia, Greece, and Romania. Households and corporation deposits have the highest share in total banks' liabilities in all observed countries. In Greece, European Central Bank borrowing makes a significant part of total liabilities (28% in 2015). Regarding the structure of loans, the loans to nonfinancial corporations and households have the largest share. In the structure of deposits, retail deposits have the largest share in all of the countries.

Regarding maturity, there is a significant mismatch between deposits and loans in all of the countries, except Albania and Turkey. In all observed countries, except these two, there is a high share of short-term deposits in total deposits and long-term loans in total loans. Romania has the highest share of short-term deposits (93% in 2015), while Croatia has the highest share of long-term loans (86.1% in 2015). In Albania and Turkey, respective shares of long-term deposits and loans in total deposits and loans were similar (approximately 70%).

Regarding the basic features of banking sector in 2015, capital adequacy ratio varied between 15 and 21 in the observed countries, the highest being recorded in Croatia and Serbia.

In 2015, Turkey recorded the lowest ratio of nonperforming to total loans (3.2%) and Greece had the highest ratio (43.9%).

Most of the countries are in the process of adopting Basel III standards. In Greece, as it is a member of the Eurozone, the regulatory framework is based on and incorporates the Basel III framework and related provisions. Croatia started the implementation of Basel III in Croatian legislation in 2013, with the adoption of the new Credit Institutions Act transposing CRD IV, following its accession to the EU. Consequently, there have been some changes in the regulatory framework regarding the coverage of risky assets in capital and capital buffers.

### 4.2. Insurance sector

As far as the insurance sector of the SEE region is concerned, it mostly stagnated in recent years, with a slight increase in most of the countries except Greece, where premium volume significantly decreased. What is interesting is that for the countries that recently joined the EU, stagnation of the sector is more evident than in the nonmember countries where the premium volume is increasing. It looks like the first group of countries reached the limits considering their economies, while the second group of countries still has potential for growth. Reasons for such situation are different, and they can be caused by some internal changes on the market, like introduction of Motor Third Party Liability (MTPL) market liberalization in the newest EU member Croatia, or by some other macroeconomic issues.

Similar to banking, ownership structure of the insurers on the markets shows dominance of insurers with majority of foreign ownership. They cover more than half of the total premium on the local markets, excluding Slovenia, where foreign insurers cover 11.1% of the market. There are also many insurers from one SEE country (such as Slovenia and Croatia) operating on the markets of other countries of the region. The largest market share of the foreign insurers is registered in Montenegro and Macedonia, where foreign insurers cover more than 90% of the markets.

http://dx.doi.org/10.5772/intechopen.70605

The total premium of the SEE insurance market in 2015 reached the amount of EUR 21.018 million with a decrease in total during last 5 years in spite of a slight increase in the premium in almost all countries. A decrease of the premium was noticed in the region's most developed markets of Greece, Slovenia, and Croatia, caused by financial crisis or by MTPL market liberalization (in case of Croatia). For this reason, if we exclude data for Turkey, we can observe that insurance premiums of the region have decreased by 10% in comparison with 2010. Slovenia and Croatia stopped recording negative trends and had an increase of premium in 2015. At the same time, total insurance premium of the Insurance Europe (IE) association members in 2015 was EUR 1,199,714 million, with 4.2% of growth compared to 2010. Like in 2010, the SEE insurance market is still 1.8% of total IE premium.

The average insurance penetration for the SEE countries in 2015 was 1.78, and it decreased from 1.81% in 2010. The average insurance density was EUR 146.35, and it increased by 3.7% in 5 years (EUR 141.15 in 2010). The average penetration of IE in 2015 was 7.4%, and it decreased from 8.9% in 2010, while insurance density was EUR 2.010 and it increased by 7.0% (EUR 1.879 in 2010). These data show that insurance penetration is 4.2 and insurance density 13.7 times below the IE average for 2015. The largest insurance density in 2015 was in Slovenia (EUR 975), Greece (EUR 343), and Croatia (EUR 271), while the other countries had density below EUR 200 showing the characteristics of poor insurance markets.

Negative trends in the premium structure during last 5 years can be recognized by a decrease of share of life insurance in total premium. In 2015, the average share of life insurance in the region was 23.1% (25.8% in 2010), while the average share in the IE members was 60.9%. It shows that there was no improvement of insurance culture in the SEE market. The low share of life insurance is a result of macroeconomic instability, consumer unfamiliarity with financial transactions as well as less sophistication of the insurance products offered. The largest share of life insurance in 2015 was registered in Greece (48.4%), Croatia (33.6%), and Slovenia (29.7%), while the other countries had a share of life insurance below 25%.

Since financial system in the region is mostly bankocentric and the share of the insurance in the total financial assets is very low, the activities of the governments in the recent years were mostly focused on providing better regulation of the banks. Insurers are still waiting to be recognized as the important subject of interest in government policies.

As we previously mentioned, the total premium of EUR 21.018 million shows that the SEE insurance market is significantly behind the developed European markets, with only 1.8% of IE premium. Since the market size depends on the economy volume, we can compare this data only for the countries of similar population and economy size. Generally, Slovenia is the closest to the IE average, but there is a lot of variety among the other countries. On the other hand, common to all countries is a high potential for the market growth.

In 2015, the data for 8 countries (since we do not have complete data about the number of companies for all countries), comparing with 2010, show a slight decrease in the number of insurers in all countries except for Macedonia, Serbia, and Slovenia. Some countries still have disproportionately large number of companies in relation to market's premium (BiH, Macedonia, and Montenegro). There are expected mergers and acquisitions in these countries, especially because of the new regulatory standards that are coming into force in the future. The average premiums per company in BiH, Macedonia, and Montenegro are below EUR 15 million, which best explains a need for decrease in the number of companies. Slovenia had the largest average premium per company (EUR 91 million).

The insurers on the SEE market are mostly organized as nonlife insurers, and this type of organization of business activities is characteristic for local insurers, while the life insurance business is mostly in the hands of insurers coming from the EU countries. A lack of insurance culture is most evident if we focus on the premium insurance structure. Namely, obligatory MTPL is still the dominant insurance product. A low competition level of local companies in life business forces them to focus on less profitable, nonlife business. Data for selected countries of SEE markets show that the average share of life insurance for SEE in 2015 was 34.2%, while MTPL covered 26.2%, and other nonlife 39.6%. During the last 5 years, the share of life insurance has increased, decreasing the share of other nonlife. Greece has the largest market share of life insurance (48.4%), followed by Croatia (33.6%) and Slovenia (29.7%). From more detailed overview of MTPL shares in the countries, we can see that dominance of MTPL is most evident in BiH, Macedonia, and Montenegro, where it covers more than 40% of the total premium.

Since the last 5 years has been very important for Europe due to the introduction of new regulatory framework, Solvency 2, changes in legal and institutional framework can be observed in the context of compliance with the new regulations. Only a few countries, already EU members, followed the deadlines for implementation of the Solvency 2 regime on the national regulatory framework. There are many reasons for this delay, but the key problem is with still underdeveloped market that cannot follow the rules that exist in the developed EU insurance market. That is accompanied with poor institutional infrastructure, not capable of fully implementing even the Solvency 1 regime. There are plenty of uncertainties facing the region regarding the Solvency 2 implementation in the insurance markets, and countries try to catch the pace according to their capabilities.

Most regulations in the countries are not compliant with the Solvency 2 regime, and some countries are still striving to fully implement Solvency 1 regime. There are some good examples, like Slovenia, Croatia, and Greece that are believed to follow the EU standards, but most of the others are going to adapt to the new standards as they are approaching the EU membership. During last 5 years, they have been focused on improving the methodology of insurer reporting and the other aspects of market supervision.

## Author details

Safet Kozarevic[1*†], Emira Kozarevic[1†], Pasqualina Porretta[2†] and Fabrizio Santoboni[2†]

*Address all correspondence to: safet.kozarevic@untz.ba

1 Faculty of Economics, University of Tuzla, Tuzla, Bosnia and Herzegovina

2 Faculty of Economics, Sapienza University of Rome, Rome, Italy

† *This study is the result of a joint effort. However, sections 2 and 4.2. are attributable to Safet Kozarevic, sections 3.1, 3.2. and 4.1. are attributable to Emira Kozarevic, sections 1. and 3.4. are attributable to Pasqualina Porretta, and sections 3.3. and 3.5. are attributable to Fabrizio Santoboni.*

## References

[1] BCBS. A global regulatory framework for more resilient banks and banking systems. 2010. December (revised June 2011). Available from: http://www.bis.org/publ/bcbs189_dec2010.pdf [Accessed: 6 November 2017]

[2] Capital Requirements Regulation (EU) No. 575/2013 of The European Parliament and Council of 26 June 2013 on Prudential Requirements for Credit Institutions and Investment Firms and Amending Regulation EU No. (648)/2012. Available from: http://eur-lex.europa.eu/legal-content/en/TXT/?uri=celex%3A32013R0575 [Accessed: 6 October 2017]

[3] McNeil AJ, Frey R, Embrechts P. Quantitative Risk Management: Concepts, Techniques and Tools. Princeton, NJ: Princeton University Press; 2005 1 p

[4] Kozarević E. Analiza i upravljanje finansijskim rizicima. Tuzla: CPA; 2009. p. 105-106

[5] Ćirović M. Bankarstvo. Bridge Company: Beograd; 2001 92 p

[6] Rose P, Hudgins S. Bankarski menadžment i finansijske usluge. Data status: Beograd; 2005 494 p

[7] Ćirović M. Bankarstvo. Beograd: Bridge Company; 2001. p. 370-371

[8] Leone P, Tutino F. Prefazione. In: Santoboni F, editor. Manuale di gestione assicurativa. Intermediazione e produzione. Padova: Cedam; 2012. p. XV

[9] Marino DM, Costa S. L'attuazione in Italia della direttiva Solvency II: un quadro di sintesi. In: . Diritto Bancario; 2015. p. 1 Available from: http://www.dirittobancario.it

[10] Marchionni F. L'impresa assicurativa. Fabbrica, finanza e ruolo sociale. Milano: Il Sole 24 Ore; 2006. p. 181

[11] Porretta P, Santoboni F. Fare assicurazioni "Solvency II compliant". Implicazioni regolamentari, organizzative e di governance. Padova: Cedam; 2016. p. 6-7

[12] Schmaltz C, Pokutta S, Heidorn T, Andrae S. How to make regulators and shareholders happy under Basel III. Journal of Banking & Finance. 2014;**46**:311-313. Available from: http://dx.doi.org/10.1016/j.jbankfin.2014.05.031 Accessed: 19 April 2015

[13] EBA/GL/2014/13. Guidelines on common procedures and methodologies for the supervisory review and evaluation process (SREP). 2014. Available from: https://www.eba.europa.eu/documents/10180/935249/EBA-GL-2014-13+(Guidelines+on+SREP+methodologies+and+processes).pdf [Accessed: 11 December 2007]

[14] EBA. Guidelines on ICAAP and ILAAP information collected for SREP purposes. 2016. Available from: https://www.eba.europa.eu/documents/10180/1645611/Final+report+on+Guidelines+on+ICAAP+ILAAP+%28EBA-GL-2016-10%29.pdf/6fa080b6-059d-4b41-95c7-9c5e db8cba81 [Accessed: 11 June 2017]

[15] ECB. Supervisory Expectations on ICAAP and ILAAP and Harmonised Information Collection on ICAAP and ILAAP. 2016. Available from: https://www.ecb.europa.eu/pub/pdf/other/ 160108letter_nouy.en.pdf [Accessed: 17 June 2017]

[16] Dissanayake AD. Basel III and emerging economies: A critique: Part I. Journal of International Banking Law and Regulation. 2012;**27/9**:353-354

[17] Corbett M. 6 Steps to a Good Risk Assessment Process [Internet]. 2015. Available from: http://www.genre.com/knowledge/blog/steps-to-a-good-risk-assessment-en.html [Accessed: 16 June 2017]

[18] Piccirillo S. Solvency II: dal presidio del rischio alla governance risk based. Giornale delle assicurazioni. 2011;**4**:58

[19] Focarelli D, Nicelli A. Il sistema assicurativo italiano: sfide e opportunità di un mercato in forte evoluzione. Economia dei servizi. 2014;**2**:156

[20] Bocconi SDA, Capgemini. Osservatorio Solvency II operational transformation. In: La gestione degli impatti operative sulle aree organizzazione e IT e commerciale & comunicazione, Proposta n. R/08/2011 [Internet]; 2012. p. 4. Available from: http://www.sdabocconi.it Accessed: 20 June 2017

[21] EIOPA. Financial Stability Report. 2015; 9 December

[22] EIOPA. Final Report on Public Consultation on Preparatory Guidelines on Product Oversight and Governance Arrangements by Insurance Undertakings and Insurance Distributors. 2016; 6 April

[23] Rago G, Santoboni F. Solvency II: cambiano le regole di fare assicurazione. In: Leone P, Porretta P, Tutino F, editors. Il governo dei rischi in banca. Nuove tendenze e nuove sfide. Cedam: Padova; 2016. 235 p

[24] Rossi S. Insurance Supervisory Authority 2014 Annual Report. Remarks by the President [Internet]; 2015. p. 5-6. Available from: http://ivass.it Accessed: 20 June 2017

Section 4

# Medical Risk

Chapter 13

# Environmental Health Surveillance for Health Risk Assessment Following Radionuclide Release

Robert Wålinder

Additional information is available at the end of the chapter

http://dx.doi.org/10.5772/intechopen.73073

**Abstract**

The currently established linear nontreshold (LNT) risk model is used for radiation protection and is actually not intended for risk assessment. Also dose concepts such as effective dose are constructions used for radiation protection, focusing on the regulatory use in standards for workers but is seldom useful for members of the public. Both the LNT model, as well as use of the concept effective dose, are also not applicable in the low dose area. An alternative method for public health risk assessment and disease surveillance can be the combination of environmental radiation monitoring and health databases. For example, after the Chernobyl accident, airborne measurements of cesium-137 gamma spectrum from the ground, activity data from food samples and high quality national health registries were used for the risk assessment of cancer development.

**Keywords:** cancer, health surveillance, ionizing radiation, nuclear accident, risk assessment

## 1. Introduction

The golden standard for risk assessment of health effects from ionizing radiation are mortality data from the LSS (Life Span Study) cohort of survivors after the atomic bombings in Hiroshima and Nagasaki in 1945. Based on epidemiological data from the 93,000 survivors, the currently accepted linear nontreshold (LNT) risk model has been established. There are however several important shortcomings of this model. Firstly the LSS cohort is mainly based on mortality data. It is well-known that mortality data is inexact in diagnostic criteria, mostly lacking autopsy data. Cancer registries for cancer incidence data using histologically verified sampling have better diagnostic accuracy. Secondly the LSS cohort is based on acute exposure at the time of the bombing, but very little chronic exposure due to local fallout.

Most radioactivity was spread in the atmosphere after the bombings giving relatively little local fallout. Thirdly the LNT model is poorly verified in the so called low dose region (<100 mGy/mSv). A recent follow-up of the LSS-cohort after 52 years point at uncertainties of the shape of the dose response curve supporting a linear-quadratic model [1]. Perhaps the most debated detail about the LNT model is the introduction of a risk reduction for the low dose region. The so called "dose and dose-rate effectiveness factor", DDREF, suggest that the risk of malignancy should be lowered by a factor of 2 in the low dose region (<100 mSv) introduced by the International Commission on Radiological Protection (ICRP). The theoretical argument for this arbitrary halving of risk estimates was that cellular repair mechanisms ought to be more efficient at low doses and low dose-rates. Perhaps a corrected radiation risk model will be considered in the light of new data and based on both physical, epidemiological and biologic data. Most other biological, medical and toxicological systems have exponential or s-shaped relationships between exposure and outcome, instead of linear. Models for radiation protection are also made mainly for regulatory purposes and do not directly reflect risk of disease, especially for exposures to populations in the low dose region. Therefore, nuclear accidents, such as in Harrisburg, Chernobyl or Fukushima, differ substantially from the conditions on which the LSS cohort and the LNT model are based. Other approaches using national health databases and environmental monitoring to detect health risks might be useful.

## 2. Assessment of dose

### 2.1. Uncertainties in dose measurements

The currently used dose estimates, such as equivalent dose or effective dose, are constructions used for radiation protection, focusing on the regulatory use in standards for workers but are seldom useful for members of the public. These dose estimates make use of several weighing factor, depending on type of radiation and the organ affected.

However, environmental exposure are often complex, including multiple tissues or whole body exposure. It is also often a combination of both external and internal exposure. For that purpose the International Commission on Radiological Protection (ICRP) recommends a weighing factor for effective dose, specific for 14 different organ/tissue categories. This weighing factor is based on years of life lost and also genetic effects, rather than the biological risk of cancer development. As a consequence thyroid cancer is weighed one third of bone marrow malignancy, breast cancer and stomach cancer, which is not related to the biological risk of developing a malignancy from a certain radiation dose.

The currently used dose estimates are primarily not constructed to be used in epidemiological studies on cancer incidence, since the weighing factors are evaluating the severity of health outcomes, mortality and even genetic effects. Therefore a discussion should be introduced about what dose risk estimates might be more suitable for epidemiological studies of cancer incidence, for example using absorbed dose energy with both radiophysically and biologically based correction factors.

### 2.2. Alternative biological dosimetry techniques

There are no biological markers for the assessment of low dose or low dose-rate exposes to humans [2]. After receiving larger external doses, nail and tooth enamel magnetic resonance analysis might be used, though with a large inaccuracy of dosimetry of 30–50 mGy, high costs and advanced laboratory equipment limiting the practical use [3]. Examples of other physical and biological dosimetry techniques being evaluated, though not yet practically applicable are: protein biomarkers, hematological changes, chromosomal damages, micronuclei and thermoluminescence [4].

### 2.3. Indirect dose assessment in non-occupational populations

Personal dosimetry is mainly used for the protection of radiation workers to ensure that the exposure to ionizing radiation is kept within dose equivalent limits. When a larger population is exposed to radionuclides dosimeters are not available in sufficiently large numbers. An exception is the internal dose to the thyroid gland which can be accessed via direct thyroid scans of radio-iodine uptake. The external dose contribution and the contributions from other radionuclides are more difficult to assess, especially multi-organ or whole-body doses.

Instead environmental monitoring from both stationary and mobile dosimeters can map geographical patterns of contamination. From these environmental data indirect dose assessments can be calculated for larger populations. External radiation doses to a population can be estimated via deposition maps, meteorological modeling or distance from the radiation source, including factors such as shielding. Internal doses can be measured for a limited amount of subjects via whole-body counting or thyroid scanners, but for most of the population estimations of doses can be made based on residence, inhalation and ingestion assumptions. An example of a well-developed model for indirect dose assessment is the Radiation Effects Research Foundation (RERF) dose estimation model (DS02R1) from an atomic bomb. The model takes into account distance to the hypocenter, shielding from buildings and terrain [1].

### 2.4. An example of indirect dose assessment among Swedish hunters

Using transfer factors based on whole-body counting from the Swedish population an example of a model for the assessment of life-time (70 years) extra dose from the Chernobyl fall-out was calculated for 16,000 hunters with families in the three mostly contaminated counties in Sweden. An extra life-time dose up to 9.4 mSv was calculated, depending on the factors age, gender and habitat. About 75% of the life-time dose was from internal contamination from food [5].

If only the external dose contribution is accounted for during the first year the relative dose contribution from so called short-lived fission products was 36%, 37% for cesium-134 and 27% for cesium-137. After 70 years the proportions were 11%, 29% and 60% respectively [6].

### 2.5. Dose assessment among reindeer herders

The highest radionuclide exposure to a population outside the former Soviet Union after the Chernobyl accident was received among Nordic reindeer herders, receiving about 10–100 times higher doses than urban populations, according to Swedish whole-body counts [7]. The reindeer livelihood was severely affected by the Chernobyl fall-out. Due to radiation protection actions about 80% of the Swedish reindeer meat was destroyed the first years following the accident, and the slaughter had to be moved from winter season to summer, when browse was less contaminated. Middle aged reindeer herdsmen also received similar or even higher doses from the global fall-out during the 1950ies and 1960ies making them exposed twice [8]. According to population data from Statistics Sweden there are only about 700 reindeer herders by occupation in Sweden, which gives too low power for epidemiological analyses on cancer incidence, but a combined study from all Nordic countries might be possible.

## 3. Use of national health data registries

Although the LSS-cohort outcome is the supposedly golden standard for cancer risk there is a fundamental shortcoming due to lack of early data covering the Hiroshima and Nagasaki prefecture populations, since prefecture cancer registries were not in use until 1958. Furthermore the LSS-cohort is mainly based on mortality data although cancer incidence registries usually are based on histological sampling with higher diagnostic accuracy. This gives uncertainties to the LSS-data concerning early health effects for the cancer risk models. In japan a "National Health Promotion" law put in place in Japan in the early 2000s said that prefectures must track illnesses including cancer. This law led to the introduction of some new cancer registries in Japan. The Fukushima prefecture begun a cancer registry in 2010 using a standardized database system governed by the Japanese National Cancer Center. But the data produced in the first few years the cancer registry was of poor quality and is still being developed by the year 2017. This is a great drawback for the estimation of health outcomes, including cancer, for the population of the Fukushima prefecture following the nuclear accident in 2011.

The lack of official health data or a national health data base were even more striking in the former Soviet union at the time of the Chernobyl accident in 1986. The absence of data for researchers has made follow-up of health outcomes difficult in the former Soviet states, though national cancer registries are now built up in Belarus, Ukraine and the Baltic states.

In the Nordic countries there are national cancer registries at the individual level covering all population. In Sweden a national cancer registry is in use since 1958 [9]. Good quality cancer registries make it possible to register changes in baseline incidences following environmental changes such as radio-nuclide releases to the population, especially as a complement when dosimetry is absent or very inexact.

### 3.1. The example of detecting increased cancer incidence in South Wales around Windscale

The first population study apart from the LSS-cohort showing a possible increased risk of cancer was in South Wales. From a fuel reprocessing plant at Windscale waste was discharged into the Irish See via a pipeline and deposition occurred in the sea bottom, fish and sea weed. The fission product ruthenium-106 was taken up very efficiently and concentrated in the sea weed *Porphyra umbilicales*. It was harvested and used in laver bread, consumed mainly in South Wales. The use had to be stopped. The activity in fish was mainly from cesium-137. When an ecosystem is contaminated with radionuclides and local food is the main source of internal exposure to radiation the individual doses to the population are very difficult to assess since food habits and lifestyle differ fundamentally between individuals and regions. Whole-body counting can be made to a small sample of the population, mainly concerning gamma-radiation from gamma-emitting nuclides and indirectly from alpha-emitters with gamma-decay, but has lower sensitivity for detecting the beta-radiation, such as from ruthenium-106. Therefore health surveillance via national cancer registries was fundamental to monitor the health effects to the population with an ecosystem is contaminated by radionuclides. Several epidemiological studies have shown increased incidences of cancer of the population around Windscale [10].

### 3.2. The example of detecting increased cancer incidence for people living at the Techa river

In 1949 the Mayak Production Association, located in the Southern Urals, started production of plutonium for the Soviet Nuclear weapons program. A cohort of 30,000 residents of 40 rural villages along the Techa river or the Chelyabinsk City with low-dose and low-dose-rate exposures have been followed for more than 50 years for incident cases of cancer. Individual radiation doses were based on geographic information of residence and food habits. Calculated external exposures were due to gamma rays from contamination of the soil and the internal exposures were assessed from expected consumption of water, milk and food containing uranium fission products. All solid cancers as a group were related to stomach doses ranging from 0 to 960 mGy with a mean of 60 mGy. Dose–response between estimated radiation dose and solid cancers and leukemia were shown with an excess relative risk (ERR) after exposure to 100 mGy of 0.08 [11].

### 3.3. The example of detecting increased incidence of thyroid cancer in Ukraine after the Chernobyl accident

Chemical composition, deposition, uptake and metabolism of iodine make thyroid dosimetry complicated, but direct measurement using a gamma-meter of the thyroid gland can be made. To estimate individual thyroid absorbed doses from radioiodine in the Ukrainian population from May–June 1986, more than 150,000 individual examinations were carried out by special dosimetric teams. The collective thyroid dose was 64,000 person-Gy, which theoretically could give about 300 extra cases of thyroid cancer [12].

Another study was performed on behalf of the International Agency for Research on Cancer (IARC). A population-based case–control study was designed of thyroid cancer among young people who lived in the areas that were heavily contaminated by the Chernobyl accident, Indirect dosimetry was performed based on data of the habitat and dietary habits of 1615 cases and controls aged 0–18 y at the time of the accident. A strong dose–response relationship was observed between estimated radiation dose to the thyroid received in childhood and thyroid cancer risk [13].

### 3.4. The example of detecting increased cancer incidence in Sweden after the Chernobyl accident

Sweden received the largest deposition of radionuclides outside the former USSR, where about 4.4% of the total Chernobyl fall-out was deposited [14]. Deposition was strongly dependent on local weathering giving highest deposition in coastal areas around the Bothnian sea. A food regulation program was introduced to assure that the annual extra dose did not exceed 1 mSv in the population. In a study indirect individual doses were assessed for 734,537 persons living in the three most contaminated counties in Sweden. Personal dosimetry could not be performed 30 years after the accident, so a cumulative exposure based on measured ground activity of cesium-137 of the residence of the subjects. A cumulative exposure estimate during 5 years following the accident was used as proxy for received dose. 82,495 cases of cancer were diagnosed from 1991 to 2010 and retrieved from the Swedish national cancer registry. A non-parametric dose–response could be shown between the deposition of cesum-137 and cancer incidence [15].

## 4. Conclusion

A paradigm shift is needed from the dominance of radiation protection to a more biologically based health risk assessment from ionizing radiation. Models for radiation protection are made for regulatory purposes and do not directly reflect the risk of disease. Also dose estimates are poorly applicable for risk assessment for populations exposed in the low dose region. Only to rely on technical surveillance could be insufficient. Instead other approaches using national health databases in combination with environmental monitoring could be more efficient for the detection of health risks. Medical surveillance and health registries are good complements, especially in the absence of dosimeter data, complex environmental exposures and when large populations are exposed. When nuclear facilities are in use national health registries could be the most sensitive source for the detection of increased cancer incidence and other disease from nuclear accidents or other emission of radionuclides to the environment. Apart from nuclear power plants possible exposure could emanate from uranium mining, fuel processing, nuclear waste processing and nuclear waste repositories.

## Author details

Robert Wålinder

Address all correspondence to: robert.walinder@medsci.uu.se

Occupational and Environmental Medicine, Department of Medical Sciences, Uppsala University, Uppsala, Sweden

## References

[1] Grant EJ, Brenner A, Sugiyama H, Sakata R, Sadakane A, Utada M, et al. Solid cancer incidence among the life span study of atomic bomb survivors: 1958-2009. Radiation Research. 2017 May;**187**(5):513-537

[2] Boice JD Jr. The linear nonthreshold (LNT) model as used in radiation protection: An NCRP update. International Journal of Radiation Biology. 2017 Oct 3;**93**(10):1079-1092

[3] Fattibene P, Callens F. EPR dosimetry with tooth enamel: A review. Applied Radiation and Isotopes. 2010 Nov 1;**68**(11):2033-2116

[4] Ainsbury EA, Bakhanova E, Barquinero JF, Brai M, Chumak V, Correcher V, et al. Review of retrospective dosimetry techniques for external ionising radiation exposures. Radiation Protection Dosimetry. 2011 Nov;**147**(4):573-592

[5] Tondel M, Rääf C, Wålinder R, Mamour A, Isaksson M. Estimated lifetime effective dose to hunters and their families in the three most contaminated counties in Sweden after the Chernobyl nuclear power plant accident in 1986 – A pilot study. Journal of Environmental Radioactivity. 2017 Oct 1;**177**:241-249

[6] Jönsson M, Tondel M, Isaksson M, Finck R, Wålinder R, Mamour A, et al. Modelling the external radiation exposure from the Chernobyl fallout using data from the Swedish municipality measurement system. Journal of Environmental Radioactivity. 2017 Nov 1;**178**:16-27

[7] Rääf CL, Hubbard L, Falk R, Agren G, Vesanen R. Transfer of 137Cs from Chernobyl debris and nuclear weapons fallout to different Swedish population groups. Science of the Total Environment. 2006 Aug 15;**367**(1):324-340

[8] Skuterud L, Thørring H. Averted doses to Norwegian Sámi reindeer herders after the Chernobyl accident. Health Physics. 2012 Feb;**102**(2):208-216

[9] Barlow L, Westergren K, Holmberg L, Talbäck M. The completeness of the Swedish Cancer Register: A sample survey for year 1998. Acta oncologica (Stockholm, Sweden). 2009;**48**(1):27-33

[10] Gardner MJ. Review of reported increases of childhood cancer rates in the vicinity of nuclear installations in the UK. Journal of the Royal Statistical Society. Series A, Statistics in Society. 1989;**152**(3):307-325

[11] Davis FG, KL Y, Preston D, Epifanova S, Degteva M, Akleyev AV. Solid cancer incidence in the Techa River incidence cohort: 1956-2007. Radiation Research. 2015 Jul;**184**(1):56-65

[12] Likhtarev IA, Shandala NK, Gulko GM, Kairo IA, Chepurny NI. Ukrainian thyroid doses after the Chernobyl accident. Health Physics. 1993 Jun;**64**(6):594-599

[13] Cardis E, Kesminiene A, Ivanov V, Malakhova I, Shibata Y, Khrouch V, et al. Risk of thyroid cancer after exposure to 131I in childhood. Journal of the National Cancer Institute. 2005 May 18;**97**(10):724-732

[14] Izrael YA, De Cort M, Jones AR, Nazarov IM, Fridman SD, Kvasnikova EV, et al. The atlas of Caesium-137 contamination of Europe after the Chernobyl accident. In: Karaoglu A, Desmet G, Kelly GN, Menzel HG, editors. The Radiological Consequences of the Chernobyl Accident. ECSC-EC-EAEC, Brussels; 1997. pp. 1-10

[15] Alinaghizadeh H, Wålinder R, Vingård E, Tondel M. Total cancer incidence in relation to 137Cs fallout in the most contaminated counties in Sweden after the Chernobyl nuclear power plant accident: A register-based study. BMJ Open. 2016 Dec 20;**6**(12):e011924

Chapter 14

# Challenges and Perspectives of the Risk Assessment of the Genetic Susceptibility to Cancer in the Next-Generation Sequencing Era

Israel Gomy

Additional information is available at the end of the chapter

http://dx.doi.org/10.5772/intechopen.72379

**Abstract**

The risk assessment of the genetic susceptibility to cancer is the process of addressing and communicating the genetic risks to individuals and families with cancer. The recent breakthroughs of the next-generation sequencing era are adding new challenges to the precision clinical care.

**Keywords:** susceptibility, next-generation sequencing, cancer genetics

## 1. Introduction

New molecular biology technologies, such as whole-exome and whole-genome sequencing have been shedding new light on the understanding of inherited cancer susceptibility. At the same time, translational oncology researches on somatic and germline mutations in actionable genes have been opening new dilemmas of the next-generation sequencing era. A critical issue of the so-called precision medicine is the genetic counseling of individuals with cancer susceptibility.

Susceptibility to cancer depends on the penetrance of germline variants or inherited alleles, which may be classified into three groups such as highly penetrant, moderately penetrant and lowly penetrant alleles.

Alleles with high penetrance have the highest lifetime risk of cancer, frequently more than 10 times the relative risk, dramatically affecting the quality of life and decreasing its expectancy. More than 50 rare Mendelian cancer syndromes are caused by germline mutations affecting either tumor suppressor genes, DNA repair genes or proto-oncogenes, mostly with autosomal dominant inheritance (**Table 1**).

| Syndrome | Gene | Mutation status | Penetrance | Tumors |
|---|---|---|---|---|
| Hereditary breast and/or ovarian cancer | *BRCA1* | Heterozygous | High | Breast cancer |
| | *BRCA2* | | | Ovarian cancer |
| | *RAD51 (B,C,D)* | | Moderate | Pancreatic cancer |
| | *ATM* | | Moderate | Prostate cancer |
| | *CHEK2* | | Moderate | Colorectal cancer |
| Lynch syndrome | *MLH1* | Heterozygous | High | Colorectal cancer |
| | *MSH2* | | | Endometrial cancer |
| | *MSH6* | | | Ovarian cancer |
| | *PMS2* | | | Gastric cancer |
| | *EPCAM* | | | Leukemia, lymphoma |
| MMR cancer syndrome | MMR genes | Homozygous | High | Rhabdomyosarcoma |
| Familial adenomatous polyposis | *APC* | Heterozygous | High | Gastrointestinal adenomas |
| | | | | Colorectal cancer |
| | | | | Duodenal cancer |
| MYH-associated polyposis | *MUTYH* | Homozygous | High | Colorectal cancer |
| Polymerase proofreading-associated polyposis | *POLE* | Heterozygous | high | Colorectal cancer |
| | *POLD1* | | | Endometrial cancer |
| Bloom syndrome | *BLM1* | Homozygous | High | Leukemia |
| | | | | Colorectal cancer |
| | | | | Wilms tumor |
| Nijmegen syndrome | *NBS1* | Homozygous | High | Lymphoma |
| | | | | Medulloblastoma |
| | | | | Rhabdomyosarcoma |
| Fanconi anemia | *FANC* genes (includes *BRCA2, PALB2, BRIP1*) | Homozygous | High | Leukemia |
| | | | | Medulloblastoma |
| | | | | Wilms tumor |
| Li-Fraumeni syndrome | *TP53* | Heterozygous | High | Breast cancer |
| Li-Fraumeni-like syndrome | *CHEK2* | | Moderate | Sarcoma |
| | | | | Adrenocortical cancer |
| | | | | Brain tumor |
| Cowden syndrome | *PTEN* | Heterozygous | High | Hamartomatous polyps |
| | | | | Skin tumors |
| | | | | Breast cancer |
| | | | | Thyroid cancer |
| | | | | Endometrial cancer |

http://dx.doi.org/10.5772/intechopen.72379

| Syndrome | Gene | Mutation status | Penetrance | Tumors |
|---|---|---|---|---|
| Hereditary diffuse gastric cancer | *CDH1* | Heterozygous | High | Gastric cancer (diffuse) |
| | | | | Breast cancer (lobular) |
| Peutz-Jeghers syndrome | *STK11* | Heterozygous | High | Hamartomatous polyps |
| | | | | Colorectal |
| | | | | Small bowel |
| | | | | Breast cancer |
| | | | | Pancreatic cancer |
| Juvenile polyposis | *SMAD4* | Heterozygous | High | Hamartomatous polyps |
| | *BMPR1A* | | | Colorectal cancer |
| | | | | Pancreatic cancer |
| Melanoma syndromes | *CDKN2A* | Heterozygous | High | Malignant melanoma |
| | *CDK4* | | | Pancreatic cancer |
| Neurofibromatosis | *NF1* | Heterozygous | High | Vestibular schwannoma |
| | *NF2* | | | Meningioma |
| | | | | Neurofibroma |
| | | | | Optic glioma |
| Tuberous sclerosis | *TSC1* | Heterozygous | High | Renal angiomyolipoma |
| | *TSC2* | | | Subependymoma |
| | | | | Giant cell astrocytoma |
| Von Hippel-Lindau syndrome | *VHL* | Heterozygous | High | Hemangioblastomas |
| | | | | Renal cell cancer |
| | | | | Pheochromocytoma |
| Chuvash polycythemia | | Homozygous | High | Vertebral angiomas |
| Birt-Hogg-Dubè syndrome | *FLCN* | Heterozygous | High | Renal cell cancer |
| | | | | Skin tumors |
| Papillary renal cancer syndromes | *FH* | Heterozygous | High | Renal cell cancer |
| | *MET* | | | |
| Retinoblastoma | *RB1* | Heterozygous | High | Retinoblastoma |
| Hereditary Paraganglioma | *SDH (A, B, C, D)* | Heterozygous | High | Paraganglioma |
| | | | | Pheochromocytoma |
| Multiple endocrine neoplasia 1 | *MEN1* | Heterozygous | High | Pituitary adenoma |
| Multiple endocrine neoplasia 2 | *RET* | | | Parathyroid adenoma |
| | | | | Medullar thyroid cancer |
| | | | | Pheochromocytoma |

**Table 1.** Hereditary cancer syndromes.

Alleles with moderate or intermediate penetrance increase the relative risk of about two to five times. Although they are rare in most populations, they may be frequently found in populations with consanguineous families due to founder effects. Affected relatives can be often identified, but the reduced penetrance of the alleles may skip generations and jeopardizes the family history.

Lowly penetrant alleles were discovered by genome-wide association studies (GWAS) and may put individuals to risk of cancer at slightly higher rates than those of the general population. This is due to a polygenic model, in which several alleles, mainly single nucleotide polymorphisms (SNPs), each one carrying a low risk, combine additively or multiplicatively to confer a range of risks in the population. In this model, individuals with few alleles would be at a reduced risk, whereas those with many alleles might suffer a lifetime risk as high as 50% [1]. It is estimated that more than 100 common variants with low risk may contribute to cancer susceptibility. Actually, they explain part of the excess familial risk, and the so-called "missing heritability" remains largely unknown [2]. Thus, it is very important to identify lowly penetrant alleles responsible for cancer genetic susceptibility. Most of these alleles are intergenic—lie between genes—and many neighbor tumor suppressor genes and proto-oncogenes, possibly affecting their expression. Nowadays, with the advance of next-generation sequencing and genotyping assays, more variants have been identified, shedding new light on the genomic architecture of the inherited susceptibility of cancer.

## 2. Risk assessment of the genetic susceptibility to cancer

The risk assessment of the genetic susceptibility to cancer (RAGSC) is a process to evaluate a personal risk of carrying a germline variant that is associated to the cancer development. RAGSC may be performed through statistical models that incorporate factors such as personal and familial history of tumors, ethnic background, and so on [3]. The advent of new sequencing technologies and bioinformatics has led to improvements of estimating more precisely risks of germline variants in many genes and assessing empiric risks of cancer.

Being part of this dynamic process [4], genetic counseling involves the analysis of pedigrees and risk assessment models to determine whether a family history is suggestive of sporadic, familial or hereditary cancer [5]. The main goal of genetic counseling is to inform susceptible individuals about their chances of developing cancer, helping them to make decisions about genetic testing, screening, prevention and treatments. Pretest and posttest genetic counseling are essential for the efficacy of implementing evidence-based protocols, in terms of reducing mortality rates [6].

**Table 2** summarizes the RAGSC process. Three main risk categories can be derived on the basis of patient and family genetic information. In the low-risk category (near-population risk), management is based on population screening, and genetic tests are generally not cost-effective; in the moderate-risk group, genetic counseling, genetic testing and management are individual-based; in the high-risk group, genetic counseling, testing and management are evidence-based and improve survival [7].

| Average risk | High | Moderate/intermediate | Low/populational |
|---|---|---|---|
| Personal/family history | Mendelian syndromes | Familial aggregation | Sporadic |
| Genetic testing | Single gene sequencing/NGS panels/WGS/WES | NGS panels/WGS/WES | DTC[&]/WGS/SNP genotyping |
| Genetic counseling | Mandatory | Advisable | Available |
| Management | Evidence-based | Individual-based[1] | Not validated |

DTC: direct-to-consumer tests; WGS: whole-genome sequencing; WES: whole-exome sequencing.
[1]Some evidence-based screening recommendations exist for breast and colorectal cancers.
[&]Restricted by the US Food and Drug Administration.

**Table 2.** Overview of the risk assessment of the genetic susceptibility to cancer.

# 3. Referrals for RAGSC

Besides sex and age, familial history is the main unmodifiable risk factor of developing cancer.

Assessing the risk factors of cancer in an individual or family is complex and raises psychological, social and ethical issues. It requires the understanding of areas of medical genetics

| | |
|---|
| **Personal history** |
| Early onset of cancer diagnosis (e.g., breast cancer <45 years, colorectal cancer <50 years) |
| Multiple associated primary cancers: breast/ovary, colorectal/endometrium |
| Male breast cancer |
| Ovarian, fallopian tube, primary peritoneal cancer |
| Breast cancer and thyroid, sarcoma, adrenocortical carcinoma |
| Multiple colon polyps (>10 cumulative) |
| Colorectal or endometrial cancer with microsatellite instability and/or lack of expression of mismatch repair protein(s) by immunohistochemistry |
| **Family history** |
| Three close relatives (same side of family) with cancer of the same or syndromically related type (breast/ovary, colorectal/endometrium) |
| Two close relatives (same side of family) with cancer of the same or related type with at least one affected under 50 years |
| One first-degree relative with early onset cancer (breast <45 years, colorectal <50 years) |
| One fist-degree relative with multiple primary cancers |
| Two or more relatives with uncommon cancers (sarcoma, glioma, hemangioblastoma, etc.) |
| Relatives of patients with known *BRCA*, *APC*, *MYH*, Lynch syndrome mutations |
| Many relatives with cancer but no criteria for testing |

**Table 3.** Referrals for hereditary cancer risk assessment.

and oncology, besides the ability of communication, and it demands more time than just a regular consultation. The American Society of Clinical Oncology (ASCO), the National Society of Genetic Counselors (NSGC) and the Oncology Nursing Society (ONS) have published guidelines for the practice of genetic counseling, risk assessment and genetic testing [6, 8]. Moreover, it includes management of at-risk individuals so that they can make informed choices about cancer screening, prevention and targeted therapies [9]. In **Table 3**, there are some indications of referral for RAGSC.

## 4. Next-generation sequencing

In 2013, at first, Roberts and Klein reported the use of next-generation sequencing (NGS) to identify a hereditary cancer syndrome. They found pathogenic germline variants in the *ATM* gene of six pancreatic cancer relatives from two different kindreds [10]. Jaeger et al. used whole-genome sequencing for the description of hereditary mixed polyposis syndrome [11].

More recently, multigene NGS panels have been used to analyze many highly and moderately penetrant variants. Although they use the same NGS technology, there is less information on predefined genes. In comparison with single-gene sequencing, panels are more time- and cost-efficient in many cases such as (1) when there is genetic or locus heterogeneity, (2) when there are actionable mutations in several genes and (3) when phenotype or family history is too unspecific or noninformative (e.g., adoption) [12].

One advantage of NGS is the possibility of including multiple genes in panels tailored to a certain familial aggregation of tumors such as breast or colon cancer. However, because of its economic viability, NGS has shifted the phenotype-driven hypothesis approach that is based on the characteristics of the syndrome. Slavin et al. found some interesting results about multigene panels. When they included only high-risk genes, the results were seldom positive, and there were more variants of unknown significance (VUS), probably because of the inclusion of more genes in the so-called "off-phenotype" pan-cancer panels [13]. Recently, evidence-based guidelines have included the utilization of multigene testing for hereditary breast and ovarian cancer risk assessment [14].

An important disadvantage of NGS is the probability of disclosing inconclusive or undetermined results. The interpretation of a VUS based on phenotype and genotype data is a difficult task and often jeopardizes the genetic counseling process. Choosing a panel with limited genes of high clinical utility specifically driven to the phenotype instead of pan-cancer panels with many low-risk genes can diminish the chances of finding variants with stressful interpretation [13]. Moreover, databases of variants with high and moderate risks are often not population-specific and may lead to misinterpretation of results.

Some ethical challenges are critical for implementing NGS in the clinics.

In March 2013, the American College of Medical Genetics and Genomics (ACMG) published recommendations on the reporting of incidental or secondary findings from NGS. The ACMG suggested the identification of 56 genes whose variants result in a high risk of developing a severe disease. Germline mutations of 16 of these genes cause hereditary cancer syndromes (**Table 4**) [15].

| Syndrome | Gene |
|---|---|
| Li-Fraumeni | *TP53* |
| Peutz-Jeghers | *STK11* |
| Familial adenomatous polyposis | *APC* |
| Von-Hippel Lindau | *VHL* |
| Multiple endocrine neoplasia | *MEN1* (type 1); *RET* (type 2) |
| Hamartomatosis | *PTEN* |
| Retinoblastoma | *RB* |
| Paraganglioma-pheochromocytoma | *SDHAF2, SDHB, SDHC, SDHD* |
| Tuberous sclerosis complex | *TSC1, TSC2* |
| Neurofibromatosis type 2 | *NF2* |
| WT1-related Wilms tumor | *WT1* |

**Table 4.** ACMG list of hereditary cancer syndromes.

In 2015, the ACMG reviewed it based on the consensus that patients could opt out of the analysis of secondary findings. This decision must be made during the process of informed consent, before testing. As some of these cancer syndromes may have the onset during childhood, these guidelines may also be applied to children, whose parents should make the decision whether or not to opt out [16].

A recent review showed that following the recommendations of international human genetic societies, parents and their children must be previously informed by a written consent about which findings should be reported. The ordering clinician must discuss with the children's parents all the possibilities of results, including the reporting of incidental findings, the "right not to know," the risks and the benefits, as well is responsible to obtain the informed consent and to provide pre- and posttest genetic counseling [17].

## 5. Conclusions

Inevitably, more challenges will arise with the application of NGS in RAGSC.

First, pretest counseling and informed consent models need to be redesigned to address the multiplex testing. Novel approaches must be developed to ensure that individuals understand the risks and benefits of choices regarding these tests. Second, the clinical management of carriers of moderately penetrant variants is still poorly defined, although some evidence-based guidelines may include them [14]. Third, finding VUS is always a potential risk, and such identification complicates data interpretation and often requires further investigation and variant reclassification. In addition, management of patients with VUS is unclear. Finally, many hereditary cancer syndromes have locus heterogeneity, incomplete penetrance and may represent phenocopies, adding difficulty in RAGSC.

In summary, the biggest challenge in counseling families with cancer is conferring precise information regarding genetic susceptibilities because it allows a better informed decision-making process about risk management, clinical surveillance, targeted therapies and preventive measures.

## Author details

Israel Gomy

Address all correspondence to: isgomy@gmail.com

Institute of Hematology and Oncology, Hospital de Clínicas da Universidade Federal do Paraná, Curitiba, Brazil

## References

[1] Pharoah PD, Antoniou A, Bobrow M, Zimmern RL, Easton DF, Ponder BA. Polygenic susceptibility to breast cancer and implications for prevention. Nature Genetics. 2002;**31**: 33-36

[2] Stadler ZK, Tom P, Robson ME, Weitzel JN, Kauff ND, Hurley KE, Devlin V, Gold B, Klein RJ, Offit K. Genome-wide association studies of cancer. Journal of Clinical Oncology. 2010;**28**:4255-4267

[3] Riley BD, Culver JO, Skrzynia C, Senter LA, Peters JA, Costalas JW, Callif-Daley F, Grumet SC, Hunt KS, Nagy RS, et al. Essential elements of genetic cancer risk assessment, counseling, and testing: Updated recommendations of the National Society of Genetic Counselors. Journal of Genetic Counseling. 2012;**21**:151-161

[4] Peters JA, Stopfer JE. Role of the genetic counselor in familial cancer. Oncology. 1996;**10**: 159-166 discussion, 176-176, 178

[5] Schneider K, Garber J. Counseling About Cancer: Strategies for Genetic Counselors. 2nd ed. New York: Wiley; 2001. p. 1904-1909

[6] Trepanier A, Ahrens M, McKinnon W, Peters J, Stopfer J, Grumet SC, Manley S, Culver JO, Acton R, Larsen-Haidle J, et al. Genetic cancer risk assessment and counseling: Recommendations of the National Society of Genetic Counselors. Journal of Genetic Counseling. 2004;**13**:83-114

[7] Schwartz GF, Hughes KS, Lynch HT, Fabian CJ, Fentiman IS, Robson ME, Domchek SM, Hartmann LC, Holland R, Winchester DJ, The Consensus Conference Committee. (2009) Proceedings of the international consensus conference on breast cancer risk, genetics & risk management. The Breast Journal. April, 2007;**15**:4-16

[8] Robson ME, Storm CD, Weitzel J, Wollins DS, Offit K. American Society of Clinical Oncology policy statement update: Genetic and genomic testing for cancer susceptibility. Journal of Clinical Oncology. 2010;**28**:893-901

[9] Weitzel JN, Blazer KR, Mac Donald DJ, Culver JO, Offit K. Genetics, genomics and risk assessment. State of the art and future directions in the era of personalized medicine. CA: A Cancer Journal for Clinicians. 2011;**61**:327-359

[10] Roberts NJ, Klein AP. Genome-wide sequencing to identify the cause of hereditary cancer syndromes: With examples from familial pancreatic cancer. Cancer Letters. 2013;**340**:227-233

[11] Jaeger E, Leedham S, Lewis A, Segditsas S, Becker M, Cuadrado PR, Davis H, Kaur K, Heinimann K, Howarth K, et al. Hereditary mixed polyposis syndrome is caused by a 40-kb upstream duplication that leads to increase and ectopic expression of the BMP antagonist GREM1. Nature Genetics. 2012;**44**:699-703

[12] Domchek SM, Bradbury A, Garber JE, Offit K, Robson ME. Multiplex genetic testing for cancer susceptibility: Out on the high wire without a net? Journal of Clinical Oncology. 2013;**31**:1267-1270

[13] Slavin TP, Niell-Swiller M, Solomon I, Nehoray B, Rybak C, Blazer KR, Weitzel JN. Clinical application of multigene panels: Challenges of next-generation counseling and cancer risk management. Frontiers in Oncology. 2015;**5**:208

[14] NCCN. NCCN Clinical Practice Guidelines in Oncology V.1.2018: Genetic/Familial High-risk Assessment: Breast and Ovarian. NCCN Clinical Practice Guidelines [Internet]. 2017. Available from: http://www.nccn.org/professionals/physician_gls/pdf/genetics_screening.pdf

[15] Greem RC, Berg JS, Grody WW, Kalia SS, Br K, Martin CL, McGuire AL, Nussbaum RL, O'Daniel JM, Ormond KE, et al. ACMG recommendations for reporting of incidental findings in clinical exome and genome sequencing. Genetics in Medicine. 2013;**15**:565-574

[16] ACMG Board of Directors. ACMG policy statement: Updated recommendations regarding analysis and reporting of secondary findings in clinical genome-scale sequencing. Genetics in Medicine. 2015;**17**:68-69

[17] Kuhlen M, Borkhardt A. Cancer susceptibility syndromes in children in the area of broad clinical use of massive parallel sequencing. European Journal of Pediatrics. 2015;**174**:987-997

Section 5

# Chemical Risk

Chapter 15

# Risk Assessment of Heavy Metals Pollution in Urban Environment

Gevorg Tepanosyan, Lilit Sahakyan,
David Pipoyan and Armen Saghatelyan

Additional information is available at the end of the chapter

http://dx.doi.org/10.5772/intechopen.70798

**Abstract**

This chapter summarizes the results of heavy metal's human health and ecological risk assessment of multipurpose ecogeochemical studies performed by the Center for Ecological-Noosphere Studies of the National Academy of Sciences of the Republic of Armenia in the young industrial cities of Yerevan and Gyumri and in an old mining region of the city of Kajaran. According to the results children non-carcinogenic risk values were greater than permissible limit of 1 indicating the possibility of an adverse health effect in the whole area of all studied cities. Among all studied elements, the riskiest were those previously identified as primary pollutants. It has also been shown that in biogeochemical provinces, where mining activities and agricultural land of rural communities are spatially juxtaposed, health risk assessment should include all possible exposure pathways. Otherwise, underestimation of possible health risk will take place. Heavy metals in soils of Yerevan and Gyumri are also an ecological risk factor and the riskiest elements having significant contribution to the overall risk and are those (Hg, Cd, and Pb) with the high level of toxicity.

**Keywords:** urban environment, heavy metals, pollution, soil, dust, risk assessment

## 1. Introduction

Soils and dust of urbanized and industrialized areas are a basis of environmental quality. Nevertheless, various pollutants of the environment, especially heavy metals, migrate linked to the complexes of dust particles [1] and finally accumulate in the soil layer. Moreover, heavy metals are known to be an ecological risk factor [2–4] and cause different disorders when entering into the human organism [1, 5].

In Armenia, risks estimation associated with the pollution of cities environment by heavy metals was included in the framework of environmental complex ecogeochemical studies, which were done since 1989 by The Center for Ecological-Noosphere Studies (CENS) of the National Academy of Sciences [6].

The results of the studies [7] performed by CENS showed that in the cities of Armenia, man-made activities lead to the formation of anthropogenically polluted areas, which were mainly localized in old mining regions (i.e., city of Kajaran) and relatively young industrial cities (i.e., Yerevan and Gyumri). In both cases, the differences of geochemical peculiarities and anthropogenic sources of pollution are conditioning the uniqueness of heavy metal's quantitative and qualitative features. In the city of Kajaran [8], which is the biggest mining center of country and houses the Zangezur Copper Molybdenum Combine (ZCMC), high contents of heavy metals are the result of the superposition of geogenic and anthropogenic components, whereas in the biggest industrial center of Yerevan and postindustrial city of Gyumri [9, 10], a significant input of heavy metals is mainly from anthropogenic sources of pollution. Although primary pollutants and the levels of anthropogenic contribution differ from city to city, the increased contents of heavy metals become a risk factor to urban ecosystems and human health.

The linking of monoelemental and multielemental pollution by heavy metals to the overall index of population prevalence, the rate of children's chronic illnesses, gestosis, and to the number of premature birth [11–13] were done in the end of 1990 through the collation of monoelemental and multielemental pollution levels spatial distribution maps [14] with the disease incidences. Later on, studies [8, 15, 16] targeted the sampling of biosubstrate and evaluation of the microelemental status of the organism among identified risk groups.

Nowadays, the most common and widely used human health risk assessment method is developed by the US Environmental Protection Agency [5, 17, 18]. The method is based on four basic steps, including hazard identification, exposure assessment, dose-response assessment, and risk characterization [18]. In the case of ecological risk from heavy metals, method developed by the Hakanson [4] was used repeatedly [2, 3, 9].

In this chapter, the results of human health and ecological risk assessment of heavy metals contents in Yerevan, Gyumri, and Kajaran environment are summarized.

## 2. Materials and methods

### 2.1. Study sites

Cities presented in this study are spatially located in different parts of Armenia (**Figure 1**). Particularly, the capital and industrial center of the country in the city of Yerevan (40°10′39.53″N and 44°30′45.10″E) is situated in the central part, whereas the cities of Gyumri (40°47′6.84″N and 43°50′29.97″E) and Kajaran (39°9′5.20″N and 46°9′12.02″E) in north-western and southern parts, respectively.

#### *2.1.1. The city of Yerevan*

Yerevan has a total area of 223 $km^2$ and 1.06 million population (4782 persons per square km) [19]. The city is located in the intermountain trough, and the natural landscape of city territory is mainly semidesert, arid steppe, and steppe. Yerevan's area is dominated by tuffs, volcanic

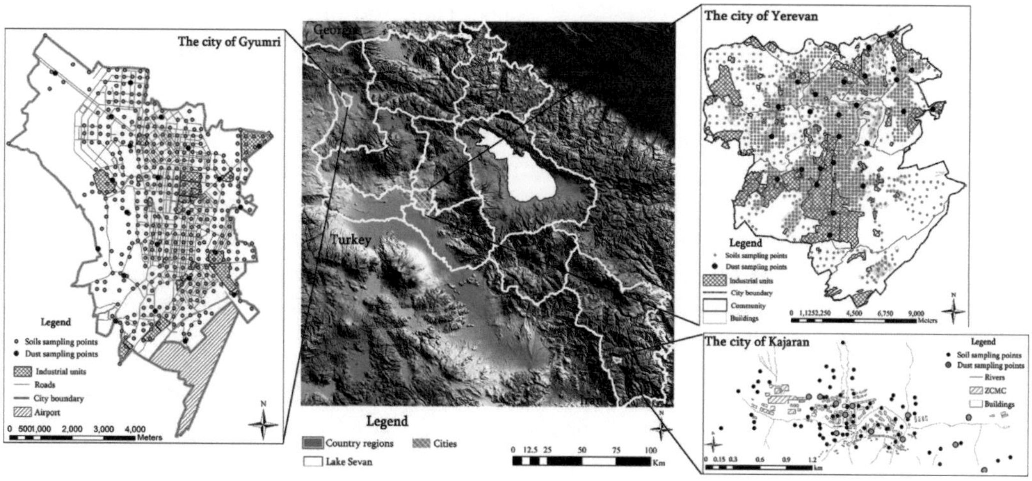


**Figure 1.** Spatial location of cities of Yerevan, Gyumri, and Kajaran and spatial distribution of soil and dust sampling points in each city.

lavas, and quaternary sediments, and the relief of the city is represented by plains, foothills, plateaus, and the River Hrazdan Canyon. The soil (mostly brown semidesert) profile of Yerevan is rich in carbonates, and at the lower horizon, the presence of gypsum is conditioning the lack of chemical element washout, thus creating a favorable environment for heavy metal accumulation on soil profiles [7].

Pollution with heavy metals in the city environment has been observed for many decades. Particularly, heavy metals were detected during the soil surveys conducted in 1979, 1989 [7], 2002 [7, 20], and 2012 [9, 21, 22], with ecogeochemical investigations of city snow cover and leaf dust [23, 24], Hrazdan river waters [25, 26], and homegrown vegetables [27, 28].

During the Soviet Union, the main sources [7, 24, 29] of heavy metal pollution in Yerevan were enterprises such as an electric bulb plant, the aluminum plant, the Car and Worsted complex, the experimental plant of milling machines, the polygraphic complex, and typography, as well as vehicular emission.

After the collapse of the Soviet Union, the socioeconomic transformations in 1990 lead to the changes in heavy metal geochemical streams' quantitative and qualitative features as many of the abovementioned industrial plants were closed. Moreover, in 2001, leaded gasoline ceased to be used in Armenia.

Nowadays, the potential sources [9] of heavy metals in Yerevan territory are urban transport and industrial units including molybdenum concentrate smelting and processing plant, Ferro-concrete constructions plant, accumulator's production, mechanical reconstruction plants, and industrial complex of metallic covers and corks, etc.

### 2.1.2. *The city of Gyumri*

Gyumri has a total area of 44.4 $km^2$ and 117.7 thousand population (2651 persons per square km) [16]. In the city, arid steppe and mountain steppe landscapes dominated and the city territory was characterized by accumulative relief of plains, lake, and alluvial-diluvial sedimentation, sometimes mixed with lavas and tuffs. Brown and mountain steppe chernozem soils dominated in Gyumri area.

During the Soviet Union period, the potential sources of heavy metals in Gyumri were forge-and-press, universal grinding machines, instrument engineering, electrotechnical, household electrical appliances, refrigerator compressors and ferro-concrete constructions plants, microelectromotor "Strommashina" plant, and foundry of machine-tool construction plant [30], which were operated till the devastating earthquake of 1988 and did not resume after the collapse of the Soviet Union. Unfortunately, there is a lack of information about the heavy metal emission from the abovementioned plants in city territory.

Nowadays, the Gyumri and its industrial sector are in reconstruction stage and there are no significant potential sources of heavy metals. In the polluted areas identified during 2013, Gyumri ecogeochemical complex investigations [10] were mainly linked to the historical pollution.

#### 2.1.3. *The city of Kajaran*

The city of Kajaran has a total area of 2.74 km$^2$, 8.4 thousand population (3066 persons per square km) [16], and is located in the valley of river Voghchi, where two types of the erosion landforms are distinguished: U-shaped river valleys in the middle and lower course of the river and V-shaped river valleys in the riverheads. Up to 1800 m, brown soils and 1800–2400 m chestnut soils predominated. The northern slope of Kajaran territory is covered with the gray mountain-forest skeletal soils [31]. The geological base of Kajaran includes volcanogenic sedimentary and intrusive rocks of the tertiary period, particularly monzonites and porphyry granites. The Kajaran sulfide copper-molybdenum deposit is timed to the monzonites, and the main ore minerals are molybdenite and chalcopyrite and the accessory minerals are pyrite, magnetite, hematite, sphalerite, tetrahedrite, bismuthine, wulfenite, vanadinite, galena, as well as native Te and Au. Besides, ore contains Re, Se, and Ag [8].

The main pollution source of Kajaran is ZCMC, including Cu-Mo opencast mine. ZCMC complex also includes ore crushing and milling, as well as ore dressing plants and active Artsvanic tailing repository. In addition, abandoned tailing repositories of Voghchi, Darazami, and Pkhrut are also significant sources of dust and heavy metals in it [8].

### 2.2. Soils, dust, and food sampling and analysis

Soil, dust, and food sampling and pretreatment were done according to the SOPs developed in compliance with methodological guidelines [32–34], international ISO [35–38] standards, and US EPA [39] guidelines. Totally, 1356, 443, and 76 soils and 25, 22, and 15 dust have been collected in Yerevan, Gyumri, and Kajaran, respectively.

Food sampling was done and 68 samples were collected from the agricultural lands of Kajaran and rural communities located near ZCMC Artsvanic tailing repository. Soils, dust, and food samples have been placed in special clean bags for transportation and storaging purposes. Prior to the analysis, samples laboratory pretreatment was done.

The total contents of heavy metals (**Table 1**) were determined using X-ray fluorescence spectrometry (Innov-X 5000, USA) [40] and atomic absorption spectrometry (AAnalyst 800 AAS PE, USA).

The analysis was done in the environmental geochemistry department and at the Central Analytical Laboratory of CENS, accredited by ISO-IEC 17025.

Detailed information concerning Yerevan's, Gyumri's, and Kajaran's soils, dust, and food sampling, samples' pretreatment, and analysis can be found in a number of manuscripts [7–10, 20–23].

### 2.3. Health risk assessment

Human health risk assessment [5] was done based on the contents of HM in soils and dust of city Yerevan, Gyumri, and Kajaran. In the case of Kajaran, health risks arising from the HM content in the food products grown near the city, ZCMC query, and its tailing storages were also studied. Health risk assessment model proposed by US EPA was used. As a preferential exposure pathway of HM for humans, soil and dust ingestion was chosen.

Noncarcinogenic health effects from the soils, dust, and food heavy metals contents was assessed using the following Eqs. [5, 17, 18].

$$CDI_{ing}\left(\frac{mg}{kg\ day}\right) = \frac{C * IngR * EF * ED * 10^{-6}}{AT * BW}, \tag{1}$$

$$HQ_{ing} = \frac{CDI_{ing}}{RfD_{ing}}, \tag{2}$$

$$HI = \sum HQ. \tag{3}$$

where CDI is the chronic daily intake of metal, C is the element concentration in studied medium (mg/kg), EF is the exposure frequency: 350 day/year for soil and dust, ED is the

| City | Medium | Heavy metals |
|---|---|---|
| Yerevan | Soil | Hg, Pb, As, Ni, Cu, Zn, Mo, Cr, Co, Mn, Ba, and V |
| | Dust | Hg, Pb, Mo, Cd, Zn, Cu, Ni, Ag, Co, Cr, and As |
| Gyumri | Soil | Hg, Pb, As, Cd, Cu, Zn, Mo, Fe, Co, Mn, and Ba |
| | Dust | Pb, Cr, Ni, Zn, Cu, and Mo |
| Kajaran | Soil | Ti, Mn, Fe, Co, Cu, Zn, Mo, and Pb |
| | Dust | Cu, Pb, Mo, Mn, Ni, Cr, V, Zn, and Sn |
| | Food | Cu, Mo, Ni, Cr, Zn, Pb, Hg, As, and Cd |

**Table 1.** Heavy metals determined in soils, dust, and food.

exposure duration: 30 years for adult [17] and 6 years for children [5], IngR is the ingestion rate: 100 mg/day$^{-1}$ for adults and 200 mg/day$^{-1}$ for children average time (AT) (AT = 365 × ED) [5], and average body weight (BW, kg): 70 kg for adults [17] and 15 kg in the case of children [5].

Taking into consideration the fact that unlike Yerevan and Gyumri where there is no local food production and consumption, in Kajaran, mining region's contribution of local plant-origin food in overall diet is significantly higher. Therefore, dietary intakes of heavy metals via consumption of selected vegetables and fruits may also be a risk factor to health.

Noncarcinogenic risk of heavy metals in food was assessed by the abovementioned formulae (1)–(3) using the following parameters: EF: 183 days/year for all investigated fruits and vegetables, except potato (365 days/year). ED was set to 63.6 for males and 69.7 for females based on the average life expectancy, starting from 8 years of age. IRS: food consumption rate was evaluated based on the result of standardized food frequency questionnaires filled by 200 males and females residing in Kajaran mining impact area. According to our polling survey in studied region, BW for males and females were considered to be 70 and 60 kg, respectively.

The reference doses (RfDs) of studied heavy metals were taken from RAIS and US EPA Human health risk assessment guidance [5, 17]. Only the RfD of Pb was taken from the WHO guideline [41]. Hazard index (HI-multielement) is the sum of all HQ (monoelement). When HI and/or HQ is less than one, there is no harmful effect to the health, whereas when HI and/or HQ values are greater than one, there is a possibility of adverse health effects.

To get overall adults health risk ($HI_{sum}$) from soils, dust ingestion, and food consumption in Kajaran, the obtained mean values of HI were summed.

### 2.4. Potential ecological risk assessment

Potential ecological risk assessment (PERI) was performed using the method proposed by Hakanson [4]. From the studied elements, only Hg, Cd, As, Pb, Cu, Ni, Cr, and Zn have "toxic-response" factors 40, 30, 10, 5, 5, 5, 2, and 1, respectively. Taking into consideration the fact that soils are the sink of city pollutants, ecological risk assessment was done based on the contents of heavy metals in soils. As the city of Kajaran is spatially located within the biogeochemical province, high contents of heavy metals are intrinsic to the city environment. Here, ecosystems have their own distinctive features and there is a deviation from common environmental patterns. Therefore, the city of Kajaran was excluded from the ecological risk assessment. RI was calculated using (4)–(6) formulas:

$$C_r^i = C_{topsoil}^i / C_n^i, \quad (4)$$

$$E_r^i = T_r^i * C_r^i, \quad (5)$$

$$\mathrm{PERI} = \sum_{i=1}^{n} E_i^r, \quad (6)$$

where PERI is potential ecological risk index, $E_r^i$ is PERI of single element, $T_r^i$ is "toxic-response" factor for the selected element (i.e., Hg = 40, As = 10, Pb = Cu = Ni = 5, Cr = 2, and

Zn = 1), $C_r^i$ is the pollution factor of the element, $C_{soil}^i$ is the concentration of element in the topsoil, and $C_n^i$ is the reference value of the selected element (local background [9, 10]). The PERI levels are classified as low (<150), moderate (150–300), considerable (300–600), and very high (>600) [4].

## 3. Results

Health noncarcinogenic risk assessment of adults and children was performed based on the contents of studied heavy metals (**Table 1**) in soils and dust of the city of Yerevan and Gyumri and in soils, dust, and food in the city of Kajaran.

### 3.1. Noncarcinogenic risk in Yerevan

The results obtained showed that in the case of Yerevan soils, monoelemental risk to adults was detected only for the contents of Pb in two sampling sites.

Multielemental noncarcinogenic risk range from 0.12 to 2.37 with the mean of 0.25, and risk was observed in four sampling sites (**Figure 2**). Monoelemental noncarcinogenic risk from dust heavy metals was observed in a single sampling site and is associated with the high contents of Mo. Multielemental risk ranges from 0.02 to 1.87 with the mean of 0.2, and risk was observed in one sampling site (**Figure 2**) situated in the southern part of the city. For

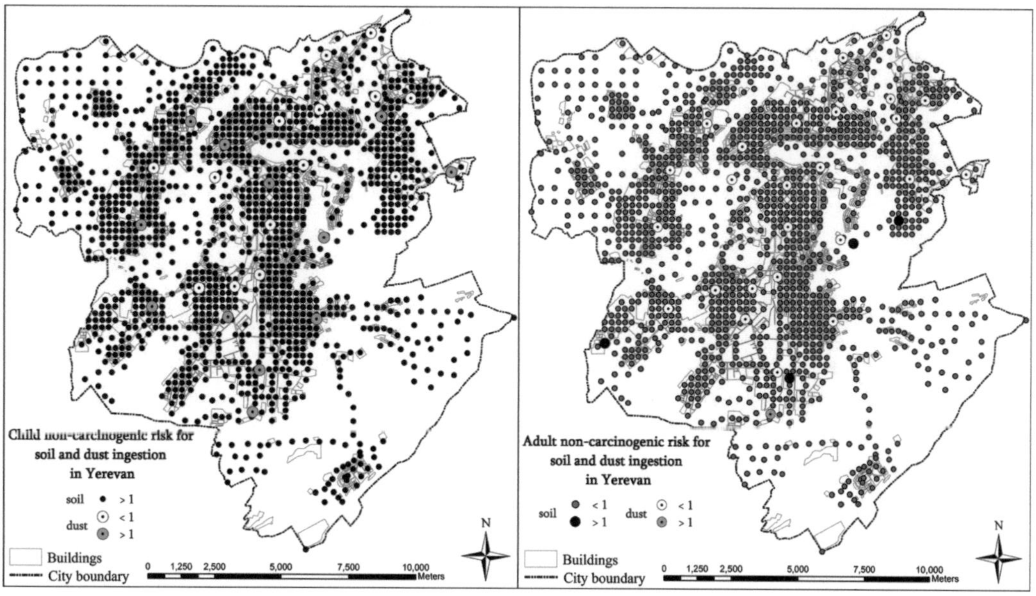


**Figure 2.** Spatial distribution of soils and dust noncarcinogenic risk to children and adults health in Yerevan.

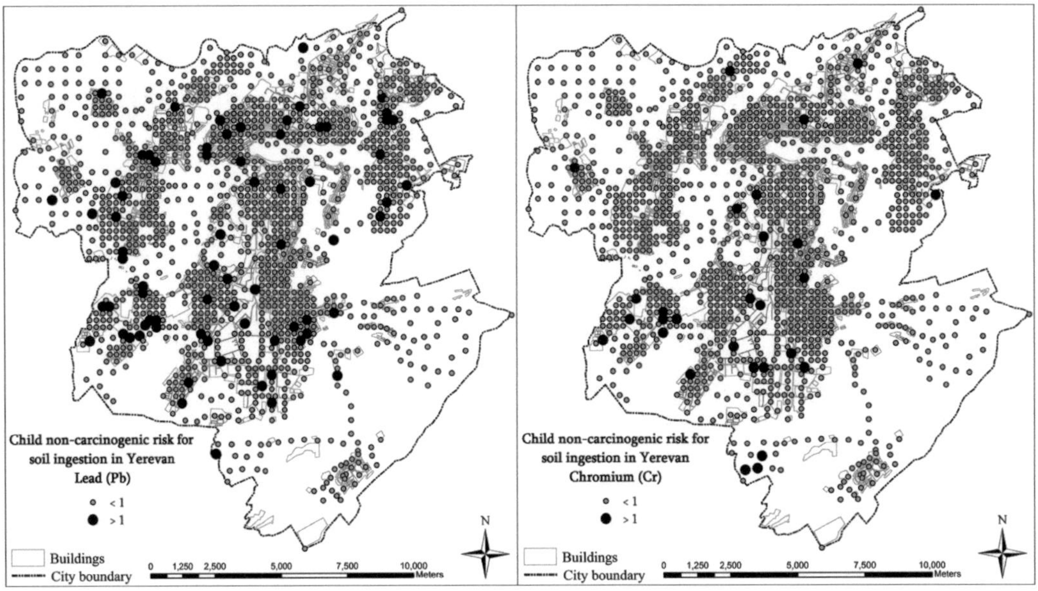


**Figure 3.** Spatial distribution of soils Pb and Cr noncarcinogenic risk to children in Yerevan.

both soils and dust, the observed risky sites are spatially allocated in or near the industrial units of Yerevan (**Figure 1**).

Children monoelement noncarcinogenic risk from soils detected for Ni, Cu, Zn, Mo, Co, and Mn in a single sampling point while for Cr and Pb risk was observed in 28 and 72 sampling sites (**Figure 3**), respectively. The study revealed [21] that riskiest contents of Pb in Yerevan are the result of the redistribution of historically polluted soils. HI values of soil's heavy metal contents range from 1.1 to 22.1 with the mean of 2.31, indicating an adverse health effect to children (**Figure 2**) in whole territory of the city. In case of dust, HQ values greater than 1 were observed from Mo, Cd, Co, and As in 1, 1, 2, and 1 sampling sites, correspondingly. Dust HI values (**Figure 2**) range from 0.25 to 17.45 with the mean of 1.82, and risk was detected in 12 sampling sites located in Yerevan's residential areas and near the industrial units (**Figures 1** and **2**).

### 3.2. Noncarcinogenic risk in Gyumri

Noncarcinogenic risk assessment showed that in Gyumri's territory, soils and dust heavy metal's HQ and HI values were less than 1, suggesting the absence of adverse health effects to adults. Risk from the dust heavy metal contents was also not detected in case of children. Soil's heavy metal HQ values greater than 1 were detected for Cu and Pb contents in 1 and 17 sampling sites, respectively. Moreover, Pb risky sites are spatially located in residential parts of the city and near its industrial units (**Figures 1** and **4**). Soil's heavy metal multielemental risk in Gyumri range from 0.85 to 7.42 with the mean of 1.56, and risk was observed in 439 of 443 sampling locations (**Figure 4**).

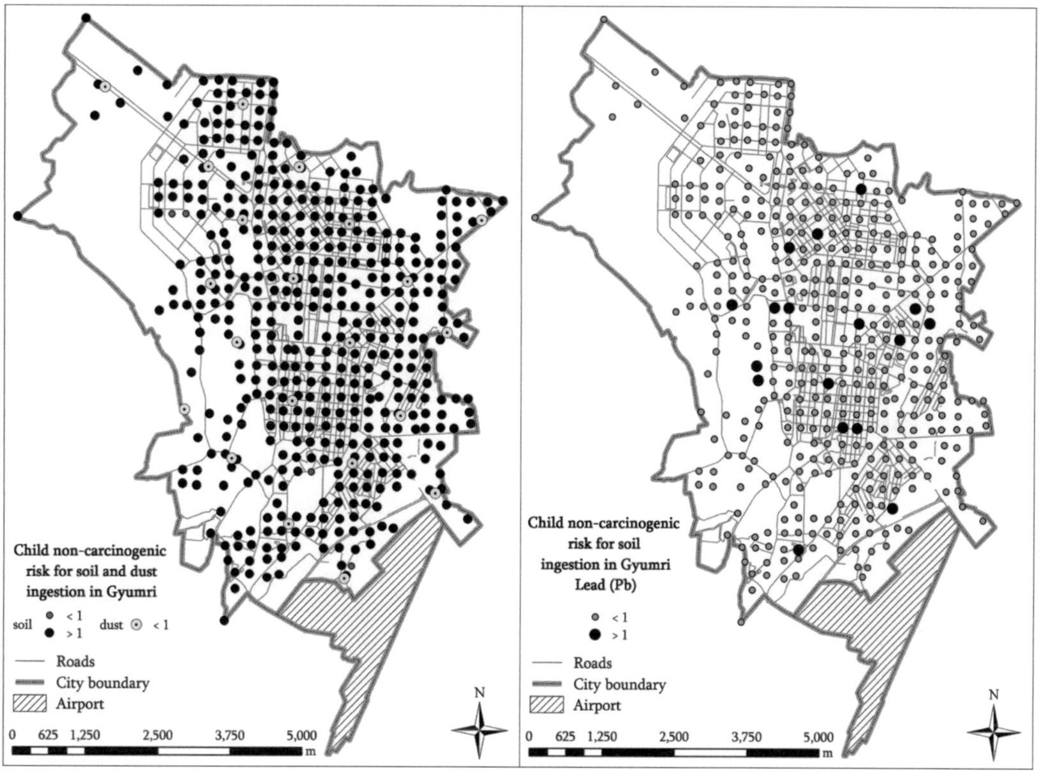


**Figure 4.** Spatial distribution of soils, dust, and soil Pb contents noncarcinogenic risk to children in Gyumri.

### 3.3. Noncarcinogenic risk in Kajaran

Noncarcinogenic risk assessment based on the detected contents of heavy metals in soils and dust of Kajaran territory showed that the HQ values of adults greater than one were detected only in four soil sampling sites for the contents of Mo. HI values of soil heavy metals range from 0.23 to 5.46 with the mean of 0.64 and risk was observed in seven sampling sites (**Figure 5**), whereas HI values of dust were all less than 1.

In the case of children, noncarcinogenic risk observed Mn, Fe, Co, Pb, Cu, and Mo in 6, 49, 18, 1, 2 and 34 sampling sites out of the 76, respectively. Soils HI values range from 2.11 to 51.0 with the mean of 5.94 and suggested an adverse health effect to children in whole area of the city. For both Fe and Mo (**Figure 6**), the risky sites are spatially located in the residential part of Kajaran and near the ZCMC ore crushing, milling, and ore dressing plants. Moreover, in the same areas of city, Mo poses a noncarcinogenic risk to children (7 of 15 dust samples).

Health risk assessment of food product consumption showed that HQ for Cu was more than 1 in maize, potato, and bean both for males and females, whereas for Mo, HQ range from 0.05 to 5.79 for males and 0.05 to 8.63 for females. Particularly, in carrot, potato, and

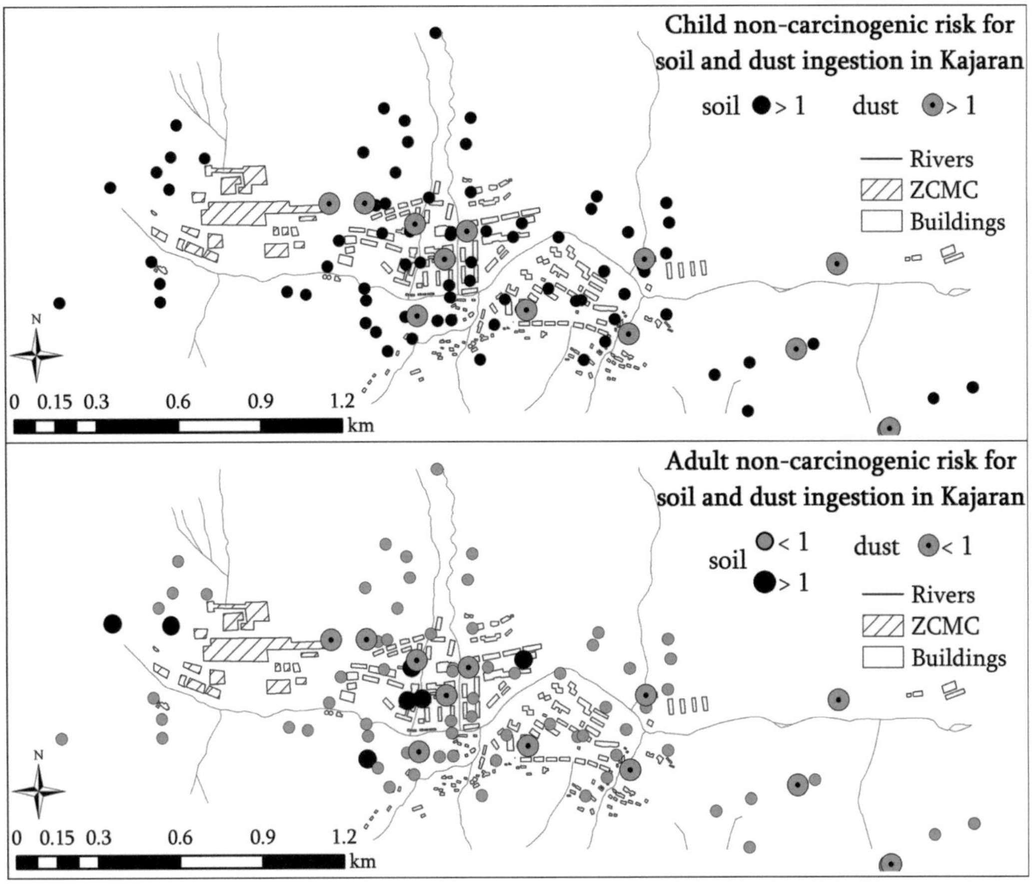


**Figure 5.** Spatial distribution of soils and dust noncarcinogenic risk to children and adults health in Kajaran.

onion leaf, HQ value is more than 5, which proves that risks are obvious. For maize consumption, the HQ is higher than 1 for males and females (3.94 and 4.40, respectively). None of the studied vegetables and fruits has a HQ > 1 for Ni, Cr, Zn, Pb, As, and Cd beside the case of Ni in maize for females. In case of Hg, beet and grape indicated HQ more than 1 both for males and females. From all studied elements, only Mo HI values from all studied vegetables and fruits were greater than 1, indicating an adverse health effect both for males and females.

The results of health risk assessment in Kajaran showed that $HI_{sum}$ were greater than 1, indicating an adverse health effect to adults from soils, dust ingestion, and food consumption. Therefore, it should be highlighted that in biogeochemical provinces where industrial activities are closely related to the agricultural lands, the risk assessment including only environmental abiotic mediums may lead to the underestimation of risk level.

Overall, heavy metals in the Yerevan, Gyumri, and Kajaran environment are a primary concern to children health. Moreover, risk assessment showed that the riskiest elements in the cities environments are those previously identified as primary pollutants.

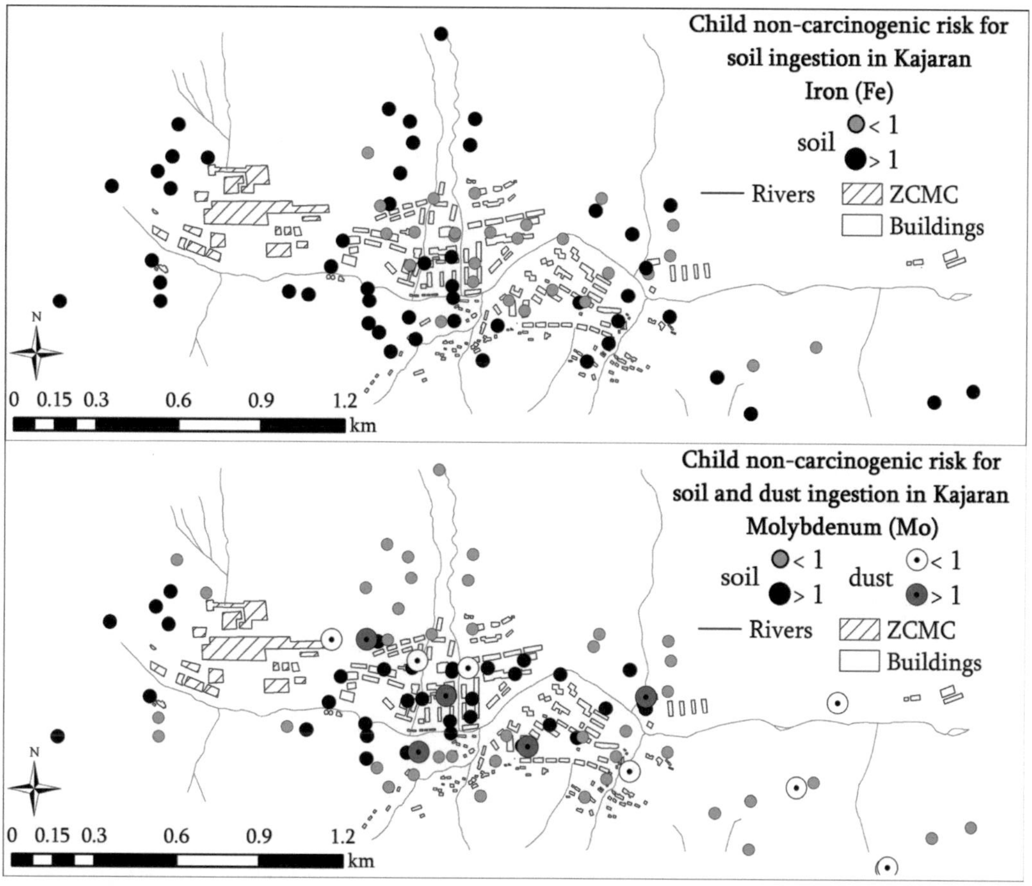


**Figure 6.** Spatial distribution of soil Fe and soils and dust Mo noncarcinogenic risk to children in Kajaran.

### 3.4. Potential ecological risk in Yerevan and Gyumri

In Yerevan, PERI was evaluated based on the contents of Hg, As, Pb, Cu, Ni, Cr, and Zn and the mean values of single ecological risk indices decreased in the following order: Hg> > Pb> > Cu > As > Ni > Cr > Zn. The results of Yerevan's soils potential ecological risk assessment showed (**Figure 7**) that PERI ranges from 53 to 5793.2 with the mean value of 425.3. The latter belongs to the considerable risk level, which was also observed in 1068 (78.8% of all samples) sampling sites. The low level (**Figure 7**) of ecological risk was detected in 38 (2.8% of all samples) and the moderate level in 155 (11.4% of all samples) sampling sites. The very high level of ecological risk was detected in 95 (7.0% of all samples) sampling sites. From all elements included in Yerevan soil's ecological risk assessment, significant contribution to the considerable and very high levels of PERI was mainly from the single ecological risk indices of Pb and Hg.

In the case of the city of Gyumri, PERI was evaluated based on the contents of Hg, Cd, As, Pb, Cu, and Zn, and the mean values of single ecological risk indices decreased in the following

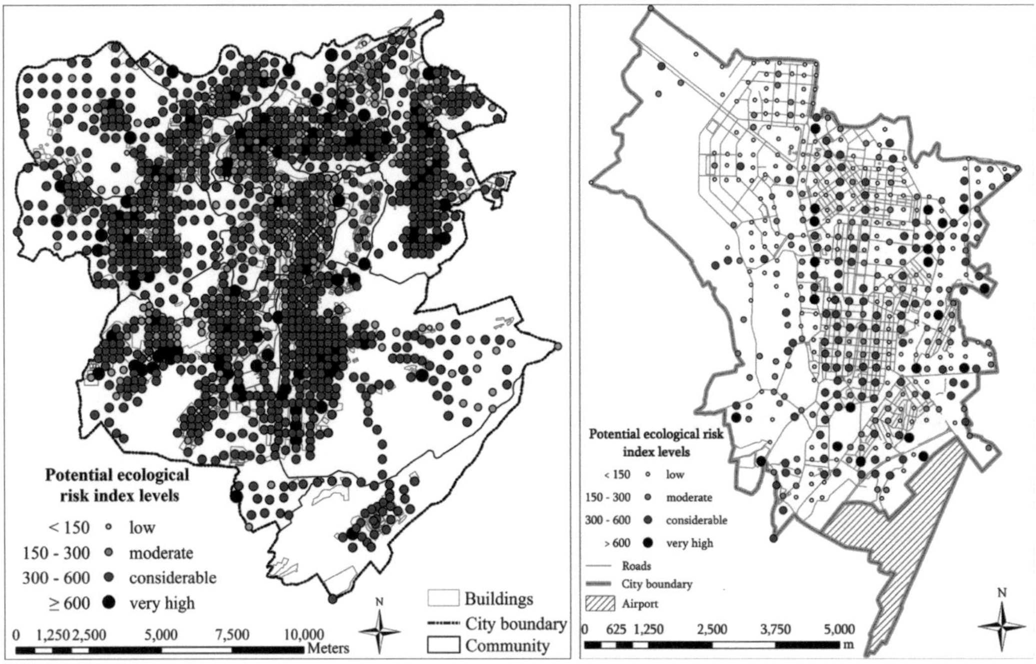


**Figure 7.** Spatial distribution of potential ecological risk levels in Yerevan and Gyumri.

order: Cd> > Hg > Pb > As > Cu > Zn. PERI ranges from 48.2 to 1892 with the mean of 252, which belongs to the moderate ecological risk level. The latter was also observed in 128 (28.9% of all samples) sampling sites. The low level (**Figure 7**) was detected in 183 (41.3% of all samples), considerable level in 111 (25.1% of all samples), and very high level of ecological risk in 21 (4.7% of all samples) sampling sites. In Gyumri, significant contribution to the very high levels of PERI was mainly from the single ecological risk indices of Cd, Pb, and Hg.

## 4. Conclusions

The result of human health risk assessment showed that soils multielemental noncarcinogenic risk (HI > 1) to adults observed in a few sampling sites both for Yerevan and Kajaran, while in Gyumri HI < 1. For children, noncarcinogenic risk values indicated possible adverse health effects approximately in the whole area of all studied cities. Also for dust, risks have been detected mainly for children in the cities of Yerevan and Kajaran. In Kajaran, risk assessment showed possible adverse health effects for the population from food, as well. The riskiest elements were Pb and Cr for Yerevan, Pb for Gyumri, and Mo for Kajaran. It should be stated that unlike anthropogenic contents of Pb in Yerevan and Gyumri, the high Mo concentrations in Kajaran can be the result of geogenic input as well. According to the results of PERI in cities of Yerevan and Gyumri, considerable and very high levels of ecological risk were observed and the riskiest elements were those (Pb, Hg, and Cd) included in the first group of toxicity. Both human health and ecological risk assessment results highlight the need for further detailed studies, especially in those areas with the highest level of identified risk.

## Author details

Gevorg Tepanosyan*, Lilit Sahakyan, David Pipoyan and Armen Saghatelyan

*Address all correspondence to: gevorg.tepanosyan@cens.am

The Center for Ecological-Noosphere Studies, NAS RA, Yerevan, Armenia

## References

[1] Bradl H, Kim C, Kramar U, Stiiben D. Interactions of heavy metals. In: Bradl HB, editor. Heavy Metals in the Environment: Origin, Interaction and Remediation. London: Elsevier; 2005. p. 28-164. DOI: 10.1016/S1573-4285(05)80021-3

[2] Sun Y, Zhou Q, Xie X, Liu R. Spatial, sources and risk assessment of heavy metal contamination of urban soils in typical regions of Shenyang, China. Journal of Hazardous Materials. 2010;**174**(1–3):455-462. DOI: https://doi.org/10.1016/j.jhazmat.2009.09.074

[3] Yuan GL, Sun TH, Han P, Li J, Lang XX. Source identification and ecological risk assessment of heavy metals in topsoil using environmental geochemical mapping: Typical urban renewal area in Beijing, China. Journal of Geochemical Exploration. 2014;**136**:40-47. DOI: https://doi.org/10.1016/j.gexplo.2013.10.002

[4] Håkanson L. An ecological risk index for aquatic pollution control: A sedimentological approach. Water Research. 1980;**14**(8):975-1001. DOI: https://doi.org/10.1016/0043-1354(80)90143-8

[5] The Risk Assessment Information System. RAIS [Internet]. [Updated: 1 January 2017]. Available from: https://rais.ornl.gov/ [Accessed: 20/08/2017]

[6] Saghatelyan AK, Sahakyan LV, Belyaeva OA, Tepanosyan GO, Pipoyan DA. Evolution of ecologo-geochemical investigations in Armenia. Izvestiya Rossiiskoi Akademii Nauk. Seriya Geograficheskaya. 2016;(4):199-127. DOI: 10.15356/0373-2444-2016-4-119-127

[7] Saghatelyan AK. Peculiarities of Heavy Metal Distribution on the Territory of. 1st ed. CENS NAS RA: Yerevan; 2005. 157 p

[8] Saghatelyan AK, Gevorgyan VS, Arevshatyan SH, Sahakyan LV. Ecological Geochemical Assessment of Environmental State of the City of Kajaran. 1st ed. CENS NAS RA: Yerevan; 2008. 200 p

[9] Tepanosyan G, Sahakyan L, Belyaeva O, Saghatelyan A. Origin identification and potential ecological risk assessment of potentially. Journal of Geochemical Exploration. 2016;**167**:1-11. DOI: http://dx.doi.org/10.1016/j.gexplo.2016.04.006

[10] Sahakyan LV, Tepanosyan GH, Kafyan MO, Saghatelyan AK. Geochemical survey and ecological risk assessment of heavy metals in soils of city of Gyumri (Armenia). Electronic Journal of NAS RA. 2017;**1**(28):16-20

[11] Saghatelyan AK, Kukulyan MA, Khachatryan TS. Assessment of the risk of perinatal mortality under the pollution of environment by heavy metals. In: Pollution of the Environment by Heavy Metals. Yerevan: Yerevan: CENS NAS RA; 1996. p. 69-70

[12] Saghatelyan AK, Babayan EA, Vardazaryan LD. The contents of lead in environmental compartment and in the blood of children and adults. In: Pollution of the Environment by Heavy Metals. Yerevan: Yerevan: CENS NAS RA; 1996. p. 68-69

[13] Babayan EA, Saghatelyan AK. Some health indices under condition of lead contamination of the environment and industrial areas. Medical Science of Armenia. 1999;**4**(39):141-151

[14] Movsesyan HS, Aroutounian RM, Sahakyan LV, Asmaryan SG, Galstyan HM, Bazikyan GK, nMartirosyan HA, Saghatelyan AK. Study of the cancer incidence in kids and juveniles in Yerevan in relation to air and soil pollution on their dwelling sites. Biological Journal of Armenia. 2013;**65**(3):69-74

[15] Saghatelyan AK, Sahakyan LV, Mikayelyan MG, Belyaeva OA. Ecological and geochemical analysis of risks of the impact of mining production upon sustainable development of Armenia. Vol. 5. Izvestiya Rossiiskoi Akademii Nauk. Seriya Geograficheskaya; 2010. p. 87-93

[16] Saghatelyan AK, Sahakyan LV, Menchinskaya OV, Zangiyeva TD, Kajtukov MZ, Uzdenova ZH. Medical geology in connection. In: Selinus O, Finkelman RB, Centeno JA, editors. Medical Geology. A Regional Synthesis. 1st ed. Netherlands: Springer Netherlands; 2010. p. 392. DOI: 10.1007/978-90-481-3430-4

[17] United States Environmental Protection Agency. US EPA [Internet]. 1989. Available from: https://www.epa.gov/risk/risk-assessment-guidance-superfund-rags-part [Accessed: 20/08/2017]

[18] Pepper I, Gerba C, Brusseau M. Environmental and Pollution Science. 2nd ed. United States of America: Elsevier Inc.; 2011. 552 p

[19] National Statistical Service of the Republic of Armenia. NSS RA [Internet]. 2016 [Updated: 2016]. Available from: http://armstat.am/en/ [Accessed: 20/08/2017]

[20] Sahakyan LV. Peculiarities of the dynamics of Yerevan soil pollution with heavy metals (Ag, Pb, Cu, Zn, Ni, Co, Cr, Mo) [dissertation]. CENS NAS RA: Yerevan; 2008. 148 p. Available from: http://cens.am/research/environmental-geochemistry/environmental-geochemistry-activities

[21] Tepanosyan G, Sahakyan L, Belyaeva O, Maghakyan N, Saghatelyan A. Human health risk assessment and riskiest heavy metal origin. Chemosphere 2017;**184**:1230-1240. DOI: http://dx.doi.org/10.1016/j.chemosphere.2017.06.108

[22] Tepanosyan G, Maghakyan N, Sahakyan L, Saghatelyan A. Heavy metals pollution levels and children health risk assessment of Yerevan. Ecotoxicology and Environmental Safety. 2017;**142**:257-265. DOI: http://dx.doi.org/10.1016/j.ecoenv.2017.04.013

[23] Saghatelyan A, Sahakyan L, Belyaeva O, Maghakyan N. Studying atmospheric dust and heavy metals on urban sites through synchronous use of different methods. Journal of Atmospheric Pollution. 2014;**2**(1):12-16

[24] Sahakyan LV. The assessment of heavy metal stream in the air basin of Yerevan. Chinese Journal of Geochemistry. 2006;**25**:95-96

[25] Nalbandyan MA, Saghatelyan AK. Peculiarities of heavy metal pollution of water bodies in the bounds of Yerevan conglomeration. In: XXI century: Ecological Science in Armenia. Yerevan: CENS NAS RA; 2003. p. 219-220

[26] Saghatelyan A, Nalbandyan M. A complex assessment of ecological status of River Hrazdan. In: Problems of River Monitoring and Ecological Safety of the South Caucasus. Yerevan: CENS NAS RA; 2005. p. 83-87

[27] Hovhannisyan HA. Ecological monitoring of vegetables grown in urban sites. Electronic Journal of NAS RA. 2004;**2**(3):10-13

[28] Hovhannisyan HA, Nersisyan GS, Saghatelyan AK. Ecological assessment of heavy metal pollution of vegetable in Yerevan (Armenia). In: XVI International Eco Conference. Novi Sad, Serbia. Novi Sad: Ecological Movement of Novi Sad; 2012, 2012. p. 331-336

[29] Saghatelyan A, Arevshatyan S, Sahakyan L. Ecological-geochemical assessment of heavy metal pollution of the territory of Yerevan. Electronic Journal of NAS RA. 2003;**1**(1):36-41

[30] Danielyan KS. Environmental Analysis of the City of Gyumri. Yerevan: For Sustainable Human Development; 2009. 76 p

[31] Klopotovski BA. Soil-geographical feature of Armenia. Proceedings of National Academy of Sciences, Arm SSR, Natural Sciences. 1947;**7**:3-26

[32] Darnley AG, Bjorklunz A, Bolviken B, Gustavsson N, Koval PV, Steenfel A, Tauchid M, Xuejing X. A Global Geochemical Database for Environmental and Resource Management. France: UNESCO Publishing; 1995. p. 122

[33] Revich BA, Smirnova RS, Sorokina ES. Methodological Guidance for Geochemical Assessment of Polluted Sites by Chemical Elements. 1st ed. IMGRE: Moscow; 1982. 112 p

[34] Saet Y, Revich B, Yanin E. Environmental Geochemistry. 1st ed. NEDRA: Moscow; 1190. 335 p

[35] Fomin GS, Fomin AG. Soil: Inspection of Quality and Ecological Safety According to International Standards. 1st ed. Moscow: Protektor; 2001. 304 p

[36] International Organization for Standardization. ISO [Internet]. 2005. Available from: https://www.iso.org/standard/32427.html [Accessed: 20/08/2017]

[37] ISO 7002:1986 Agricultural food products. Layout for a standard method of sampling from a lot. International Organization for Standardization [Internet]. 1986. Available from: https://www.iso.org/standard/13569.html [Accessed: 20/08/2016]

[38] ISO 874:1980 Fresh fruits and vegetables – Sampling. International Organization for Standardization [Internet]. 1980. Available from: https://www.iso.org/standard/5259.html [Accessed: 20/08/2017]

[39] US EPA. pdf document [Internet]. 1999. Available from: [Accessed: 20/08/2017]

[40] Method 6200. Field Portable X-Ray Fluorescence Spectrometry for the Determination of Elemental Concentrations in Soil and Sediment. US EPA [Internet]. 2007. Available from: https://www.epa.gov/hw-sw846/sw-846-test-method-6200-field-portable-x-ray-fluorescence-spectrometry-determination [Accessed: 20/08/2017]

[41] World Health Organization. Guidelines for Drinking-water Quality—Volume 1: Recommendations. 3rd ed. Geneva: World Health Organization; 2008. 668 p

Chapter 16

# Pharmaceuticals and Personal Care Products: Risks, Challenges, and Solutions

Zakiya Hoyett

Additional information is available at the end of the chapter

http://dx.doi.org/10.5772/intechopen.70799

**Abstract**

Pharmaceuticals and personal care products (PPCPs) encompass a large class of chemical contaminants that can originate from human usage and excretions and veterinary applications. These pollutants have captured the attention of scientists, governments, and the public as several studies across the globe reveal their widespread occurrence in low-level concentrations in wastewater and the aquatic environment. Most of the research on PPCPs has been generated from efforts in highly developed countries, primarily North America and Europe, although investigations and reports are emerging from Southeast Asia and China. With the increased concern of potential threats triggered by the occurrence of these chemicals in the environment, environmental risk assessment (ERA) strategies for such compounds have considerably evolved over the past decade. Regulations are in effect or planned in several western nations, however, there is no global standard for conducting ERAs. As the scope of the problem evolves, substantial research will be imperative to address these contaminants and their occurrence in the environment. This chapter will discuss the evolution of the risk associated with the occurrence of PPCPs in the environment, the challenges faced by their existence here, and the colloquy about solutions to address this escalating issue.

**Keywords:** pharmaceuticals and personal care products (PPCPs), pharmaceuticals, contaminants of emerging concern (CECs), environmental risk assessment (ERA), aquatic environment

## 1. Introduction

Anthropogenic pollutants enter surface and ground waters via a multitude of processes. Commercial activities such as manufacturing emissions, waste disposal, and accidental releases are a few examples [1]. Other practices include deliberate introduction such as sewage sludge application to land, groundwater recharge, and consumer activity which involves both the excretion and purposeful disposal of a wide range of naturally occurring and anthropogenic chemicals [1, 2]. During the last few decades, the impact of chemical pollution in the water has focused almost exclusively on the conventional "priority pollutants" [3]. Priority pollutants are a group of chemicals regulated under legislation such as the Clean Water Act (CWA) of 1972 by the United States Environmental Protection Agency (US EPA) and the Water Framework Directive (2000/60/EC) (WFD), an updated version of Council Directive 76/464/EEC, by the Environment Directorate General of the European Commission (DG Environment) for the European Union (EU) [4, 5]. These pollutants are chemicals that have specific effects on organisms, comprised mainly of agricultural and industrial chemicals and their synthesis by-products [4–6]. The prioritized lists of 126 pollutants and 33 substances in the US and EU, respectively, currently include chemicals that were selected primarily because of their toxicity, persistence, and degradability, among other factors [4, 7, 8]. Chemical production rates and the frequency of occurrence in waters was also considered [4, 5].

Pharmaceuticals and personal care products (PPCPs) are among a group of chemicals termed "contaminants of emerging concern" (CECs). CECs are not necessarily new pollutants as they may have been present in the environment for several years, but their presence and significance are only now being evaluated [3]. Due to their medical properties, PPCPs have an inherent biological effect; furthermore, they behave as persistent pollutants because of their continual infusion into the aquatic ecosystem [9–11].

## 2. Risks

Considering scientific literature dating as far back as the early 1900s, more than 130 million organic and inorganic substances had been indexed by the American Chemical Society in the Chemical Abstracts Service (CAS) Registry, which is updated daily with about 15 thousand new substances [12]. Over eight million chemicals are commercially available, but only 350 thousand are inventoried and/or regulated globally [4, 8, 12–14].

**Figure 1** shows that the majority of chemicals in commerce are "industrial" chemicals, a significant percentage of these chemicals fall into the categories of "cosmetics ingredients" and "pharmaceuticals". Collectively, these two categories contain several compounds that are potentially persistent and bioaccumulative [14]. Caffeine, nicotine, and aspirin are a few of the pharmaceutically active compounds that have been known for years to enter the environment [3]. Only more recently has it become evident that drugs and personal care products from a wide spectrum of therapeutic and consumer-use classes exist in the environment in low concentrations [15, 16]. Over 50 million pounds of antibiotics are produced annually in the United States, with approximately 60% for human use and 40% for animal agriculture, therefore, veterinary medicines contribute considerably to PPCP occurrence [17]. In addition to pharmaceuticals, compounds such as synthetic fragrances, detergents, disinfectants, and insect repellents are among the man-made chemicals that are now beginning to accumulate in the natural environment [18].

http://dx.doi.org/10.5772/intechopen.70799

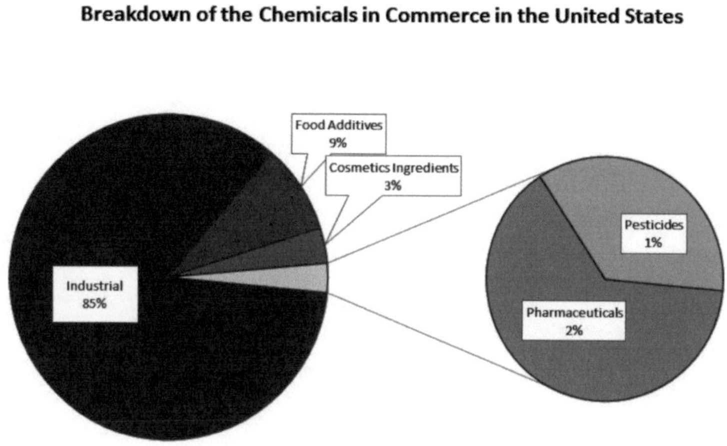


**Figure 1.** Estimated number and categories of chemicals in commerce registered for use in the United States over the past 30 years. Not all chemicals may be in current use. Similar proportions would be anticipated in other countries. Adapted with permission from [14]. Copyright 2006 American Chemical Society.

Increasing introduction to the marketplace of new pharmaceuticals is adding to the already large array of poorly understood chemical classes that each have distinct modes of biochemical action [1]. In the United States, legislation exists that requires an assessment of potential risk to the environment by new pharmaceutical products. Under this policy, the Food and Drug Administration (US FDA) is required to consider the environmental impacts of manufacture, use, and distribution of human drugs as well as investigational use and approvals of veterinary drugs [19, 20]. The European Commission recently published a Roadmap that acknowledges the Commission's effort toward developing a similar strategy that will address the manufacture, use and disposal of active pharmaceutical ingredients [21].

**Figure 2** illustrates the numerous pathways by which antibiotics and other PPCPs are introduced into the environment which can be both point and non-point sources [22, 23]. Municipal sewage, both treated and untreated, is the most likely route for human use drugs to enter the environment. Wastewater treatment processes achieve variable and often incomplete removal of antibiotics [24, 25]. Human pharmaceuticals are excreted from the body in urine and feces as unchanged parent compounds, metabolites or conjugated substances; furthermore, because of their polarity, water solubility and persistence some of these compounds may not be completely eliminated or transformed during sewage treatment [26, 27]. Therefore, residential and commercial healthcare facilities, specifically hospitals, are known contributors of antibiotics to municipal wastewater [2, 19, 28–30]. Additionally, the incorrect disposal of expired or unwanted medicines in the sink, toilet, or in household solid waste that is then taken to landfills contribute to the occurrence of pharmaceuticals in wastewater [31–33]. Another possible pathway begins with the disposal of unwanted illicit drugs, synthesis byproducts, raw products and intermediates into domestic sewage systems by clandestine drug operations [3, 32, 34]. Other probable entries include leakage from pipelines, tanks, waste ponds or landfills, and atmospheric deposition [35].

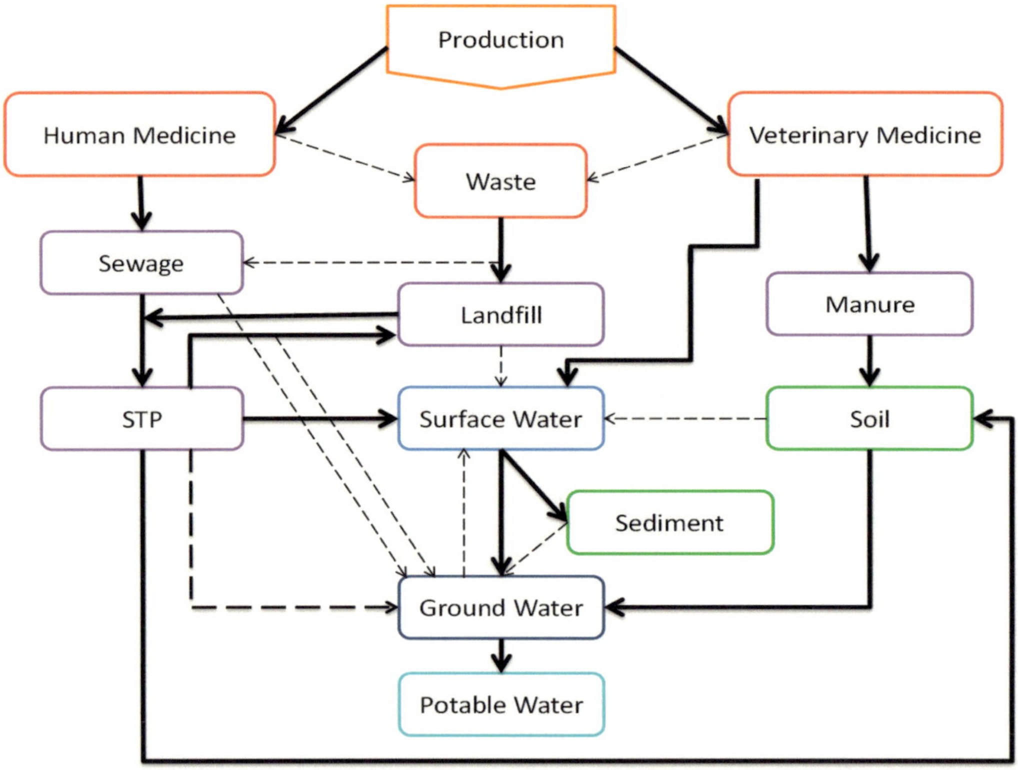


**Figure 2.** Source, fate, and distribution of PPCPs in the environment.

Veterinary medicines may enter the environment through a number of pathways, with terrestrial runoff from concentrated animal feeding operations (CAFOs) and wind-borne drift of agriculturally-applied antimicrobials to crops being the primary sources [32, 34, 36]. After administration, the substances may be metabolized in the animal which changes their physical, chemical and eco-toxicological properties, but even metabolites may be reconverted to their parent compounds after excretion [37, 38]. Accidental leakage or leaching from animal waste storage can also be a source. Still another major channel by which veterinary antibiotics are released into the environment is through application of manure or slurry to agricultural fields as fertilizer [34, 36, 39].

Dependent upon the chemical properties and structures of PPCPs, several processes can affect the fate and transport of these compounds in the environment. These include, but are not limited to, sorption, biotic transformation, and abiotic transformation [7, 24, 27]. Most PPCPs are water soluble and have a low volatility, although there are few that may strongly adsorb to soils and are somewhat persistent. These characteristics allow them to be easily transported and omnipresent in various aquatic environments [7, 19]. Because PPCPs can be introduced on a continual basis to the aquatic environment, they are ubiquitously present in waters; their removal or transformation by biodegradation, hydrolysis, photolysis, and other processes is continually countered by their replenishment [3].

With concentrations typically ranging from the low parts per trillion (ppt) and parts per billion (ppb) levels, several individual PPCPs or their metabolites from a variety of therapeutic classes (**Table 1**) have been detected in environmental samples from all over the world [3]. More than 80 pharmaceuticals and their metabolites have been detected in almost

| Therapeutic class | Examples of generic names | Examples of brand names |
|---|---|---|
| Analgesics/non-steroidal anti-inflammatories (NSAIDs) | Acetaminophen (analgesic)<br>Diclofenac<br>Ibuprofen<br>Ketoprofen<br>Naproxen | Tylenol<br>Voltaren<br>Advil<br>Oruvail<br>Naprosyn |
| Antimicrobials/antibiotics | e.g., sulfonamides, fluoroquinolones | Many |
| Antiepileptics | Carbamazepine | Tegretal |
| Antihypertensives (betablockers, beta-adrenergic receptor inhibitors) | Bisoprolol<br>Metoprolol | Concor<br>Lopressor |
| Antineoplastic | Cyclophosphamide<br>Ifosfamide | Cycloblastin<br>Holoxan |
| Antiseptics | Triclosan | Irgasan DP 300 |
| Contraceptives | β-Estradiol<br>17a-Ethinyl estradiol | Diogyn<br>Oradiol |
| Hormonally active agents<br>Androgens<br>Anti-acne agents adrenocorticosteroids inhalable Steroids<br>Estrogen antagonists | Fluoxymesterone<br>Isotretinoin<br>Tretinoin<br>Prednisone<br>Triamcinolone<br>Fluticasone<br>Tamoxifen | Accutane<br>Retin-A<br>Flovent<br>Nolvadex |
| $\beta_2$-Sympathomimetics (bronchodilators) | Albuterol | Ventolin |
| Lipid regulators (anti-lipidemics; cholesterol-reducing agents; and their bioactive metabolites) | Clofibrate (active metabolite: clofibric acid)<br>Gemfibrozil | Atromid-S<br>Lopoid |
| Musks (synthetic) | Nitromusks<br>Polycyclic musks<br>Reduced metabolites of nitromusks | Musk xylene<br>Celestolide<br>Substituted amino nitrobenzenes |
| Anti-anxiety/hypnotic agents | Diazepam | Valium |
| Sun screen agents | Methylbenzylidene camphor<br>avobenzene<br>Octyl methoxycinnamate | Eusolex 6300<br>Parsol A<br>Parsol MOX |
| X-ray contrast agents | Diatrizoate | Hypaque |



**Table 1.** Chemical classes (and members) of PPCPs detected in environmental samples.

| Chemical class | Location | Concentration range (ng/L) | References |
|---|---|---|---|
| Multiple pharmaceuticals | North America<br>*U.S.* | ND – 72 | [51] |
| | East Asia<br>*China*<br>*Japan*<br>*Korea* | ND – 5911 | [52–57] |
| | Europe<br>*Finland*<br>*Norway*<br>*Portugal*<br>*U.K.* | ND – 126,000 | [58–61] |
| Antimicrobials/antibiotics | North America<br>*U.S.* | 90–320 | [24, 62] |
| | East Asia<br>*China*<br>*Korea* | ND – 21,278 | [52, 63–72] |
| | Europe<br>*Finland*<br>*Sweden*<br>*U.K.* | ND – 3052 | [61, 73, 74] |
| Hormonally active agents | North America<br>*Canada* | 0.2–96 | [75] |
| | East Asia<br>*China*<br>*Japan*<br>*Korea* | ND – 253.8 | [52, 53, 76–83] |
| | Europe<br>*Portugal* | ND – 25 | [60] |
| Antiepileptics | East Asia<br>*China* | 230–1110 | [84] |
| Antiseptics | Europe<br>*Norway* | 160–480 | [59] |
| Musks (synthetic) | North America<br>*U.S.* | 495–3730 | [85] |
| | East Asia<br>*China*<br>*Japan* | <4–2050 | [86–89] |
| | Europe<br>*Portugal* | 1–889 | [60] |
| Sun screen agents | East Asia<br>*China* | 21–1287 | [90] |
| | Europe<br>*U.K.* | <2–6325 | [61] |

ND: not detected.

**Table 2.** Representation of the global occurrence of PPCPs in WWTP effluents.

every aquatic environment in North America and Europe surface waters [33, 40–44]. A national reconnaissance study on the occurrence of pharmaceuticals, hormones, and other organic wastewater contaminants (OWCs) in United States streams found that one or more OWCs were found in 80% of the stream samples, with 82 compounds of the 95 analyzed for detected during the study [40]. In another project, source water, finished drinking water, and distribution system (tap) water from 19 United States drinking-water treatment (DWT) plants was analyzed for 51 pharmaceuticals and pharmaceutical metabolites. Targeted compounds were detected most frequently in source water with at least one compound being detected in all 19 source waters; they were also found in approximately 89% of finished drinking waters and 87% of distribution systems [45]. In yet another study conducted by the United States Geological Survey (USGS) and the Centers for Disease Control and Prevention (CDC), several compounds that were frequently detected in samples of stream water and raw-water supplies were also detected in samples collected throughout the DWT facility, indicating that these compounds resist removal through conventional water-treatment processes [46].

PPCPS have been reported in hospital wastewaters, wastewater treatment plant (WWTP) effluents, WWTP biosolids, soil, surface waters, groundwaters, sediments, biota, and drinking water [33, 40, 47–50]. Since WWTPs are considered a major source of these pollutants, several investigations of environmental loads of PPCPs examine WWTP effluents (**Table 2**) [28]. There is less documented research of PPCP occurrence in coastal or marine ecosystems. A wide distribution of clofibric acid, caffeine, and DEET in concentrations up to 19, 16, and 1.1 ng/L, respectively, was measured throughout the North Sea and along European coasts [91]. Sulfamethoxazole, carbamazepine, tamoxifen, and indomethacin were discovered in China in the Yangtze River Estuary at levels ranging from 4.2 to 159 ng/L [92]. In the United States, sulfamethoxazole was detected in at least four bays ranging in concentrations from 4.8 to 65 ng/L, while trimethoprim was found at a maximum concentration of 72.2 ng/L in Jamaica Bay, New York and 2.1 ng/L off the coast of California [93–95].

## 3. Challenges

An ecological or environmental risk assessment (ERA) is defined as the means of evaluating the probabilities and magnitudes of adverse effects to human health or ecological receptors, directly or indirectly, as a result of exposure to pollutants and other anthropogenic activities [96]. ERAs are employed to estimate any potential harm that could emerge from environmental contaminants, with a known degree of certainty, using scientific methodologies. The innovation of ERAs has become necessary as improved research reveals chemicals in the environment at levels that are potentially toxic to humans and/or our valuable natural resources [11]. The specific methodology for carrying out an ERA may vary depending on the chemical being assessed, but the core principles and the key stages of the process are fundamentally the same in each case (**Figure 3**).

ERAs can be used to predict the likelihood of future adverse effects, prospective, or to evaluate the likelihood that effects are caused by past exposure to stressors, retrospective [97].

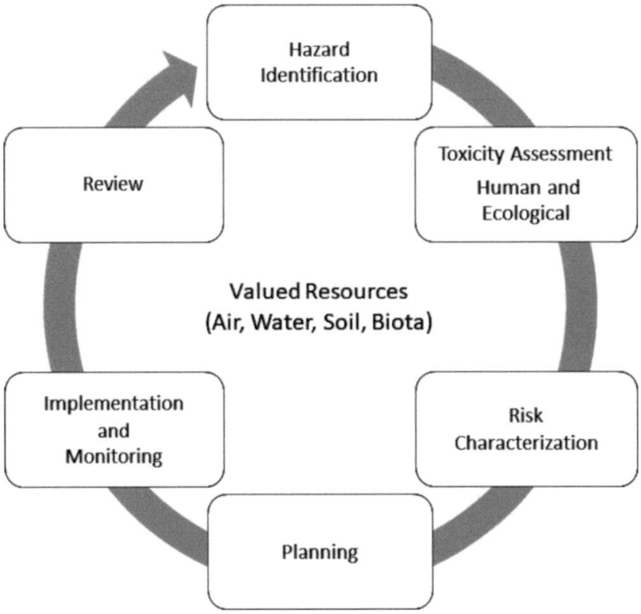


**Figure 3.** Flow chart for a general ERA process.

Examples of prospective uses include establishing drinking water goals or wastewater discharge limits. Federal and state regulatory programs also utilize prospective ERAs to reduce toxic tort liabilities and improved public relations. The government may use retrospective ERAs as a decision making tool, for example, when determining Comprehensive Environmental Response, Compensation, and Liability Act – CERCLA or Superfund – projects [11, 98, 99]. In many cases, both approaches are included in a single risk assessment. Combined retrospective and prospective risk assessments tend to be beneficial in situations where ecosystems have a history of previous impacts and/or the potential for future effects from multiple chemical, physical, or biological stressors [97].

Although the concentrations of these PPCPs generally range from the low ppt- to ppb-levels, there is increasing evidence that PPCPs may have significant impacts on natural biotic communities. There are two major concerns with the presence of low-level concentrations of pharmaceuticals in the aquatic environment: the potential toxicity of these compounds to aquatic organisms and the exposure to humans through drinking water [23, 31, 100]. Some PPCPs, such as antidepressants, birth control drugs, and other medications have been detected in fish tissue and were identified as the cause of neurological, biochemical, and physiological changes [100, 101]. Because pharmaceuticals are designed to target specific metabolic and molecular pathways in humans and animals, it is assumed that they may affect the same pathways in animals with identical or similar target organs, tissues, cells or biomolecules. Certain receptors in lower vertebrates resemble those in humans, while others are different or lacking; in these cases, dissimilar modes of actions may occur in the lower animals [102, 103].

http://dx.doi.org/10.5772/intechopen.70799

Acute toxicity studies typically show that the concentrations of PPCPs to produce effects such as death in half of the exposed organisms ($EC_{50}$) range of 25 to ≥500 mg/L; one particular example found that the chronic toxicity or median lethal dose ($LC_{50}$) of furazolidone, which is largely used in medicated fish feed, at 40 mg/kg in the mosquito larvae, *Culex pipiens* [31]. In a study that tested the effects of tylosin and oxytetracycline on three species of soil fauna, neither of the substances had any effect at environmentally relevant concentrations; however, as soil ecosystems are built up by complex and linked food webs, the study concluded that it is not yet possible to exclude that indirect effects on soil fauna driven by changes in the microbial community and alteration of the decomposer system may occur [38].

Since antibiotics are specifically designed to control bacteria in plants and animals of economic interest, this obviously makes them hazardous to bacteria and other micro-organisms in the environment. There is growing concern that low level concentrations of antibiotics in the environment contribute to the emergence of strains of disease-causing bacteria that are resistant to even high doses of these drugs [23]. Current evidence supports that feeding low doses of antibiotics to livestock in an attempt to improve production efficiency has produced resistant strains of certain microorganisms. Bacterial strains evolve and become resistant to multiple antibiotics if they are continually exposed to low doses of antibiotics in the environment since the three mechanisms of gene transfer – conjugation, transduction, and transformation – all occur in the aquatic environment [104].

Streams and rivers that receive low levels of chronic antibiotic exposure can be viewed as a source and a reservoir of resistant genes as well as a means for their dispersion. In addition, if non-target organisms, such as cyanobacteria, are over-exposed to antibiotics, they may be negatively affected, which will disturb the aquatic food chain [6]. Increased bacterial resistance has been seen in waste effluent from hospitals and pharmaceutical plants indicating that the ultimate disposal of antibiotics may be a serious public health issue [23, 29]. Furthermore, individual compounds may interact synergistically or antagonistically with other chemicals present in the environment [6, 15, 16].

## 4. Solutions

The production and usage of most pharmaceutical and personal care products will either stabilize or increase. It is probable that the environmental load of these chemicals will follow the same trend. Although remedying this issue seems unfeasible, it cannot be regarded as a terminal quandary. Instead, tactics should be implemented to minimize their impact on the environment.

There are four major factors that determine the concentrations of drug residues reported in environmental samples: (1) frequency of use, (2) excretion of un-metabolized drugs, (3) persistence on biodegradation, and (4) the analytical method used [105]. Due to the consequential concern resulting from the detection of PPCPs in the aquatic environment, sensitive analytical techniques have been developed to investigate this new class of environmental pollutants; techniques that will have to continue to evolve in order to improve method accuracy and sensitivity [105]. Likewise, methodology must be designed to analyze compounds in combination [93].

Perhaps reform should begin with production of PPCPs, specifically pharmaceuticals. Medicinal drugs are intended to be metabolized by organisms, yet, approximately 20% or more of these compounds are excreted in their parent form or as metabolites [26, 105]. After excretion, these compounds could possibly mix with other chemicals already present in the environment or biodegradation and transformation may occur: circumstances which could produce other metabolites or by-products, conceivably leading to a substance that may be far more toxic than the parent compounds [105]. Production of pharmaceuticals that are fully absorbed or completely metabolized by the organism would be ideal; this, however, may be impractical. The responsibility then shifts from the pharmaceutical industry to the medical industry. By purposefully managing prescriptions with deep scrutiny, doctors may begin to begin to alleviate the issue through reduction of input [26].

Effective regulation of PPCPs is implausible without a global colloquy giving great consideration to the creation and installation of a well-developed, universal ERA procedure for these contaminants. Existing protocols must be expanded to adapt to the gravity of the potential impacts of these unique compounds in the environment. Implementation of a retrospective aspect to the protocol may also be necessary in the near future [93].

## Author details

Zakiya Hoyett

Address all correspondence to: zakiya.hoyett@gmail.com

College of Science and Technology, Florida Agricultural and Mechanical University, Tallahassee, FL, USA

## References

[1] Daughton CG. Pharmaceuticals and personal care products (PPCPs) as environmental pollutants: Pollution from personal actions. In: California Bay-Delta Authority Contaminant Stressor Workshop; Sacramento, CA; 2004

[2] Daughton CG. Chemicals from the practice of healthcare: Challenges and unknowns posed by residues in the environment. Environmental Toxicology and Chemistry. 2009;**28**(12):2490-2494

[3] Daughton CG. Pharmaceuticals and personal care products in the environment: Overarching issues and overview. In: Pharmaceuticals and Care Products in the Environment. Washington, DC: American Chemical Society; 2001. p. 2-38

[4] Clean Water Act of 1972. 33 U.S.C. § 1251 et seq. 2002

[5] European Union. EUR-Lex [Internet]. [Updated: 2017]. Available from: http://eur-lex.europa.eu/oj/direct-access.html [Accessed: 19 July 2017]

[6] Jones OAH, Voulvoulis N, Lester JN. Potential impact of pharmaceuticals on environmental health. Bulletin of the World Health Organization. 2003;**81**(10):768-769

[7] Ellis JB. Pharmaceutical and personal care products (PPCPs) in urban waters. Environmental Pollution. 2006;**144**:184-189

[8] Lee GF, Jones-Lee A. Unrecognized environmental pollutants. In: Water Encyclopedia: Surface and Agricultural Water. Hoboken, NJ: Wiley; 2005. p. 371-373

[9] Harada A, Komori K, Nakada N, Kitamura K, Suzuki Y. Biological effects of PPCPs on aquatic lives and evaluation of river waters affected by different wastewater treatment levels. Water Science and Technology. 2008;**58**(8):1541-1546

[10] Ferrari B, Mons R, Vollat B, Fraysse B, Paxēaus N, Giudice RL, et al. Environmental risk assessment of six human pharmaceuticals: Are the current environmental risk assessment procedures sufficient for the protection of the aquatic environment? Environmental Toxicology and Chemistry. 2004;**23**(5):1344-1354

[11] Van der Oost R, Beyer J, Vermeulen NPE. Fish bioaccumulation and biomarkers in environmental risk assessment: A review. Environmental Toxicology and Pharmacology. 2003;**13**(2):57-149

[12] Chemical Abstracts Service. CAS Content. Retrieved from CAS | A Division of the American Chemical Society; 2017, July 17. http://www.cas.org/content

[13] Hooper K. Lessons from the PBDEs—The value of monitoring community body burdens using breast milk. In: 6th Biennial State of the Estuary Conference; Oakland, CA; 2003

[14] Muir DCG, Howard PH. Are there other persistent organic pollutants? A challenge for environmental chemists. Environmental Science & Technology. 2006;**40**(23):7157-7166

[15] Jones OAH, Voulvoulis N, Lester JN. Human pharmaceuticals in the aquatic environment a review. Environmental Technology. 2001;**22**(12):1383-1394

[16] Hansen H, De Rosa CT, Pohl H, Fay M, Mumtaz MM. Public health challenges posed by chemical mixtures. Environmental Health Perspectives. 1998;**106**(Suppl 6):1271-1280

[17] Levy SB. The challenge of antibiotic resistance. Scientific American. 1998;**278**(3):32-39

[18] Nilsen EB, Rosenbauer RR, Furlong ET, Burkhardt MR, Werner SL, Greaser L, et al. Pharmaceuticals, Personal Care Products and Anthropogenic Waste Indicators Detected in Streambed Sediments of the Lower Columbia River and Selected Tributaries. National Ground Water Association. 2007; Paper 4483, p. 15

[19] Breton R, Boxall A. Pharmaceuticals and personal care products in the environment: Regulatory drivers and research needs. QSAR & Combinatorial Science. 2003;**22**(3): 399-409

[20] National Environmental Policy Act of 1969. 42 U.S.C. § 4332 et seq. 1994

[21] European Commission (EC), Evaluation Road map, available at: http://ec.europa.eu/smart-regulation/roadmaps/docs/2016_move_059_ev_accident_investigation_en.pdf

[22] Díaz-Cruz MS. Environmental behavior and analysis of veterinary and human drugs in soils, sediments and sludge. TrAC Trends in Analytical Chemistry. 2003;**22**(6):340-351

[23] Yang S, Carlson K. Routine monitoring of antibiotics in water and wastewater with a radioimmunoassay technique. Water Research. 2004;**38**(14-15):3155-3166

[24] Brown KD, Kulis J, Thomson B, Chapman TH, Mawhinney DB. Occurrence of antibiotics in hospital, residential, and dairy effluent, municipal wastewater, and the Rio Grande in New Mexico. Science of the Total Environment. 2006;**366**(2-3):772-783

[25] Deblonde T, Cossu-Leguille C, Hartemann P. Emerging pollutants in wastewater: A review of the literature. International Journal of Hygiene and Environmental Health. 2001;**214**:442-448

[26] Kümmerer K, Al-Ahmad A. Estimation of the cancer risk in humans resulting from the presence of cyclophosphamide and ifosfamide in surface water. Environmental Science and Pollution Research. 2010;**17**:486-496

[27] Monteiro SC, Boxall ABA. Occurrence and fate of human pharmaceuticals in the environment. In: Reviews of Environmental Contamination and Toxicology. New York, NY: Springer New York; 2010

[28] Liu J, Wong M. Pharmaceuticals and personal care products (PPCPs): A review on environmental contamination in China. Environment International. 2013;**59**:208-224

[29] Guardabassi L, Petersen A, Olsen JE, Dalsgaard A. Antibiotic resistance in *Acinetobacter* spp. isolated from sewers receiving waste effluent from a hospital and a pharmaceutical plant. Applied and Environmental Microbiology. 1998;**64**(9):3499-3502

[30] Hartmann A, Alder AC, Koller T, Widmer RM. Identification of fluoroquinolone antibiotics as the main source of umuC genotoxicity in native hospital wastewater. Environmental Toxicology and Chemistry. 1998;**17**(3):377-382

[31] Halling-Sørensen B, Nielsen SN, Lanzky PF, Ingerslev F, Lützhøft HCH, Jørgensen SE. Occurrence, fate and effects of pharmaceutical substances in the environment—A review. Chemosphere. 1998;**36**(2):357-393

[32] Hirsch R, Ternes T, Haberer K, Kratz KL. Occurrence of antibiotics in the aquatic environment. The Science of the Total Environment. 1999;**225**(1-2):109-118

[33] Bound JP, Voulvoulis N. Household disposal of pharmaceuticals as a pathway for aquatic contamination in the United Kingdom. Environmental Health Perspectives. 2005;**113**(12):1705

[34] Campagnolo ER, Johnson KR, Karpati A, Rubin CS, Kolpin DW, Meyer MT, et al. Antimicrobial residues in animal waste and water resources proximal to large-scale swine and poultry feeding operations. The Science of the Total Environment. 2002;**299**(1-3): 89-95

[35] Pepper IL, Gerba CP, Brusseau ML, editors. Pollution Science. San Diego, CA: Academic Press; 1996

[36] Howard P, Muir DCG. Identifying new persistent and bioaccumulative organics among chemicals in commerce II: Pharmaceuticals. Environmental Science & Technology. 2011;**45**:6938-6946

[37] Gibson GG, Skett P. Introduction to Drug Metabolism. Cheltenham, Great Britain: Nelson Thornes; 2001

[38] Baguer AJ, Jensen J, Krogh PH. Effects of the antibiotics oxytetracycline and tylosin on soil fauna. Chemosphere. 2000;**40**(7):751-757

[39] Kim SC, Davis JG, Truman CC, Ascough JC II, Carlson K. Simulated rainfall study for transport of veterinary antibiotics-mass balance analysis. Journal of Hazardous Materials. 2010;**175**(1-3):836-843

[40] Kolpin DW, Furlong ET, Meyer MT, Thurman EM, Zaugg SD, Barber LB, et al. Pharmaceuticals, hormones, and other organic wastewater contaminants in US streams, 1999-2000: A national reconnaissance. Environmental Science & Technology. 2002;**36**(6): 1202-1211

[41] Ternes TA. Occurrence of drugs in German sewage treatment plants and rivers. Water Research. 1998;**32**(11):3245-3260

[42] Wiegel S, Aulinger A, Brockmeyer R, Harms H, Löffler J, Reincke H, et al. Pharmaceuticals in the river Elbe and its tributaries. Chemosphere. 2004;**57**(2):107-126

[43] Sacher F, Lange FT, Brauch H-J, Blankenhorn I. Pharmaceuticals in groundwaters: Analytical methods and results of a monitoring program in Baden-Württemberg, Germany. Journal of Chromatography A. 2001;**938**(1-2):199-210

[44] Brown AFM, Dortch Q, Van Dolah FM, Leighfield TA, Morrison W, Thessen AE, et al. Effect of salinity on the distribution, growth, and toxicity of *Karenia* spp. Harmful Algae. 2006;**5**(2):199-212

[45] Benotti MJ, Trenholm RA, Vanderford BJ, Holady JC, Stanford BD, Snyder SA. Pharmaceuticals and endocrine disrupting compounds in U.S. drinking water. Environmental Science & Technology. 2008;**43**(3):597-603

[46] Stackelberg PE, Furlong ET, Meyer MT, Zaugg SD, Henderson AK, Reissman DB. Persistence of pharmaceutical compounds and other organic wastewater contaminants in a conventional drinking-water-treatment plant. Science of the Total Environment. 2004;**329**(1-3):99-113

[47] Watkinson AJ, Murby EJ, Kolpin DW, Costanzo SD. The occurrence of antibiotics in an urban watershed: From wastewater to drinking water. Science of the Total Environment. 2009;**407**(8):2711-2723

[48] Kümmerer K, Henninger A. Promoting resistance by the emission of antibiotics from hospitals and households into effluent. Clinical Microbiology and Infection. 2003;**9**(12):1203-1214

[49] Costanzo SD, Murby J, Bates J. Ecosystem response to antibiotics entering the aquatic environment. Marine Pollution Bulletin. 2005;**51**(1-4):218-223

[50] Kim SC, Carlson K. Quantification of human and veterinary antibiotics in water and sediment using SPE/LC/MS/MS. Analytical and Bioanalytical Chemistry. 2007;**387**(4): 1301-1315

[51] Thomas PM, Foster GD. Tracking acidic pharmaceuticals, caffeine, and triclosan through the wastewater treatment process. Environmental Toxicology and Chemistry. 2005;**24**(1):25-30

[52] Behera SK, Kim HW, Oh JE, Park HS. Occurrence and removal of antibiotics, hormones and several other pharmaceuticals in wastewater treatment plants of the largest industrial city of Korea. Science of the Total Environment. 2011;**409**(20):4351-4360

[53] Nakada N, Tanishima T, Shinohara H, Kiri K, Takadaet H. Pharmaceutical chemicals and endocrine disrupters in municipal wastewater in Tokyo and their removal during activated sludge treatment. Water Research. 2006;**40**(17):3297-3303

[54] Zhao JL, Ying GG, Liu YS, Chen F, Yang JF, Wang L, et al. Occurrence and a screening-level risk assessment of human pharmaceuticals in the Pearl River system, South China. Environmental Toxicology and Chemistry. 2010;**29**(6):1377-1384

[55] Zhou HD, Wu CY, Huang X, Gao MJ, Wen XH, Tsuno H, et al. Occurrence of selected pharmaceuticals and caffeine in sewage treatment plants and receiving rivers in Beijing, China. Water Environment Research. 2010;**82**(11):2239-2248

[56] Sui Q, Huang J, Deng SB, Yu G, Fan Q. Occurrence and removal of pharmaceuticals, caffeine and DEET in wastewater treatment plants of Beijing, China. Water Research. 2010;**44**(2):417-426

[57] Sui Q, Huang J, Deng SB, Chen WW, Yu G. Seasonal variation in the occurrence and removal of pharmaceuticals and personal care products in different biological wastewater treatment processes. Environmental Science & Technology. 2011;**45**(8):3341-3348

[58] Vieno NM, Tuhkanen T, Kronberg L. Analysis of neutral and basic pharmaceuticals in sewage treatment plants and in recipient rivers using solid phase extraction and liquid chromatography-tandem mass spectrometry detection. Journal of Chromatography A. 2006;**1134**(1-2):101-111

[59] Weigel S, Berger U, Jensen E, Kallenborn R, Thoresen H, Hühnerfuss H. Determination of selected pharmaceuticals and caffeine in sewage and seawater from Tromsø/Norway with emphasis on ibuprofen and its metabolites. Chemosphere. 2004;**56**(6):583-592

[60] Salgado R, Noronha JP, Oehmen A, Carvalho G, Reis MAM. Analysis of 65 pharmaceuticals and personal care products in 5 wastewater treatment plants in Portugal using a simplified analytical methodology. Water Science & Technology. 2010;**62**(12): 2862-2871

[61] Kasprzyk-Hordern B, Dinsdale RM, Guwy AJ. The removal of pharmaceuticals, personal care products, endocrine disruptors and illicit drugs during wastewater treatment and its impact on the quality of receiving waters. Water Research. 2009;**43**(2):363-380

[62] Yang SW, Carlson K. Evolution of antibiotic occurrence in a river through pristine, urban and agricultural landscapes. Water Research. 2003;**37**(19):4645-4656

[63] Peng XZ, Wang ZD, Kuang WX, Tan JH, Li K. A preliminary study on the occurrence and behavior of sulfonamides, ofloxacin and chloramphenicol antimicrobials in wastewaters of two sewage treatment plants in Guangzhou, China. Science of the Total Environment. 2006;**371**(1-3):314-322

[64] Peng XZ, Tan JH, Tang CM, Yu YY, Wang ZD. Multiresidue determination of fluoroquinolone, sulfonamide, trimethoprim, and chloramphenicol antibiotics in urban waters in China. Environmental Toxicology and Chemistry. 2008;**27**(1):73-79

[65] Minh TB, Leung HW, Loi IH, Chan WH, So MK, Mao JQ, et al. Antibiotics in the Hong Kong metropolitan area: Ubiquitous distribution and fate in Victoria Harbour. Marine Pollution Bulletin. 2009;**58**(7):1052-1062

[66] Li B, Zhang T. Mass flows and removal of antibiotics in two municipal wastewater treatment plants. Chemosphere. 2001;**83**(9):1284-1289

[67] Gulkowska A, Leung HW, So MK, Taniyasu S, Yamashita N, Yeung WYL, et al. Removal of antibiotics from wastewater by sewage treatment facilities in Hong Kong and Shenzhen, China. Water Research. 2008;**42**(1-2):395-403

[68] Tong CL, Zhuo XJ, Guo Y. Occurrence and risk assessment of four typical fluoroquinolone antibiotics in raw and treated sewage and in receiving waters in Hangzhou, China. Journal of Agricultural and Food Chemistry. 2001;**59**(13):7303-7309

[69] Gao LH, Shi YL, Li WH, Niu HY, Liu JM, Cai YQ. Occurrence of antibiotics in eight sewage treatment plants in Beijing, China. Chemosphere. 2012;**86**(6):665-671

[70] Chang H, Hu JY, Wang LZ, Shao B. Occurrence of sulfonamide antibiotics in sewage treatment plants. Chinese Science Bulletin. 2008;**53**(4):514-520

[71] Xiao Y, Chang H, Jia A, Hu JY. Trace analysis of quinolone and fluoroquinolone antibiotics from wastewaters by liquid chromatography-electrospray tandem mass spectrometry. Journal of Chromatography A. 2008;**1214**(1-2):100-108

[72] Zhao JL, Ying GG, Liu YS, Chen F, Yang JF, Wang L. Occurrence and risks of triclosan and triclocarban in the Pearl River system, South China: From source to the receiving environment. Journal of Hazardous Materials. 2010;**179**(1-3):215-222

[73] Vieno NM, Tuhkanen T, Kronberg L. Elimination of pharmaceuticals in sewage treatment plants in Finland. Water Research. 2007;**41**(5):1001-1012

[74] Lindberg RH, Wennberg P, Johansson MI, Tysklind M, Andersson BAV. Screening of human antibiotic substances and determination of weekly mass flows in five sewage treatment plants in Sweden. Environmental Science & Technology. 2005;**39**(10):3421-3429

[75] Servos MR, Bennie DT, Burnison BK, Jurkovic A, McInnis R, Neheli T, et al. Distribution of estrogens, 17beta-estradiol and estrone in Canadian municipal wastewater treatment plants. Science of the Total Environment. 2005;**336**(1-3):155-170

[76] Liu S, Ying GG, Zhao JL, Zhou LJ, Yang B, Chen ZF, et al. Occurrence and fate of androgens, estrogens, glucocorticoids and progestagens in two different types of municipal wastewater treatment plants. Journal of Environmental Monitoring. 2012;**14**(2):482-491

[77] Liu S, Ying GG, Zhao JL, Chen F, Yang B, Zhou LJ, et al. Trace analysis of 28 steroids in surface water, wastewater and sludge samples by rapid resolution liquid chromatography-electrospray ionization tandem mass spectrometry. Journal of Chromatography A. 2011;**1218**(10):1367-1378

[78] Jin SW, Yang FX, Liao T, Hui Y, Xu Y. Seasonal variations of estrogenic compounds and their estrogenicities in influent and effluent from a municipal sewage treatment plant in China. Environmental Toxicology and Chemistry. 2008;**27**(1):146-153

[79] Zhou YQ, Zha JM, Wang ZJ. Occurrence and fate of steroid estrogens in the largest wastewater treatment plant in Beijing, China. Environmenal Monitoring and Assessment. 2011;**184**(11):6799-6813

[80] Zhou HD, Huang X, Wang XL, Zhi XH, Yang CD, Wen XH, et al. Behaviour of selected endocrine-disrupting chemicals in three sewage treatment plants of Beijing, China. Environmenal Monitoring and Assessment. 2009;**161**(1-4):107-121

[81] Chang H, Wan Y, Wu SM, Fan ZL, Hu JY. Occurrence of androgens and progestogens in wastewater treatment plants and receiving river waters: Comparison to estrogens. Water Research. 2011;**45**(2):732-740

[82] Chang H, Hu JY, Shao B. Occurrence of natural and synthetic glucocorticoids in sewage treatment plants and receiving river waters. Environmental Science & Technology. 2007;**41**(10):3462-3468

[83] Wang YQ, Hu W, Cao ZH, Fu XQ, Zhu T. Occurrence of endocrine-disrupting compounds in reclaimed water from Tianjin, China. Analytical and Bioanalytical Chemistry. 2005;**383**(5):857-863

[84] Zhou XF, Dai CM, Zhang YL, Surampalli RY, Zhang TC. A preliminary study on the occurrence and behavior of carbamazepine (CBZ) in aquatic environment of Yangtze River Delta, China. Environmenal Monitoring and Assessment. 2011;**173**(1-4):45-53

[85] Reiner JL, Berset JD, Kannan K. Mass flow of polycyclic musks in two wastewater treatment plants. Archives of Environmental Contamination and Toxicology. 2007;**52**(4):451-457

[86] Zeng XY, Sheng GY, Gui HY, Chen DH, Shao WL, Fu JM. Preliminary study on the occurrence and distribution of polycyclic musks in a wastewater treatment plant in Guangdong, China. Chemosphere. 2007;**69**(8):1305-1311

[87] Zhang XL, Yao Y, Zeng XY, Qian GR, Guo YW, Wu MH, et al. Synthetic musks in the aquatic environment and personal care products in Shanghai, China. Chemosphere. 2008;**72**(10):1553-1558

[88] Zhou HD, Huang X, Gao MJ, Wang XL, Wen XH. Distribution and elimination of polycyclic musks in three sewage treatment plants of Beijing, China. Journal of Environemntal Sciences. 2009;**21**(5):561-567

[89] Nakata H, Shinohara R. Concentration of benzotriazole UV stabilizers and polycyclic musks in wastewater treatment plant samples in Japan. In: Interdisciplinary Studies on Environmental Chemistry—Environmental Specimen Bank. Setagaya, Tokyo: Terrapub; 2010. p. 51-59

[90] Li WH, Ma YM, Guo CS, Hu W, Liu KM, Wang YQ, et al. Occurrence and behavior of four of the most used sunscreen UV filters in a wastewater reclamation plant. Water Research. 2007;**41**(15):3506-3512

[91] Weigel S, Bester K, Hühnerfuss H. New method for rapid solid-phase extraction of large-volume water samples and its application to non-target screening of North Sea water for organic contaminants by gas chromatography-mass spectrometry. Journal of Chromatography A. 2001;**912**(1):151-161

[92] Yang Y, Fu J, Peng H, Hou L, Liu M, Zhou JL. Occurrence and phase distribution of selected pharmaceuticals in the Yangtze Estuary and its coastal zone. Journal of Hazardous Materials. 2011;**190**(1):588-596

[93] Hoyett Z, Owens MA, Clark CJ II, Abazinge A. A comparative evaluation of environmental risk assessment strategies for pharmaceuticals and personal care products. Ocean & Coastal Management. 2016;**127**:74-80

[94] Benotti MJ, Brownawell BJ. Distributions of pharmaceuticals in an urban estuary during both dry- and wet-weather conditions. Environmental Science & Technology. 2007;**41**(16):5795-5802

[95] Vidal-Dorsch DE, Bay SM, Maruya K, Snyder SA, Trenholm RA, Vanderford BJ. Contaminants of emerging concern in municipal wastewater effluents and marine receiving water. Environmental Toxicology and Chemistry. 2012;**31**(12):1-9

[96] United States Environmental Protection Agency. Methods for Measuring the Acute Toxicity of Effluents and Receiving Waters to Freshwater and Marine Organisms. Washington, DC: EPA/600/4-90/027F; 1991

[97] United States Environmental Protection Agency. Ecological Effects Test Guidelines; OPPTS 850.1025 Oyster Acute Toxicity Test (Shell Deposition). Washington, DC: EPA 712-C-96-115; 1996

[98] Suter GW. Ecological Risk Assessment. Boca Raton, FL: CRC Press; 2007

[99] Comprehensive Environmental Response Compensation and Liability Act of 1980. 42 U.S.C. § 9601 et seq. (1986)

[100] Daughton CG, Ternes TA. Pharmaceuticals and personal care products in the environment: Agents of subtle change? Environmental Health Perspectives. 1999;**107**(Suppl 3):907-938

[101] Krock B, Pitcher GC, Ntuli J, Cembella AD. Confirmed identification of gymnodimine in oysters from the west coast of South Africa by liquid chromatography-tandem mass spectrometry. African Journal of Marine Science. 2009;**31**(1):113-118

[102] Sumpter JP, Johnson AC. Lessons from endocrine disruption and their application to other issues concerning trace organics in the aquatic environment. Environmental Science & Technology. 2005;**39**(12):4321-4332

[103] Fent K, Weston AA, Caminada D. Ecotoxicology of human pharmaceuticals. Aquatic Toxicology. 2006;**76**(2):122-159

[104] Migliore L, Brambilla G, Cozzolino S, Gaudio L. Effect on plants of sulphadimethoxine used in intensive farming (*Panicum miliaceum, Pisum sativum* and *Zea mays*). Agriculture, Ecosystems & Environment. 1995;**52**:103-110

[105] Kostopoulou M, Nikolaou A. Analytical problems and the need for sample preparation in the determination of pharmaceuticals and their metabolites in aqueous environmental matrices. Trends in Analytical Chemsitry. 2008;**27**(11):1023-1035

Chapter 17

# Estimation of PM2.5 Trajectory Using Atmospheric Dispersion Models and GIS in the Tokyo Metropolitan Area

Kayoko Yamamoto and Zhaoxin Yang

Additional information is available at the end of the chapter

http://dx.doi.org/10.5772/intechopen.70608

**Abstract**

The present study aims to use atmospheric dispersion models and geographical information system (GIS) to make estimations of the trajectory of PM2.5 (particulate matter) discharged from specific generation sources, by grasping the atmospheric concentration within the Tokyo metropolitan area in Japan. It is expected that such estimation results should contribute to the risk assessment concerning the influences of PM2.5 on human health and ecosystem. Using ADMER in the first stage, estimations of the atmospheric concentration distribution of PM2.5 throughout the entire Tokyo metropolitan area from 2009 to 2014 were conducted. As a result, areas with high atmospheric concentration of PM2.5 focused in the same area each year, and it was revealed that the entire Tokyo and Saitama had high atmospheric concentrations. Additionally, as a result of setting Tokyo the detail estimation range, it was grasped that the atmospheric concentrations are high in Shinjuku ward and Tachikawa city in Tokyo. Based on the results in the first stage, using METI-LIS in the second stage, estimations of the trajectory of PM2.5 discharged from specific generation sources were conducted in Tachikawa city. As a result, it was made clear that PM2.5 had spread within 500 m of the specific generation sources, and the atmospheric concentrations were intensively high.

**Keywords:** PM2.5 (particulate matter), estimation of trajectory, risk assessment, air pollution, atmospheric dispersion model, geographic information systems (GIS)

## 1. Introduction

Humans produce a variety of waste due to production and consumption activities. If the amount of such waste is not so large, it can be processed through the natural purification effect. However, in addition to waste disposal increasing beyond the natural processing capacity, new types of waste materials that cannot be naturally processed are also being produced.

Such pollution of the natural environment has progressed, and this has led to the disruption of ecosystem, damage to human health, as well as various pollution issues including air pollution. Worldwide, especially in Asian countries which have achieved rapid industrialization, the amount of air pollutant discharged has rapidly increased along with the expanding scale of economic activities. Though air pollutants can be lessened by exchanging fuel used by main generation sources such as automobiles and plants with fuel that is less of a burden on the environment, there have been few reduction measures. It is necessary to accurately grasp the actual condition of air pollution and take appropriate measures to handle this issue.

Based on the background mentioned above, the present study aims to use atmospheric dispersion models and geographical information system (GIS) to make estimations of the trajectory of air pollutants discharged from specific generation sources by grasping the atmospheric concentration within the Tokyo metropolitan area in Japan. As a target air pollutant for estimations in the present study, PM2.5 (particulate matter), which has been a serious concern to human health, will be discussed. The estimations in the present study will be conducted using two types of atmospheric dispersion models in two stages. In the first stage, wide-range and long-term estimations will be conducted in the entire estimation target area. By means of the estimation results, the areas with high atmospheric concentration of PM2.5 will be selected, its generation source will be investigated, and the PM2.5 trajectory will be estimated. In the second stage, by means of the estimation results of the first stage, detailed estimations in smaller areas surrounding specific generation sources with high atmospheric concentration of PM2.5 will be conducted, and its trajectory will be estimated in detail. Based on such estimation results, the information concerning the measures to reduce PM2.5 that is more effective than before can be provided, and the estimation methods of PM2.5 trajectory proposed in the present study can be used for other air pollutants as well as in other areas. Additionally, it is expected that such estimation results should contribute to the risk assessment concerning the influences of PM2.5 on human health and ecosystem.

## 2. Related work

Regarding studies that grasped the behavior of air pollutants using atmospheric dispersion model, there have been many with dioxin as its subject. Some of the representative studies in recent years include that of Sasaki et al. [1], Teshima et al. [2], Hoa [3], Viel [4], Ripamonti et al. [5], Ashworth et al. [6], Ishii and Yamamoto [7], Sun et al. [8], and Zhang et al. [9], in which simulations of the behavior of dioxins in the atmosphere were conducted with incinerators as its generation source. Maantay et al. [10], Chen et al. [11], Kawashima et al. [12], Onofrio et al. [13], Zhou et al. [14], and Chandra et al. [15] conducted simulations of dioxins in the atmosphere. Additionally, Armitage et al. [16], Huang and Liang [17], and Zhou [18] conducted simulations of the behavior of dioxins underwater in places including the sea, canals, and lakes.

Among the related studies above, Maantay et al. [10] and Viel [4] demonstrated the effectiveness to combine atmospheric dispersion model and GIS to estimate the behavior of dioxins in the atmosphere. Additionally, Ishii et al. [7] combined two types of atmospheric dispersion models and GIS to grasp the dispersion conditions of dioxins in both wide-rage and small-range areas with high concentration and proposed a method to evaluate environmental risks.

In a similar manner, representative studies in recent years with PM2.5 as its subject include that of Mueller et al. [19], Saide et al. [20], Chen et al. [21], Solazzo et al. [22], Lee et al. [23], Saraswat et al. [24], and Rizza et al. [25], in which simulations of the atmospheric behavior of PM2.5 were conducted. Simulations of the atmospheric behavior of PM2.5 were conducted with the generation source being traffic by Lang et al. [26], daily activities by Louge et al. [27], and incinerators by Kodros et al. [28]. However, studies on Japan are very rare, whereas most focus their target on China, North America, and Europe.

In contrast with the studies mentioned above, the present study will focus on PM2.5 discharged from specific generation sources that has been seldom targeted in Japan and demonstrate the originality by proposing detailed trajectory estimation methods using both two types of atmospheric dispersion models and GIS. Additionally, by means of two-stage estimations targeting wide areas and narrow areas with high atmospheric concentration, the atmospheric concentration distribution of PM2.5 can be accurately grasped. Moreover, using both atmospheric dispersion models and GIS for the two-stage estimation method in the present study, the effectiveness is demonstrated by quantitatively and spatially grasping the dispersion conditions of PM2.5 discharged from specific generation sources. More specifically, the areas with high atmospheric concentration will be extracted by estimating the atmospheric concentration distribution of PM2.5 throughout the entire estimation target area using atmospheric dispersion models. Furthermore, in these areas, reflecting the land use by means of GIS, the trajectory of PM2.5 discharged from specific generation sources will be estimated in detail.

## 3. Estimation method

### 3.1. Overview of atmospheric dispersion method and GIS

With the present study, since estimations of PM2.5 trajectory are made in two stages, two types of atmospheric dispersion models will be used. In the wide-area estimations involving the entire estimation target area of the first stage, the explosion risk evaluation atmospheric dispersion model (AIST-ADMER Ver.3) by the National Institute of Advanced Industrial Science and Technology [29–31] is used. This is an atmospheric dispersion model suitable for estimating wide-ranged and long-term atmospheric concentration distribution of chemicals according to the amount of PM2.5 discharged from generation sources as well as meteorological conditions.

In the localized and detailed estimations with selected specific generation sources as the target in the second stage, the low-rise industrial source dispersion model (METI-LIS Ver.3.2.1) by the Ministry of Economy, Trade, and Industry [32–34] is used. This model estimates the atmospheric concentration of chemicals surrounding specific generation sources. Additionally, this model takes into consideration the downwash that occurs when the air current is disturbed due to buildings surrounding the specific generation sources. By entering the data on height of the buildings near the specific generation sources, the model takes into consideration the buildings' influence on chemical dispersion, and estimations of small-ranged and detailed atmospheric concentration dispersions can be made.

Additionally, the ArcGIS Ver.10.2 of ESRI will be used as GIS. Upon the estimations of PM2.5 trajectory, using GIS, an overlay analysis with the estimation results which are obtained from two types of atmospheric dispersion models and digital map data, as well as statistical processing will be conducted.

### 3.2. Overview of estimation method

The flow of the estimation method in the present study is as shown in **Figure 1**, and the details will be explained below.

1. In the present study, the estimation target area is selected in the beginning, the data concerning PM2.5 discharged from generation sources is gathered and processed, and the generation source data is prepared.
2. The above data concerning PM2.5 discharged from generation sources and meteorological data are entered into the atmospheric dispersion model, ADMER in the first stage, and the atmospheric concentration of PM2.5 in the entire estimation target area is estimated.

   Based on the estimation results, the areas with high atmospheric concentration are selected, and detailed estimation range, where a detailed estimation of atmospheric concentrations is conducted, is set. By using ADMER in this way, the PM2.5 trajectory of the entire estimation target area can be grasped on a macro scale, and areas with high atmospheric concentration can be set as a detailed estimation range to be confirmed.
3. Based on the estimation results of the first stage, the estimation target area is selected for the second stage. The meteorological data and digital map data concerning the estimation target area of the second stage are gathered and processed, and entered into the atmospheric dispersion model, METI-LIS. In the second stage, as downwash and stack-tip downwash due to buildings are taken into consideration, the PM2.5 trajectory can be accurately estimated. Additionally, the trajectory of PM2.5 discharged especially from specific generation sources is estimated.

### 3.3. Selection of estimation target area

For the present study, the Tokyo metropolitan area (Ibaraki, Tochigi, Gunma, Kanagawa, Saitama, Chiba, and Tokyo) was selected as the target area for the estimations of PM2.5 trajectory. **Figure 2** shows the Tokyo metropolitan area. In the estimations of the entire target area using ADMER in the first stage, the entire Tokyo metropolitan area will be the target area. In the second stage using METI-LIS for estimations of the atmospheric concentration of chemicals surrounding specific generation sources, areas with high atmospheric concentration of PM2.5 will be selected from the entire Tokyo metropolitan area based on the estimation results of the previous stage. Furthermore, the cause of high atmospheric concentrations in those areas will be considered, and the trajectory of PM2.5 discharged from specific generation sources will be estimated in detail.

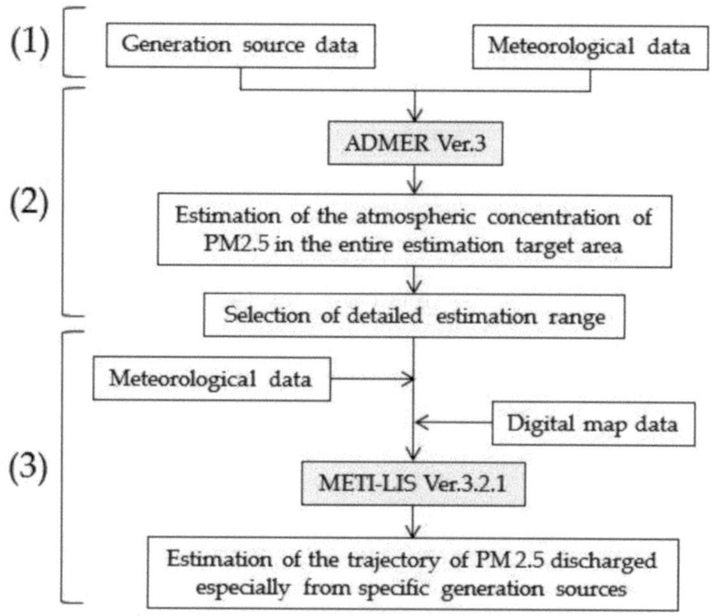


**Figure 1.** Flow of the estimation method.

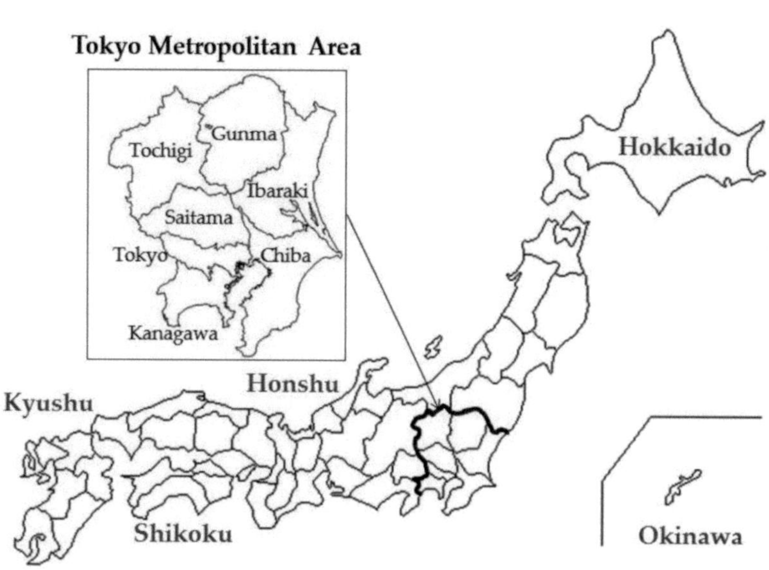


**Figure 2.** Tokyo metropolitan area as estimation target area.

# 4. Gathering and processing data

Data shown in **Table 1** are used in the present study. Generation source data and meteorological data will be entered into the atmospheric dispersion models, while measured data of

| Type | Name | Source |
|---|---|---|
| Generation source data | Data of the amount of discharged PM2.5 (2009–2014) | Ministry of Trade, Economy, and Industry |
| Meteorological data | AMeDas and rainfall data for ADMER (2009–2014) | National Institute of Advanced Industrial Science and Technology |
| | Tokyo weather data (2014) | Japan Meteorological Agency |
| Digital map data | Administrative division data (2012) | Ministry of Land, Infrastructure, Transport, and Tourism |
| | Base map information (scale level of 2500) | |
| Measured data of PM2.5 | Monthly and annual data of the atmospheric environment (2014) | National Institute of Environmental Studies |

**Table 1.** List of data used.

PM2.5 will be processed into GIS data and used for spatial analysis. For data of the amount of discharged PM2.5, the data concerning the amount of discharged chemicals announced on the basis of the Pollutant Release and Transfer Register Law (PRTR Law, enacted in 2001) will be used in the present study.

According to the PRTR Law, it is necessary for the businesses themselves to grasp the amount of hazardous chemicals discharged from businesses into the environment (atmosphere, water, and soil) and included in waste substances to be released outside business facilities, and report it to the national government. Additionally, based on the above reported data and statistics, it is essential for the national government to tally and announce the amount of chemicals discharged and transferred. As businesses with the responsibility of notification according to the PRTR Law are restricted by category of business and plant, employee scale and transaction volume, the amount of chemicals discharged from the generation sources exempt from the law are estimated and announced by the national government.

## 5. Estimation in the entire estimation target area

### 5.1. Estimation targets

With the entire Tokyo metropolitan area as the estimation target area in the first stage, the atmospheric concentrations of PM2.5 will be estimated using ADMER. By setting the estimation target area to 34° 50′ 00″ - 37° 12′ 30″ north latitude and 138° 18′ 45″ - 140° 56′ 15″ east longitude in accordance with the Tokyo metropolitan area, the grid will be set to 5*5 of the tertiary grid square (1 km grid) meaning a 5 km grid square units of area, and the grid number for the entire Tokyo metropolitan area will be 42*57 (2,394). Moreover, the estimation target period for the first stage is 6 years from 2009 to 2014. This is because a period with available data concerning the estimation target area was selected.

http://dx.doi.org/10.5772/intechopen.70608

## 5.2. Estimation results

### *5.2.1. Results for the entire estimation target area*

The atmospheric concentration distribution of PM2.5 was estimated using ADMER, and the estimation results for each year are shown in **Figure 3**. As shown in the figure, the areas with high

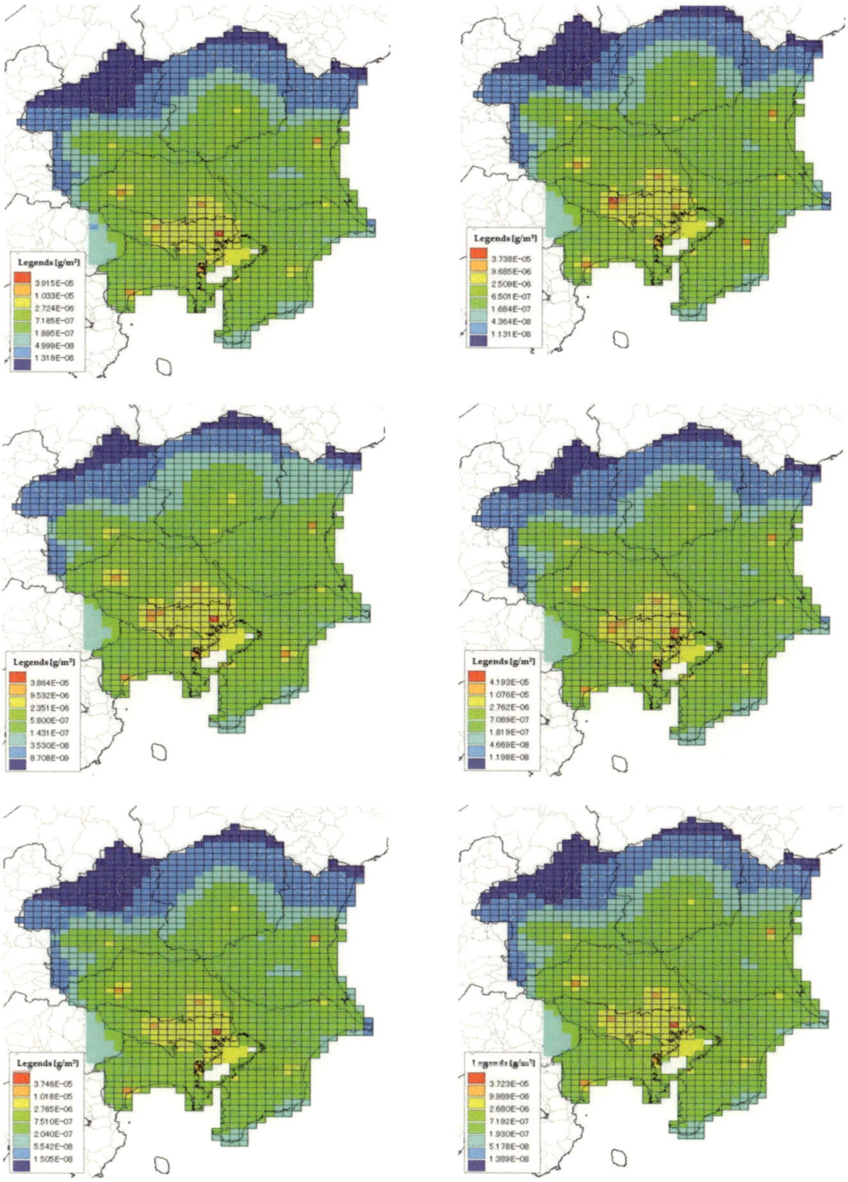


**Figure 3.** Atmospheric concentration distribution of PM2.5 in the Tokyo metropolitan area (2009–2014).

atmospheric concentration were focused each year, and the entire Tokyo and Saitama as well as some parts of Ibaraki, Kanagawa, and Chiba had high atmospheric concentrations of PM2.5. For Ibaraki, Kanagawa, and Chiba, an investigation of generation sources was conducted. As areas with high atmospheric concentration of PM2.5 are located in urban central parts, main generation sources are considered to be transportation including automobiles and railroad vehicles.

### *5.2.2. Estimation results for detailed estimation range*

Based on the estimation results in the previous section, because it was made clear that Tokyo and Saitama had high atmospheric concentrations of PM2.5, Tokyo was set as the detailed estimation range, and the atmospheric concentration distribution was estimated. One reason for this is that there are many measurement stations of PM2.5, and there is an abundance of data to aid the grasping of the atmospheric concentration distribution of PM2.5. The second reason is that the atmospheric concentration of PM2.5 in Tokyo has been decreasing every year making great improvements (55% reduction of atmospheric concentration in 10 years from 2001 to 2011). However, the rate for meeting environmental standards is low, and the atmospheric concentrations of PM2.5 in Tokyo are slightly above the environmental standards (annual average of below 15 μg/m$^3$).

After setting the detailed estimation range to 35° 27′ 30″ - 35° 55′ 0″ north latitude and 138° 52′ 30″ - 140° 0′ 0″ east longitude in accordance with Tokyo, a grid of 100 m*100 m is created. Then, the atmospheric concentration distribution of PM2.5 is estimated using ADMER, and the estimation results for each year are shown in **Figure 4**. From the figure, it is clear that atmospheric concentrations are high in Shinjuku ward and Tachikawa city each year.

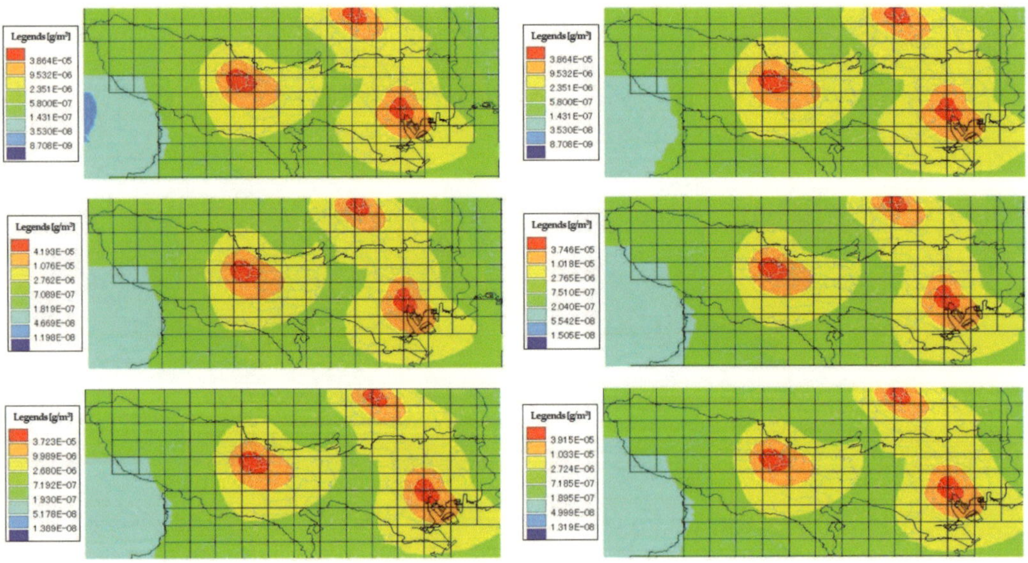


**Figure 4.** Atmospheric concentration distribution of PM2.5 in Tokyo (2009–2014).

### 5.3. Discussion

From the estimation results in this section, it is clear that the areas with high atmospheric concentration of PM2.5 focus in the same areas each year. Especially in the entire Tokyo and Saitama, as well as in certain parts of Ibaraki, Kanagawa, and Chiba, the atmospheric concentrations were high. Moreover, as shown in **Figure 5**, in order to verify the validity of the estimation results of atmospheric concentration of PM2.5 in the present study, the levels of estimated atmospheric concentration and measured atmospheric concentration for 2014 are compared. The measured atmospheric concentration is from the environment numerical database of the National Institute for Environmental Studies.

As shown in **Figure 5**, the estimation results of atmospheric concentrations of PM2.5 in the Tokyo metropolitan area showed excellent reproducibility. With the atmospheric dispersion model, the consistency reference for the ratio of estimated and measured atmospheric concentration is set to be around ½- to 2-fold. Though the estimated atmospheric concentration in Tokyo and Saitama was above the measured atmospheric concentration level, the former for Ibaraki, Tochigi, Gunma, Kanagawa, and Chiba was below the latter. For Tokyo and Saitama, the amounts of PM2.5 discharged from businesses and plants in addition to the amount from automobiles and railroad vehicles were also large. Though the amounts of PM2.5 discharged from these generation sources are fixed according to the categories in ADMER, it could be that the estimated atmospheric concentration is higher than the measured atmospheric concentration level, as there is a possibility of the actual amount of discharged PM2.5 being less. On the other hand, in Ibaraki, Tochigi, Gunma, Kanagawa, and Chiba, the main generation sources for PM2.5 are considered to be transportation such as automobiles and

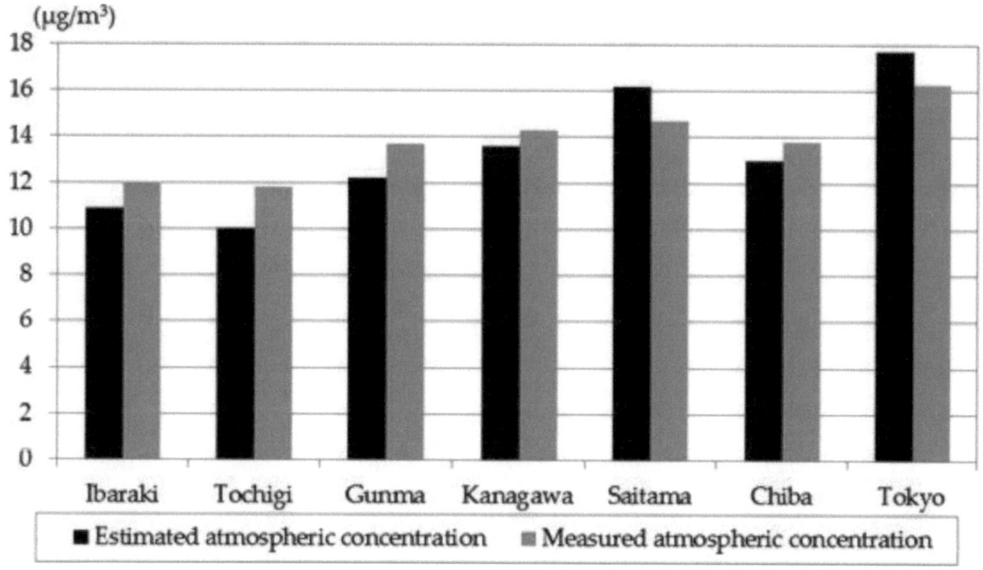


**Figure 5.** Comparison between the measured atmospheric concentration and estimated concentration in the present study in the Tokyo metropolitan area (2014).

railroad vehicles, as well as businesses and general households. Additionally, as the amounts of PM2.5 transported from other areas and naturally discharged into the environment are not taken into account, the estimated atmospheric concentration may have been lower than the measured atmospheric concentration level.

## 6. Estimations in detailed estimation target area

### 6.1. Estimation target

From the estimation results of ADMER in the previous section, it was made clear that the atmospheric concentrations of PM2.5 were especially high in Shinjuku ward and Tachikawa city. Shinjuku ward is made up of some of the most prominent busy streets in Tokyo, and there are many commuters due to the many train lines. The fuel combustion from automobiles and railroads vehicles is considered to be the main cause of the high atmospheric concentration of PM2.5 in Shinjuku ward. Additionally, as there are many high-rise buildings in Shinjuku ward and these prevent air circulation and the dispersion of PM2.5, it is thought that this results in long-term high atmospheric concentration of PM2.5. In Tachikawa city, there are many commercial facilities as well as offices that are clustered together, and there are many commuters due to the many train lines like Shinjuku ward. For this reason, the PM2.5 discharged from the combustion of fuel from automobiles and railroad vehicles is considered to be one of the causes for Tachikawa city being an area with high atmospheric concentration. Moreover, business facilities and plants are considered to be main generation sources of PM2.5.

From the reasons stated above, as the estimation target area of the second stage, Tachikawa city was selected. Because PM2.5 discharged not only from fuel combustion due to automobiles and railroad vehicles, but also from specific generation sources such as business facilities and plants. Accordingly, in order to estimate the trajectory of PM2.5 discharged from specific generation sources, two areas within Tachikawa city (area A and area B) were extracted. **Figure 6** shows the distribution of specific generation sources in the estimation targets of the second stage. 80 m grid square units is set for area A and 100 m grid square units is set for area B. Additionally, to make identifying building and road placements in the estimation target area easier, the outer peripheral lines and arterial roads downloaded from the basic map information were displayed. Moreover, the estimation target period of the second stage is 2014 in which the latest data can be obtained.

### 6.2. Estimation results

The estimation results of atmospheric concentration distribution of PM2.5 in area A and area B using METI-LIS are shown in **Figure 7**. The estimation results are shown in 80 m grid square units for area A and 100 m grid square units in area B. From the estimation results, it is clear that PM2.5 spreads within a range of about 500 m from the specific generation sources, and the atmospheric concentration distributions are higher. In such areas, a downwash occurs due to buildings, which in turn prevents PM2.5 from spreading by the wind.

**Figure 6.** Detailed estimation target area.

### 6.3. Discussion

In this section, the trajectory of PM2.5 discharged from specific generation sources was estimated. The measured atmospheric concentration levels show the total amount of PM2.5 discharged in the areas surrounding the measurement station. Therefore, as the measured atmospheric concentrations which indicate the amount of PM2.5 discharged from specific

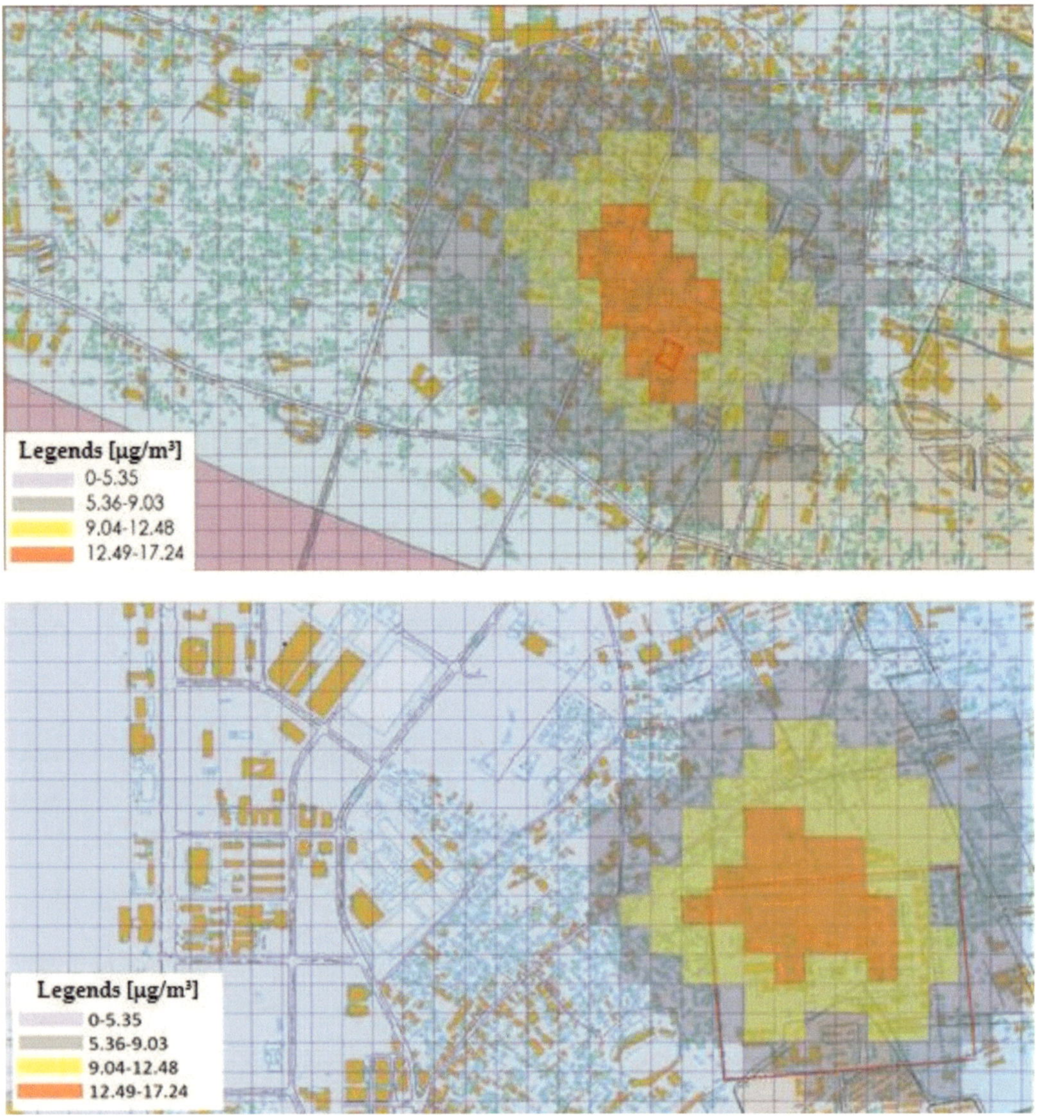


**Figure 7.** Atmospheric concentration distribution of PM2.5 in Tachikawa city (2014).

generation sources do not exist, the levels of estimated atmospheric concentration and measured atmospheric concentration cannot be compared. In the vicinity of estimation target area A, though there are many buildings, it is considered to have little effect on the dispersion of PM2.5 as most are low-rise buildings. Additionally, in estimation target area B, though there are few buildings within the premises of specific generation source, because there are many surrounding buildings, this causes a downwash in the dispersion of PM2.5 which may reduce the atmospheric concentration.

http://dx.doi.org/10.5772/intechopen.70608

## 7. Conclusion

The conclusion of the present study can be summarized into the following four points:

1. By means of the estimation method proposed in the study, estimations of the trajectory of air pollutants discharged from specific generation sources were conducted. PM2.5 was selected as a target air pollutant. Regarding its trajectory, a wide-range estimation for the entire estimation target area was conducted using ADMER in the first stage, and a narrow-range and detailed estimation using METI-LIS in areas surrounding specific generation sources was conducted in the second stage. By using two types of atmospheric dispersion models and GIS, it is possible to grasp the atmospheric concentration distribution of PM2.5 discharged from specific generation sources, and estimate its trajectory.

2. Using ADMER in the first stage, estimations of the atmospheric concentration distribution of PM2.5 throughout the entire Tokyo metropolitan area from 2009 to 2014 were conducted. As a result, areas with high atmospheric concentration focused in the same area each year, and it was revealed that the entire Tokyo and Saitama, as well as some parts of Ibaraki, Kanagawa, and Chiba had high atmospheric concentrations of PM2.5. Additionally, as a result of setting Tokyo the detailed estimation range, it was grasped that the atmospheric concentrations are high in Shinjuku ward and Tachikawa city in Tokyo.

3. Using METI-LIS in the second stage, estimations of the trajectory of PM2.5 discharged from specific generation sources were conducted in Tachikawa city, which was estimated to be a high atmospheric concentration area from the results in the first stage, with the estimation period being 2014 in which the latest data can be obtained. As a result, it was made clear that PM2.5 had spread within 500 m of the specific generation sources, and the atmospheric concentrations were intensively high. This verified the fact that a downwash occurs due to buildings and that this prevents the wind from spreading PM2.5.

4. If data concerning generation sources can be created, the estimation method proposed in the present study can be applied to any air pollutant besides PM2.5 as well as in other areas. Additionally, because the atmospheric concentration distribution of PM2.5 can be grasped using the estimation method of the present study, the information concerning the measures to reduce PM2.5 that is more effective than the past may be provided. It is expected that such estimation results should contribute to the risk assessment concerning the influences of PM2.5 on human health and ecosystem.

## Author details

Kayoko Yamamoto* and Zhaoxin Yang

*Address all correspondence to: k-yamamoto@is.uec.ac.jp

University of Electro-Communications, Tokyo, Japan

## References


[1] Sasaki H, IImura F, Tsukui T, Yoshioka H, Sasaki Y, Ando H, Kashiwagi N. Study on the composition of dioxins of environment in Tokyo. In: FY2004 Annual Report of the Tokyo Metropolitan Research Institute for Environmental Protection; 2004. p. 117-123

[2] Teshima T, Shibakawa S, Fujita Y, Matsumoto A, Takeda N, Takaoka M. Reduction of dioxin emissions by retrofitting a municipal solid waste incinerator. Journal of the Japan Society Waste Management Experts. 2006;**17**(4):281-292

[3] Hoa TT. Spatial distribution of dioxin plumes in the vicinity of an incinerator using air dispersion coupled geostatistical model [Ph.D. Dissertation in Environmental Engineering]. University of Michigan; Ann Arbor, Michigan State, US. 2009. p. 207

[4] Viel JF. GIS and atmospheric diffusion modelling for assessment of individual exposure to dioxins emitted from a municipal solid waste incinerator. In: Maantay JA, McLafferty S, editors. Geospatial Analysis of Environmental Health. Springer Verlag; Berlin and Heidelberg, Germany. 2011. pp. 443-456

[5] Ripamonti G, Lonati G, Baraldi P, Cadini F, Zio E. Uncertainty propagation in a model for the estimation of the ground level concentration of dioxin/furans emitted from a waste gasification plant. Reliability Engineering and System Safety. 2013;**111**(120):98-105

[6] Ashworth DC, Fuller GW, Toledano MB, Font A, Elliott P, Hansell AL, Hoogh K. Comparative assessment of particulate air pollution exposure from municipal solid waste incinerator emission. Journal of Environmental and Public Health. 2013. Article ID.560342: http://dx.doi.org/10.1155/2013/560342; [Assessed: July 20, 2017] p.13

[7] Ishii M, Yamamoto K. An environmental risk evaluation method employing atmospheric dispersion models and GIS. Journal of Environmental Protection. 2013;**4**(12):1392-1408

[8] Sun R, Ismail TM, Ren X, El-Salam A. Numerical simulation of gas concentration and dioxin formation for MSW combustion in a fixed bed. Journal of Environmental Management. 2015;**157**:111-117

[9] Zhang M, Buekens A, Li X. Brominated flame retardants and the formation of dioxins and furans in fires and combustion. Journal of Hazardous Materials. 2016;**304**:26-39

[10] Maantay JA, Tu J, Marokoa AR. Loose-coupling an air dispersion model and a geographic information system (GIS) for studying air pollution and asthma in the Bronx, New York City. International Journal of Environmental Health Research. 2009;**19**(1):59-79

[11] Chen Q, Sun J, Liu J. Toxicity prediction of dioxins and dioxin-like compounds based on the molecular fragments variable connectivity index. Environmental Contamination and Toxicology. 2011;**87**(2):134-137

[12] Kawashima M, Tobe T, Kaga A, Kondo A, Inoue Y, Matsumoto D, Dong Y. A multimedia model for the evaluation of environmental behavior of dioxins. In: Annual report of FY2004, The core university program between Japan Society for the Promotion of Science and Vietnamese Academy of Science and Technology; 2005. p. 59-64

[13] Onofrio M, Spataro R, Botta S. The role of a steel plant in north-west Italy to the local air concentrations of PCDD/FS. Chemosphere. 2011;**82**(5):708-717

[14] Zhou ZG, Zhao B, Qi L, Xu PJ, Ren Y, Li N, Zheng S, Zhao H, Fan S, Zhang T, Liu AM, Huang YR, Shen L. Distribution of polychlorinated dibenzo-p-dioxins and dibenzofurans in the atmosphere of Beijing, China. Aerosol and Air Quality Research. 2014;**14**:1269-1278

[15] Chandra SR, Lee WJ, Mutiara MPE, Mwangi JK, Wang LC, Lin NH, Chien GPC. Atmospheric deposition of polychlorinated dibenzo-p-dioxins and dibenzofurans at coastal and high mountain areas in Taiwan. Aerosol and Air Quality Research. 2015;**15**:1390-1411

[16] Armitage JM, MS ML, Wiberg K, Jonsson P. A model assessment of polychlorinated dibenzo-p-dioxin and dibenzofuran sources and fate in the Baltic Sea. Science of the Total Environment. 2009;**407**(12):3784-3792

[17] Huang CK, Liang J. Effects of basin topography and monsoon conditions on transport and occurrence of atmospheric PCDD/FS in the Taichung Basin. Environmental Science and Pollution Research. 2011;**18**(8):1305-1315

[18] Zhou F. Polychlorinated dibenzo-p-dioxins and polychlorinated dibenzofuran concentrations in Xingyun Lake sediment close to the pulp mill region of Jiangchuan (China): A typical case study. Water Science & Technology. 2014;**69**(1):69-75

[19] Mueller SF, Mallard JW. Contributions of natural emissions to Ozone and PM2.5 as simulated by the community multiscale air quality (CMAQ) model. Environmental Science & Technology. 2011;**45**:4817-4823

[20] Saide PE, Carmichael GR, Spak SN, Gallardo L, Osses AE, Mena-Carrasco MA, Pagowski M. Forecasting urban PM10 and PM2.5 pollution episodes in very stable nocturnal conditions and complex terrain using WRF-chem model CO tracer model. Atmospheric Environment. 2011;**45**:2769-2780

[21] LWA C, Lowenthal DH, Watson JG, Koracin D, Kumar N, Knipping EM, Wheeler N, Craig K, Reid S. Toward effective source apportionment using positive matrix factorization: Experiments with simulated PM2.5 data. Journal of the Air & Waste Management Association. 2017;**60**:43-54

[22] Solazzo E, Galmarini S, Bianconi R, Rao ST. Model evaluation for surface concentration of particulate matter in Europe and North America in the context of AQMEII. In: Steyn D, Builtjes P, Timmermans R, editors. Air Pollution Modelling and its Application XXII. NATO Science for Peace and Security Series C: Environmental Security. Springer; Berlin and Heidelberg, Germany. 2014. pp. 1-5

[23] Lee CL, Huang HC, Wangd CC, Sheuc CC, Wu CC, Leung SY, Lai RS, Lin CC, Weij YF, Lai IC, Jiang H, Chou WL, Chung WY, Huang MS, Huang SK. A new grid-scale model simulating the spatiotemporal distribution of PM2.5-PAHs for exposure assessment. Journal of Hazardous Materials. 2016;**314**:286-294

[24] Saraswat A, Kandlikar M, Brauer M. PM2.5 population exposure in New Delhi using a probabilistic simulation framework. Environmental Science & Technology. 2014;**50**:3174-3183

[25] Rizza U, Barnaba F, Miglietta MM, Mangia C, Liberto LD, Dionisi D, Costabile F, Grasso F, Gobbi GP. WRF-chem model simulations of a dust outbreak over the central Mediterranean and comparison with multi-sensor desert dust observations. Atmospheric Chemical and Physics. 2017;**17**:93-115

[26] Lang J, Cheng S, Li J, Chen D, Zhou Y, Wei X, Han L, Wang H. A monitoring and modelling study to investigate regional transport and characteristics of PM2.5 pollution. Aerosol and Air Quality Research. 2013;**13**:943-956

[27] Logue JM, Lunden MM, Singer BC. Development and Application of a Physics-based Simulation Model to Investigate Residential PM2.5 Composition and Size Distribution Across the US. Ernest Orlando Lawrence Berkeley National Laboratory; Berkeley, California State, US. 2014. p. 11

[28] Kodros JK, Wiedinmyer C, Ford B, Cucinotta R, Gan R, Magzamen S, Pierce JR. Global burden of mortalities due to chronic exposure to ambient PM2.5 from open combustion of domestic waste. Environmental Research Letters. 2016;**11**:1-9

[29] National Institute of Advanced Industrial Science and Technology: ADMER [Internet]. 2012. Available from: http://www.aist-riss.jp/software/admer/ja/index_ja.html [Accessed: July, 20 2017]

[30] National Institute of Advanced Industrial Science and Technology: ADMER ver.2.5 Instruction Manual; 2016. p. 171

[31] National Institute of Advanced Industrial Science and Technology: Detailed Risk Assessment Technical Guidance - Detailed Edition - No. 2. ADMER - Exposure and Risk Assessment Atmospheric Dispersion Model; 2005. p. 114

[32] Ministry of Economy, Trade and Industry. Ministry of Economy, Trade and Industry - Low-Rise Industrial Source Dispersion Model - METI-LIS Model ver.3.0.2. - Instruction Manual; 2016. p. 116

[33] Ministry of Economy, Trade and Industry. Manual for Method of Estimating Environmental Impact in the Vicinity of Sources Relating to Hazardous Air Pollutants (Ministry of Economy, Trade and Industry - Low-Rise Industrial Source Dispersion Model: METI-LIS) ver.3.0.2; 2009. p. 78

[34] Ministry of Economy, Trade and Industry. Detailed Risk Assessment Technical Guidance - Detailed Edition - No.3. Low-Rise Industrial Source Dispersion Atmospheric Level Estimation System (METI-LIS); 2006. p. 221

Chapter 18

# Risk Assessment and Prediction of Aflatoxin in Agro-Products

Peiwu Li, Xiaoxia Ding, Yizhen Bai, Linxia Wu,
Xiaofeng Yue and Liangxiao Zhang

Additional information is available at the end of the chapter

http://dx.doi.org/10.5772/intechopen.70792

**Abstract**

Aflatoxin (AFT), highly toxic and carcinogenic to humans, seriously threatens consumption safety of agro-products. It is necessary to conduct risk assessment of aflatoxin contamination in agro-food products to find out critical control points (CCPs) and develop prediction, prevention and control theories and technologies. In this chapter, risk assessment and prediction of aflatoxin contamination in peanut were taken as an example. The values under the limit of detection (LOD) were replaced by zero, 1/2 LOD or LOD according to their respective proportion, and the distribution of values higher than the LOD was fitted by @RISK software. $AFB_1$ dietary exposure was evaluated based on non-parametric probability risk assessment and margin of exposure (MOE). A risk ranking method was adopted for mycotoxins based on food risk expectation ranking. Spatial analysis of $AFB_1$ contamination was conducted using geographic information system (GIS). Average climatic conditions were calculated by Thiessen polygon method and the relationship between $AFB_1$ concentration and average pre-harvest climatic conditions was obtained through multiple regression. To fulfill the purposes of reducing cost, increasing efficiency, maximizing the role of risk assessment and prediction, and improving the quality and safety of agricultural products, we will continuously focus on developing advanced and integrated technologies and solutions.

**Keywords:** peanut, aflatoxin, dietary exposure, risk ranking, risk prediction

## 1. Introduction

Risk prediction of agro-product, especially oil and grain products, is becoming more and more important. In this chapter, risk assessment and prediction of aflatoxin (AFT) in peanuts were taken as an example. We presented the development and research progress on risk assessment and prediction of aflatoxin in agro-products in the following aspects: (1) data processing and simulation methods of peanut aflatoxin contamination (determination and simulation of highly skewed data); (2) risk assessment methods (non-parametric probability risk assessment method and margin of exposure (MOE) method); (3) risk ranking method (multi-mycotoxin risk ranking method based on the expert scoring method); (4) risk prediction technologies (large-scale aflatoxin prediction based on ArcGIS) and (5) prospect of future research.

Aflatoxin (AFT) is highly toxic and carcinogenic and has been therefore classified as a Group I carcinogen by the International Agency for Research on Cancer [1]. The most important types of aflatoxins occurring naturally in agro-products are aflatoxin $B_1$ ($AFB_1$), aflatoxin $B_2$ ($AFB_2$), aflatoxin $G_1$ ($AFG_1$) and aflatoxin $G_2$($AFG_2$) [2, 3]. The total output and output per acre of peanuts always rank first of all oil crops cultivated in China. Peanuts produced in China account for about 40% of the world's peanut trade. In addition, peanuts contribute large amounts of vegetable oil, protein and vitamin E to developed countries [3–5]. Unfortunately, aflatoxin has been detected in more than 100 kinds of agro-products, especially in peanut and maize. Aflatoxin contamination might occur during the whole process of agro-products from production, storage, processing to trade, which seriously threatens consumption safety. To control aflatoxin contamination and ensure consumption safety, it is necessary to assess the risk of aflatoxin contamination in agro-products to identify critical control points (CCPs) and develop prediction, prevention and control theories and technologies for precise control in practice.

The mechanism of aflatoxin contamination is still not unclear since it is complex and multifactor dependent. Moreover, the aflatoxin contamination processes are significantly different over several consecutive years, and the contamination shows seriously skewed distribution. Among agro-products, peanuts are most seriously contaminated by aflatoxin. Since peanuts are popular food and oilseed worldwide, the prediction and control of aflatoxin contamination in peanuts are hot issues difficult to be resolved.

## 2. Data processing and simulation methods of peanut aflatoxin contamination

### 2.1. Data processing of peanut aflatoxin contamination

According to post-harvest peanut aflatoxin data in China from 2009 to 2010, the proportion of "trace data" (below the detection limit) was over 70%, and the proportions of trace data for $AFB_1$, $AFB_2$, $AFG_1$ and $AFG_2$ were 78.3, 78.3, 98.8 and 97.2%, respectively. The aflatoxin data for the Chinese peanuts were positively skewed, and the Kolmogorov-Smirnov test proved that $AFB_1$ and the total aflatoxin did not conform to the normal distribution, with about 90% of the aflatoxin data concentrated in the range of 0–2 μg/kg, so that the aflatoxin data for Chinese peanuts were left censored data and in line with a left skewed distribution. The "trace data" were distributed between 0 and the detection limit, which could not be accurately quantified due to the accuracy limitation of available instruments or equipment or unsatisfactory detection techniques. The presence of these trace data posed some difficulties to subsequent statistical analyses and could not be simply ignored because they had influences on the results of risk assessment.

During the process of building a risk assessment model for the peanuts' aflatoxin exposure, it is important to consider how to deal with the considerable values below the detection limit.

In accordance with the previous studies, there were mainly two solutions, which were point substitution and theoretical distribution substitution.

The point substitution method has been widely used in risk assessment of chemicals, such as heavy metals or pesticides. International aflatoxin risk assessment, conducted by JECFA or EFSA, also adopted this approach. Global Environment Monitoring System-Food Contamination Monitoring and Assessment Program [6] suggested that when the proportion of non-quantified or non-detected results was greater than 60%, the value under the limit of detection (LOD) was replaced by zero or LOD to produce upper and lower boundaries; when the proportion was less than 60%, LOD/2 was chosen as the substitute to produce statistical estimates. The point substitution method was a convenient operation, but its results were relatively rough and could not be used to evaluate uncertainty and variability.

Taking the process of total aflatoxin in peanuts as an example, the values below the LOD were all replaced by 0, 1/2LOD and LOD at first to generate three data sets, respectively. Then, the percentiles were calculated and it showed the difference occurred at <65th percentile, the three alternative results of total aflatoxin approached gradually from 65th to 85th percentile, and the maximum absolute difference was reduced from 0.16 to 0.14 μg/kg and then to 0.11 μg/kg at 95th percentile. The difference was mainly from detection limit of test method for aflatoxins. So, low detection limit was the main approach for reducing the difference among three alternative methods and improving the evaluation accuracy. The optimal detection method for aflatoxin was liquid chromatography coupled with immunoaffinity chromatography, which had relatively higher sensitivity and accuracy.

The theoretical distribution substitution method was based on the characteristics of contamination data. Taking $AFB_1$ in post-harvest peanuts as research object, the method was performed in two steps as follows. First, we sorted all data and eliminated the trace values that were lower than LOD from the entire dataset. Second, we fitted the distribution function with the values that were higher than the LOD by @RISK software, and then used the Kolmogorov-Smirnov (K-S) or Anderson-Darling (A-D) method to perform statistical tests on the fitting results. Through screening and optimization, Pearson V, Inverse Gauss and log-normal distributions were suitable to aflatoxin distribution in peanuts, and the comparison of frequency distribution and probability density indicated that Pearson V for goodness of fit was the best.

### 2.2. Risk assessment methods

#### *2.2.1. Risk assessment based on non-parametric probability*

$AFB_1$ dietary exposure was evaluated based on a probability distribution of aflatoxin contamination and consumption in agro-products, and the results were standardized by human bodyweight. The Monte Carlo method was chosen to perform the entire simulation process by @RISK program and the uncertainty was described by 90% confidence interval or quartile. The risk posed by dietary exposure to $AFB_1$ was modeled by the following formula: Population risk = exposure × average potency; exposure (daily intake of $AFB_1$ expressed as ng $kg^{-1}$ bw day) = (contamination level × consumption amount)/bw; average potency = 0.3 × P + 0.01 × (1 − P), where P represents the hepatitis-B-virus surface antigen (HBsAg) prevalence rate for different age groups.

For example, $AFB_1$ risk assessment in peanuts was conducted on the basis of dietary exposure to $AFB_1$ and its potential to cause hepatic cancer. Based on the results of peanut aflatoxin survey in China conducted in 2009–2010, as well as peanut consumption data and average bodyweight in each age-gender group from the 2002 Chinese Residents Nutrition and Health Survey Report [7–9], dietary exposure to $AFB_1$ was calculated and simulated by Monte Carlo. In line with the guidelines of Global Environment Monitoring System-Food Contamination Monitoring and Assessment Program [6], the values which were less than the LOD, were estimated, assuming that the proportion of non-quantified or non-detected results was more than 60% but less than 80%, the values under the LOD were substituted by zero or the LOD, which could provide a lower or higher boundary [10].

Excess risks for liver cancer incidence per year, resulting from $AFB_1$ dietary intake through peanut consumption, were calculated from dietary exposure to $AFB_1$ multiplied by the average $AFB_1$ cancer potency. According to the $AFB_1$ risk assessment report from JECFA [11], the average cancer potency was produced by setting the individual potencies of $HBsAg^+$ and $HBsAg^-$ to 0.3 and 0.01 cancers/year/100,000/ng $kg^{-1}$ bw $day^{-1}$, respectively. In this assessment, the age-adjusted $HBsAg^+$ prevalence rate was obtained from the 2006 National Sero-epidemiological Survey report.

To evaluate potential health risk to Chinese under $AFB_1$ exposure in food, the excess risk for liver cancer in adults was estimated based on the mean and 97.5th percentile of the contamination and consumption data. The estimated $AFB_1$ intake from raw peanuts was between 0.11 and 5.66 ng $kg^{-1}$ bw $day^{-1}$ and the population risk was 0.003–0.17 cancer cases/year/100,000 from raw peanut consumption. The population risk was 0.03–2.06 cancer cases/year/100,000 from peanut oil intake of 0.84–68.8 ng $kg^{-1}$ bw $day^{-1}$. These data indicated that the risk from peanut oil was 10 times or more that from raw peanuts.

#### *2.2.2. Margin of exposure (MOE) method*

A "margin of exposure" was calculated from a chosen point of departure (POD) on a dose–response curve divided by the human dietary exposure estimate, which was obtained based on the benchmark dose (BMD) developed by EFSA [12, 13]. The PODs, employed to quantify an increased cancer risk, were summarized in **Table 1**. When the POD was determined, a smaller

| Sources | BMDL10[a] | BMDL10[b] | BMDL1[b] | T25 |
|---|---|---|---|---|
| BfR | 150 | – | – | – |
| Brien et al. [14] | 160 | – | – | 500 |
| EFSA [10] | 170–340 | 870–1100 | 78–121 | – |
| Dybing et al. [15] | 160–300 | – | – | 500 |
| Benford et al. [16] | 250 | – | – | 390 |

[a]Derived from animal carcinogenicity data.
[b]Derived from Chinese epidemiological data.

**Table 1.** Reference points/PODs derived from animal carcinogenicity and Chinese epidemiological data (ng $kg^{-1}$ bw $day^{-1}$).

MOE value represented a greater risk. Compared with a traditional low-dose extrapolation approach, the MOE value easily indicated what the risk level was, provided that the POD value had been defined. The MOE value would be smaller when exposure became greater. In general, a smaller MOE represented a greater risk.

Taking risk assessment of peanut aflatoxin exposure expressed by the MOE in China as an example, the MOEs were calculated on the basis of Chinese peanut aflatoxin exposure and PODs from the reported literature, which were developed based on Chinese epidemiological data by EFSA [10] or rodent experimental data by Benford et al. The relative PODs were summarized in **Table 1**. Here, BMDL10 (140 ng $kg^{-1}$ bw $day^{-1}$ for rodent [16] and 870 ng $kg^{-1}$ bw $day^{-1}$ for human) and BMDL1 (78 ng $kg^{-1}$ bw $day^{-1}$ for human), were introduced into the MOE calculation, which represented the 95% lower confidence limit (CL) of the BMD for a 10 or 1% increased cancer risk .

The estimated MOE values ranging from 24.1 to 1272 were higher than the results estimated by EFSA (88–483) [10] for Africa (0.2–121.4) [17]. Far lower than 10,000, would be regarded as low concern [12], and a higher MOE value implied a lower risk. The MOE values of peanuts based on the rodent data were 24.7–1272 and 2.0–167, respectively. In other words, the cancer risk, which originated from direct consumption of post-harvest peanuts or raw peanuts, was much lower than that from peanut oil. The above results were consistent with the conclusions, which were calculated on the basis of the cancer potencies of aflatoxin employed by JECFA. However changing index among two different methods was not found by now.

## 3. Risk ranking methods

The Codex Alimentarius Commission (CAC) recommended a food risk expectation ranking method. This method is usually based on literature review, authoritative data and database records of the countries that have evaluated the food hazard/consumption frequency and detected frequency information. Then the main food contamination factors are accurately identified, and the risks from different sources are compared. According to the calculated scores of the indexes, ranking of the risk will be obtained. This approach has the advantage of clear scoring criteria and that direct use of the defined scoring criteria. In 1999, Houghton et al. [18] studied the ranking of risk factors for anxiety disorders in the UK using the risk expectation approach. In 2011, the method of risk expectation was used to systematically study the ranking and change in the relative risk index of liquefied petroleum gas transportation in Mexico metropolitan area [19]. In 2014, Speybroeck et al. [20] studied the ranking of risk factors in food chain by means of sampling survey and risk expectation. Nevertheless, there are few studies on risk factor of mycotoxins in peanuts and no study on risk ranking of mycotoxins in China until now.

In order to give a reference for risk monitoring and assessment of peanut quality and safety, the risk ranking method of mycotoxins in peanuts was proposed on the basis of the food risk expectation ranking method. A total of 604 peanut samples from 8 provinces were collected. Based on the mycotoxins concentration in peanuts and maximum residue levels, hazard degrees were identified. The effective evaluation indicators were chosen, and a normalized method was established for searching, identifying and ranking peanut mycotoxin risk factors.

### 3.1. Hazard degree identification

This study referred to the risk ranking sample tool recommended by the CAC and considered human health threats caused by the risk factors. The hazard severity and probability of occurrence were considered from the qualitative and quantitative points of view. The hazard degrees of the risk factors and risk ranking score evaluation criteria (**Table 2**) were identified with their toxicity, degree of difficulty in risk control, severity, social reputation, maximum amount of detection residue and detection rate considered.

According to the basic requirements of risk identification, the modified risk identification method was developed for peanut mycotoxins after several cycles of discussion, screening and expert opinion collection (**Table 3**).

### 3.2. Risk factor analysis and ranking

According to the risk ranking score evaluation criteria (**Table 2**) and identification of peanut mycotoxin risk degrees (**Table 3**), in addition to the toxicities, degrees of difficulty in risk

| Index | Index value (score = 5) | Index value (score = 4) | Index value (score = 3) | Index value (score = 2) |
|---|---|---|---|---|
| Toxicity | High | Relatively high | Medium | Low |
| Degree of difficulty inrisk control | Difficult | Poor | Potentially poor | Capable |
| Severity | Serious | Relatively serious | Medium | Noteworthy |
| Social reputation | Serious | Relatively serious | Medium | Noteworthy |
| Maximum amount of detection residue/(μg/kg) | >5000 | 1000–5000 | 500–1000 | 0–500 |
| Detection rate/% | >10 | 8–10 | 6–8 | 4–6 |

**Table 2.** Identification of food hazards and risk ranking score evaluation criteria.

| Risk factors | Toxicity | Degree of difficulty in risk control | Severity | Social reputation |
|---|---|---|---|---|
| $AFB_1$ | High toxicity, class 1 carcinogens, damage to liver | Difficult | Serious | Serious |
| $AFB_2$ | | Poor | Relatively serious | Relatively serious |
| $AFG_1$ | | Poor | Serious | Relatively serious |
| $AFG_2$ | | Poor | Relatively serious | Relatively serious |
| OTA | Medium toxicity, class 2 possible carcinogens | Potentially poor | Medium | Medium |
| DON | Medium toxicity, class 2 possible carcinogens | Potentially poor | Medium | Medium |

**Table 3.** Identification of peanut mycotoxin risk degrees.

control, severities, social reputations, maximum amounts of detection residue and detection rates of peanut $AFB_1$, $AFB_2$, $AFG_1$, $AFG_2$, Ochratoxin A (OTA) and Deoxynivalenol (DON) in peanuts in China, the mycotoxin risk factor scores for peanuts were calculated by Formula (1).

$$s = \frac{\sum_{i=1}^{n} X_{Ai}}{n} \times \frac{\sum_{i=1}^{n} (X_{Bi} + X_{Ci} + X_{Di} + X_{Ei} + X_{Fi})}{n} = U_A \times (U_B + U_C + U_D + U_E + U_F) \quad (1)$$

S: mycotoxin risk factor scores of peanuts; $X_{Ai}$: mycotoxin toxicity score of sample i; $X_{Bi}$: mycotoxin score of the degree of difficulty in risk control of sample i; $X_{Ci}$: mycotoxin severity score of sample i; $X_{Di}$: mycotoxin social reputation score of sample i; $X_{Ei}$: mycotoxin score of the maximum amount of detection residue of sample i; $X_{Fi}$: mycotoxin detection rate score of sample i; n: number of samples; $U_A$: average score of mycotoxin toxicity; $U_B$: average score of the degree of difficulty in risk control; $U_C$: average score of severity; $U_D$: average score of social reputation; $U_E$: average score of the maximum amount of detection residue; $U_F$: average score of the detection rate.

Mycotoxin risk factor scores and ranking for peanuts in China were listed in **Table 4**. It indicated that high attention needed to be paid to $AFB_1$, relatively high attention needed to be paid to $AFG_1$, moderate attention needed to be paid to $AFB_2$ and $AFG_2$, and low attention needed to be paid to OTA and DON.

| Mycotoxin | UA | UB | UC | UD | UE | UF | S | Risk degree |
|---|---|---|---|---|---|---|---|---|
| $AFB_1$ | 5 | 5 | 5 | 5 | 5 | 4 | 120 | High |
| $AFB_2$ | 4 | 4 | 4 | 4 | 4 | 3 | 76 | Moderate |
| $AFG_1$ | 4 | 4 | 4 | 4 | 5 | 5 | 88 | Relatively high |
| $AFG_2$ | 4 | 4 | 4 | 4 | 4 | 3 | 76 | Moderate |
| OTA | 3 | 3 | 3 | 3 | 2 | 3 | 42 | Low |
| DON | 3 | 3 | 3 | 3 | 2 | 3 | 42 | Low |

**Table 4.** Mycotoxin risk ranking for peanuts in China.

## 4. Risk prediction technologies

Geographic information systems (GIS) and geostatistics can be used to describe, analyze and display spatial patterns of a wide variety of variables at any scale and, by improving resource management and revealing causal relationships among geographically variable factors, assist in real world problems [21]. Kriging, a regression technique for interpolation of spatially correlated data, is the most common geostatistical procedure for surface interpolation, can be used to locally average the weights of data from sampled locations surrounding an unsampled location based on statistical similarity to unsampled locations; it gives unbiased estimates with the estimated variance minimized. The weights are determined using semivariance analysis between sampled locations [22]. Areas in China with the highest risk of $AFB_1$ contamination were identified by geostatistical analyses and Kriging maps. According to different locations, terrain features, climatic conditions, variety distributions and cultivation systems, the peanut planting areas in China were divided into four sections: Northeast, North, Yangtze River and South [23]. Agricultural practices including crop rotation, tillage, irrigation and fertilization, as well as the planting date, genetic resistance, soil type and climatic conditions all impact AFT contamination of peanuts before harvest [24]. Nevertheless, climatic conditions significantly influence the AFT contamination level. In serious drought and/or high temperature conditions before harvest, fungus invasion and AFT accumulation become accelerated [25, 26].

### 4.1. Spatial analysis of $AFB_1$ contamination of peanuts in China

A total of 9741 peanut samples were collected from main produce area in China from 2009 to 2014 and on the $AFB_1$ content of these peanut samples were analyzed. Geostatistical analyses were performed on the annual average $AFB_1$ content to obtain the patterns of $AFB_1$ contamination throughout China. Kriging of $AFB_1$ showed that aflatoxin contamination of peanuts in China was a perennial problem presenting both temporal and spatial (regional) variations. Kriging interpolation of $AFB_1$ contamination indicated a patchy distribution, which varied with the seasons. Results showed that aflatoxin contamination was almost not found in the Northeast region during the study period. And it presented significantly temporal and spatial (regional) variations in the Yangtze River Basin region and Southeast Coast region.

### 4.2. Relationship between $AFB_1$ contamination levels in peanuts and climatic conditions before harvest

Cole *et al.* found that ripe and integral peanuts exposed to simultaneously drought and heat (25.7–31.3°C) stress became be prone to *Aspergillus flavus* invasion and AFT production in the last 4–6 weeks of the growing season [27]. A total of 2983 peanut samples were collected from 122 counties in 6 provinces of China's Yangtze River ecological region from 2009 to 2014. Based on Thiessen polygon interpolation, average precipitation and mean temperature data in 2009–2014 in the Yangtze River ecological region were calculated by the climatic conditions of 118 weather stations. In **Figure 1**, we found that there was less precipitation and higher daily mean temperature (around 25°C) during peanut growing season (June–August) in 2013, which aggravated the $AFB_1$ contamination. Taking Hunan province as an example, the

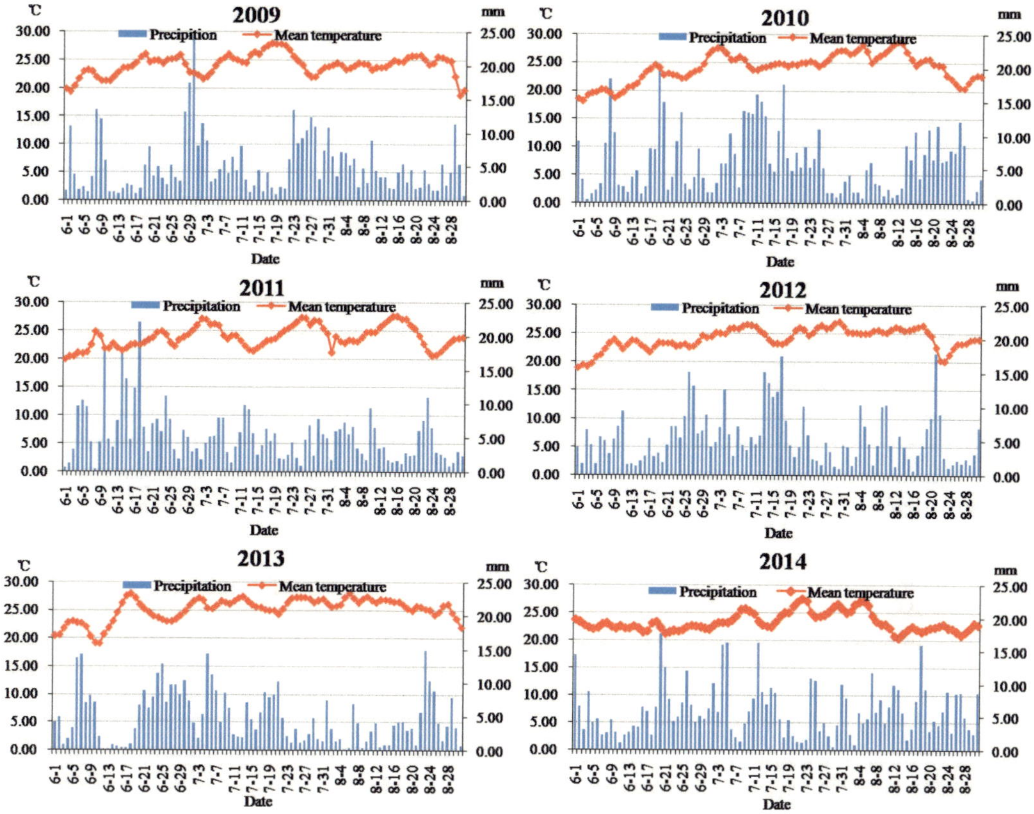


**Figure 1.** Precipitation and mean temperature of the Yangtze River ecological region during the peanuts' growing season (2009–2014).

determination coefficient ($R^2$) fitted by the $AFB_1$ content with the average climatic conditions in different pre-harvest periods was obtained by multiple regression (**Figure 2**). Results indicated that the average precipitation and mean temperature of 1 month before harvest had a significant influence on $AFB_1$ contamination. Moreover, Hunan and Jiangxi were greatly affected. Due to the annual and climatic variation of $AFB_1$ contamination level, it is necessary to build a prediction model by developing a continuous and effective $AFB_1$ monitoring program for pre-harvest peanuts during its growing season. Up to now, there had been some progress on model building in Australia and USA [28, 29].

## 5. Prospect of future research

The occurrence and control of aflatoxin contamination in agro-products are world wide hot issues difficult to resolve. Studies on risk monitoring, risk assessment and early risk prediction of aflatoxin in peanuts, maize and other agro-products have long been considered as an important premise for effective aflatoxin contamination control. Hence, our further efforts

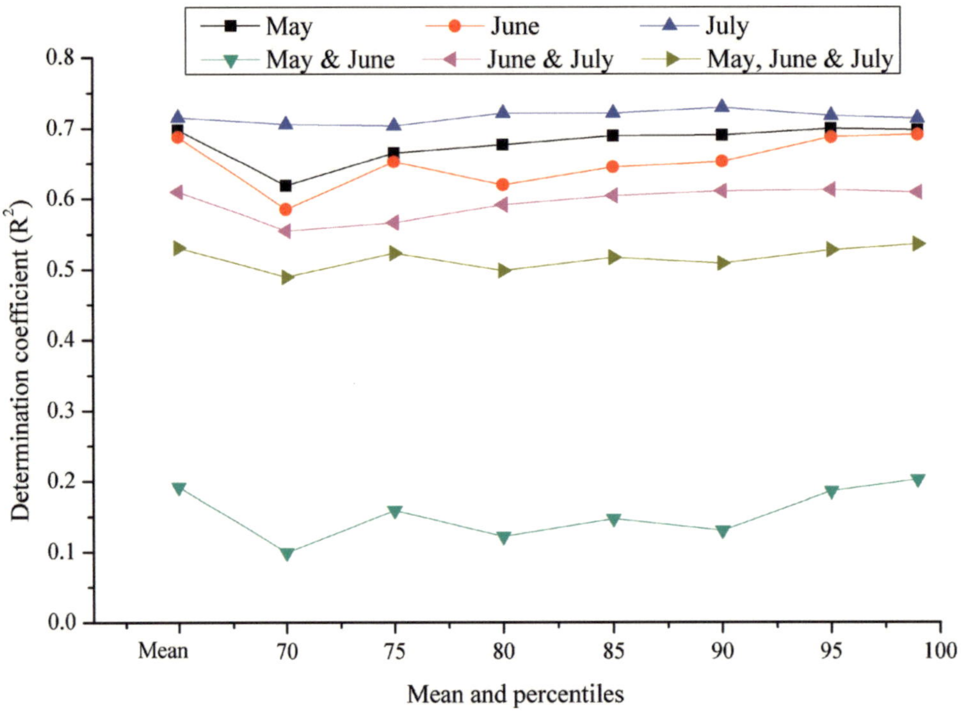


**Figure 2.** Multiple regression determination coefficient ($R^2$) fitted by the $AFB_1$ content with average precipitation and mean temperature in different periods of time (Hunan province).

should be focused on enhancing and perfecting the basic database of aflatoxin contamination, drawing geographic risk maps, developing reliable and accurate risk forecasting techniques and a forecasting system, as well as establishing a smart and low-cost platform for sustainable aflatoxin management and communication in the field and during storage, processing and transportation. Moreover, risk monitoring, simultaneous detection technologies of multi-mycotoxin contamination and their interaction mechanisms should also be taken into account.

### 5.1. Characteristics and geographic risk maps for aflatoxin contamination and main toxigenic fungal population

To boost the progress of mycotoxin risk assessment, what is crucial is to enhance the basic database construction for mycotoxin contamination, especially aflatoxin, which has the highest acute and chronic toxicity among all mycotoxins. Hence, it is necessary to conduct a continuous and effective AFT monitoring program for obtaining quantitative data from different latitudes, altitudes and ecological regions in a global level via international cooperation by sampling and detecting representative fields. Precise risk maps for aflatoxin contamination will be drawn to highlight the distribution, concentration and trend of annual occurrence. Certainly, advanced and accurate analytical techniques will be an essential part of guaranteeing the quality of monitoring and data. Meanwhile, the main toxigenic fungal population database, including *A. flavus* and *Aspergillus parasiticus*, should also be determined by collecting and identifying isolates from peanuts and soil in the corresponding areas. Geographic maps of these fungi populations will be defined. The aflatoxin production, biodiversity and phylogenetic clades of these toxigenic fungi will be revealed. Because the changes in climatic conditions will lead to a shift in the fungal population and mycotoxin patterns, more attention should be paid in a climate change scenario. What is more, multi-mycotoxin co-occurrence and interactions with different fungi, such as *A. flavus* and *Fusarium verticillioides*, have been recognized as emerging problems and cannot be ignored.

From a risk assessment perspective, to determine risk maps of aflatoxin contamination and toxigenic fungal population to a global level, which was based on a continuous and effective AFT monitoring program, is a key step in risk prediction. Additionally, aflatoxin and toxigenic fungi risk maps could be used as a communication tool for stakeholders and farmers. Moreover, the maps could be provided as a tool for scientific supervision, decision support and governments' policy-making, as well as prioritization of a more targeted approach and intervention strategies, especially in high-risk zones.

### 5.2. Building an early predictive model of aflatoxin by combining macroscopic and molecular warning technologies

In further studies, there is an urgent need to establish a precise and reliable forecasting system with advanced prediction methods to reflect actual occurrence of aflatoxin contamination so that we can make appropriate management and agronomic strategies especially in high-risk areas, minimize the risk of pre-harvest contamination and therefore protect public and animal health. Moreover, applying the early warning model can significantly reduce the detoxification cost. Until now, a lot of researches indicated that key environmental factors including temperature, humidity and precipitation significantly influenced fungus growth, infection as well as aflatoxin production [30–32]. And some efforts have been devoted to developing models to predict aflatoxin contamination in peanuts and maize with climatic data used as the main or only input, such as the Agricultural Production Systems Simulator (APSIM) and the CSM-CROPGRO-Peanut model [29, 33]. However, few models were actually applied to the field to predict the future aflatoxin risk or just a small region for validation and demonstration. Besides climatic data, the factors such as ecological zones, peanut varieties and microbial structures should not be ignored, which were also believed to be aflatoxin-related factors. In short, there is further work to develop a large-scale risk prediction model based on multi-factors and apply it to different fields.

Firstly, advanced technologies in digital and smart agriculture are essential for effectively monitoring the fields. GIS, environmental sensors or satellite systems will be used to monitor crop conditions and abiotic factors such as humidity, temperature, precipitation, wind and sunshine in real time, in order to acquire spatial and temporal distribution information of the crops and climatic data rapidly and accurately during crop-growing seasons. These local and accurate real-time data will be directly translated for researchers and other stakeholders. And then, correlation analysis will be, respectively, carried out between the aflatoxin contamination data and these real-time data, agronomic information and toxigenic fungus population to evaluate the role and contribution of these relevant factors in the field of aflatoxin contamination risks. In particular, the role of $CO_2$ should be taken into account, which is increasingly important in a climate change scenario. At last, a macro-scale predictive model will be built and operated with these "real-time" data as the input to obtain specific early predictions regarding the risk of aflatoxin. Appropriate management decisions and recommendations for farmers and other stakeholders will be formulated when the contamination risk is high, which is based on the output data.

In recent years, with rapid development of molecular biology, molecular prediction technologies have gradually become frontier for early warning of mycotoxins. The interactive conditions of $a_w$ × temperature × elevated $CO_2$ have a significant impact on aflatoxin biosynthetic gene expression, such as the structural genes *aflD* and *aflM* and regulatory genes *aflS* and *aflR*, and the production of $AFB_1$ [32, 34]. A physical model was built and used to relate gene expression to $a_w$ and temperature conditions to predict $AFB_1$ production. And its relationship with the observed $AFB_1$ production provided a good linear regression fit to the predicted production based on the model [34]. The expression data ratio of aflS/aflR has a relationship with the amount of $AFB_1$ or $AFG_1$. High ratios in the range between 17 and 30°C corresponded to the production profile of $AFG_1$ biosynthesis. A low ratio was observed at >30°C, which was related to $AFB_1$ biosynthesis [35]. We therefore believed that it is possible to predict the aflatoxin risk via the expression model of key genes or secondary metabolites. In our future work, we will devote to screening and identifying more effective molecular markers to make the prediction more reliable and build a molecular forecasting model.

Therefore, it is believed that effective integration of macro-scale, molecular, ecophysiological and secondary metabolite data sets could be critical in predicting the risk of aflatoxin contamination under different biotic and abiotic stress scenarios and agronomic strategies. Such combinative technologies will be beneficial to more accurate predictions of the aflatoxin risk in different regions and also the potential for new emerging toxin threats.

### 5.3. Developing a smart platform for aflatoxin risk communication and management

A convenient and user-friendly platform, such as a mobile app, will be developed. The platform will provide key information about the crops, contamination risks or levels, recommendations, practical solutions, problem consultation and answers to farmers and other stakeholders who require suggestions for rapid and low-cost intervention. The interpretation of the output of the predictive models and recommendations will be transformed into the platform. Thus, farmers can not only obtain the growth status of the crops in the field, but also timely and cost-effective strategies for prevention or remediation of the risks during their harvest, storage, processing and transportation.

In conclusion, it is necessary to further focus on the development of advanced and integrated technologies and solutions to achieve the purposes of reducing costs, increasing efficiency, maximizing the role of risk assessment and risk prediction, and definitely improving the quality and security of agricultural products.

## Author details

Peiwu Li[1,2,3,4,5], Xiaoxia Ding[1,2,3,5], Yizhen Bai[1,2,5], Linxia Wu[1]*, Xiaofeng Yue[1,2,3,5] and Liangxiao Zhang[1,2,3,5]

*Address all correspondence to: 1060038025@qq.com

1 Oil Crops Research Institute of Chinese Academy of Agriculture Science, Wuhan, People's Republic of China

2 Laboratory of Risk Assessment for Oilseeds Products (Wuhan), Ministry of Agriculture, Wuhan, People's Republic of China

3 Key Laboratory of Detection for Mycotoxins, Ministry of Agriculture, Wuhan, People's Republic of China

4 Key Laboratory of Oil Crop Biology and Genetic Improvement of Oil Crops, Ministry of Agriculture, Wuhan, People's Republic of China

5 Quality Inspection and Test Center for Oilseeds Products, Ministry of Agriculture, Wuhan, People's Republic of China

## References

[1] International Agency for Research on Cancer (IARC). Some naturally occurring substances: Food items and constituents, heterocyclic aromatic amines and mycotoxins. IARC Monographs on the Evaluation of Carcinogenic Risks to Humans. 1993;**56**:245-395

[2] Idris YM, Mariod AA, Elnour IA, et al. Determination of aflatoxin levels in Sudanese edible oils. Food and Chemical Toxicology. 2010;**48**(8-9):2539-2541. DOI: 10.1016/j.fct.2010.05.021

[3] Williams RJ, McDonald D. Grain molds in the tropics: Problems and importance. Annual Review of Phytopathology. 1983;**21**:153-178. DOI: 10.1146/annurev.py.21.090183.001101

[4] Chen YC, Liao CD, Lin HY, et al. Survey of aflatoxin contamination in peanut products in Taiwan from 1997 to 2011. Journal of Food and Drug Analysis. 2013;**21**(3):247-252. DOI: 10.1016/j.jfda.2013.07.001

[5] LX Wu, Ding XX, Li PW, et al. Aflatoxin contamination of peanuts at harvest in China from 2010 to 2013 and its relationship with climatic conditions. Food Control. 2016;**60**:117-123. DOI: 10.1016/j.foodcont.2015.06.029

[6] World Health Organization (WHO). GEMS/Food-EURO Workshop on Reliable Evaluation of Low-Level Contamination of Food: Report on a workshop in the frame of GEMS/Food-Euro, Kulmbach, Federal Republic of Germany, 3-5 March 1994

[7] Wang LD. 2002 Chinese Resident Nutrition and Health Survey Report. Beijing, China: China Ministry of Health; 2005

[8] Wang J, Liu X. Assessment of dietary aflatoxins exposure in Chinese residents. Chinese Journal of Food Hygiene. 2007;**19**(3):832 932, 42

[9] Zhai FY. The Prospective Study on Nutrition Transition in China. Beijing: Science Press; 2008

[10] European Food Safety Authority (EFSA). Opinion of the Scientific Panel on Contaminants in the Food Chain on a request from the Commission related to the potential increase of consumer health risk by a possible increase of the existing maximum levels for aflatoxins in almonds, hazelnuts and pistachios and derived products. The EFSA Journal. 2007;**446**:1-127

[11] Joint FAO/WHO Expert Committee on Food Additives (JECFA). Forty-ninth Meeting of the Joint FAO/WHO Expert Committee on Food Additives. Safety evaluation of certain assitives and contaminants in food: afaltoxins. Food Additives Series. 1998;**40**:359-469

[12] European Food Safety Authority (EFSA). Opinion of the Scientific Committee on a request from EFSA related to A Harmonized Approach for Risk assessment of Substances which are both Genotoxic and Carcinogenic. The EFSA Journal. 2005;**282**:1-31

[13] Barlow S, Schlatter J. Risk assessment of carcinogens in food. Toxicology and Applied Pharmacology. 2010;**243**:180-190. DOI: 10.1016/j.taap.2009.11.004

[14] O'Brien J, Renwick AG, Constable A, et al. Approaches to the risk assessment of genotoxic carcinogens in food: A critical appraisal. Food and Chemical Toxicology. 2006;**44**:1613-1635. DOI: 10.1016/j.fct.2006.07.004

[15] Dybing E, O'Brein J, Renwick AG, et al. Risk assessment of dietary exposures to compounds that are genotoxic and carcinogenic - An overview. Toxicology Letters. 2008;**180**:7-110. DOI: 10.1016/j.toxlet.2008.05.007

[16] Benford D, Leblanc JC, Setzer RW. Application of the margin of exposure (MOE) approach to substances in food that are genotoxic and carcinogenic example: Aflatoxin B1(AFB1). Food and Chemical Toxicology. 2010;**48**:534-541. DOI: 10.1016/j.fct.2009.09.039

[17] Shephard GS. Risk assessment of aflatoxins in food in Africa. Food Additives and Contamination. 2008;**25**(10):1246-1256. DOI: 10.1080/02652030802036222

[18] Houghton JR, Murray E, Ball DJ. Risk ranking by the British public: A survey of worry about a broad spectrum of risk issues. Human and Ecological Risk Assessment. 1999;**5** (3):509-526. DOI: 10.1080/10807039.1999.10518876

[19] López-Atamoros LG, Fernández-Villagómez G, Cruz-Gomez MJ, et al. Development of a relative ranking risk index for risk assessment in the transportation of liquefied petroleum gas in Mexico City's Metropolitan Area. Human and Ecological Risk Assessment. 2011;**17**:1193-1209. DOI: 10.1080/10807039.2011.618383

[20] Speybroeck N, Devleesschauwer B, Depoorter P, et al. Needs and expectations regarding risk ranking in the food chain: A pilot survey amongst decision makers and stakeholders. Food Control. 2015;**54**:135-143. DOI: 10.1016/j.foodcont.2014.12.041

[21] Jaime-Garcia R, Cotty PJ. Aflatoxin contamination of commercial cottonseed in South Texas. Phytopathology. 2003;**93**:1190. DOI: 10.1094/PHYTO.2003.93.9.1190

[22] Borgelt C, Wieda RE, Sudduth KA. Geostatistical analysis of soil chemical properties from nested grids. Journal of Physical Chemistry. 1997;**13**:477-483. DOI: 10.13031/2013.21631

[23] Ding X, Li P, Bai Y, et al. Aflatoxin B-1 in post-harvest peanuts and dietary risk in China. Food Control. 2012;**23**:143-148. DOI: 10.1016/j.foodcont.2011.06.026

[24] Torres AM, Barros GG, Palacios SA, et al. Review on pre- and post-harvest management of peanuts to minimize aflatoxin contamination. Food Research International. 2014;**62**:11-19. DOI: 10.1016/j.foodres.2014.02.023

[25] Sanders TH, Blankenship PD, Cole RJ, et al. Effect of soil temperature and drought on peanut pod and stem temperatures relative to *Aspergillus flavus* invasion and aflatoxin contamination. Mycopathologia. 1984;**86**:51-54. DOI: 10.1007/BF00437229

[26] Diener UL, Davis ND. Production of aflatoxin on peanuts under controlled environments. Journal of Stored Products Research. 1969;**5**:251-258. DOI: 10.1016/0022-474X(69)90040-X

[27] Cole RJ, Sanders TH, Hill RA, et al. Mean geocarposphere temperatures that induce preharvest aflatoxin contamination of peanuts under drought stress. Mycopathologia. 1985;**91**:41-46. DOI: 10.1007/BF00437286

[28] Henderson CE, Potter WD, McClendon RW, et al. Predicting aflatoxin contamination in peanuts: A genetic algorithm/neural network approach. Applied Intelligence. 2000;**12**: 183-192. DOI: 10.1023/A:1008310906900

[29] Chauhan YS, Wright GC, Rachaputi RCN, et al. Application of a model to assess aflatoxin risk in peanuts. Journal of Agricultural Science. 2010;**148**:341. DOI: 10.1017/S002185961000002X

[30] Van der Fels-Klerx HJ, Liu C, Battilani P. Modelling climate change impacts on mycotoxin contamination. World Mycotoxin Journal. 2016;**9**:717-726. DOI: 10.3920/WMJ2016.2066

[31] Ding XX, LX Wu, Li PW, et al. Risk assessment on dietary exposure to aflatoxin b1 in post-harvest peanuts in the Yangtze River ecological region. Toxins. 2015;**7**:4157-4174. DOI: 10.3390/toxins7104157

[32] Medina A, Rodriguez A, Magan N. Effect of climate change on *Aspergillus flavus* and aflatoxin B1 production. Frontiers in Microbiology. 2014;**5**:1-7. DOI: 10.3389/fmicb.2014.00348

[33] Boken VK, Hoogenboom G, Williams JH, et al. Monitoring peanut contamination in Mali (Africa) using the AVHRR satellite data and a crop simulation model. International Journal of Remote Sensing. 2008;**29**:117-129. DOI: 10.1080/01431160701264250

[34] Hadi AA, Heydt MS, Parra R, et al. A systems approach to model the relationship between aflatoxin gene cluster expression, environmental factors, growth and toxin production by *Aspergillus flavus*. Journal of the Royal Society Interface. 2012;**9**:757-767. DOI: 10.1098/rsif.2011.0482

[35] Heydt MS, Rüfer CE, Hadi AA, et al. The production of aflatoxin B1 or G1 by *Aspergillus parasiticus* at various combinations of temperature and water activity is related to the ratio of aflSto aflR expression. Mycotoxin Research. 2010;**26**:241-246. DOI: 10.1007/s12550-010-0062-7

Section 6

# Social Risk

Chapter 19

# Risk Management in Complex Organisations

Andrea Antonucci

Additional information is available at the end of the chapter

http://dx.doi.org/10.5772/intechopen.70762

**Abstract**

The aim of the chapter is to examine thoroughly the aspects connected with risk management in complex organisations; it starts with the consideration of theoretical models of accident's analysis in which the accident is attributable not so much to a single worker's malpractice, but rather to deficiencies in the whole organisational structure for the development and control of processes, and concludes with the concept of the 'resilient system', i.e. dynamic organisations in which the very idea of safety and its applications in the objective context evolve and adapt to the changes in situation by learning from their mistakes. With this in mind, the chapter aims to emphasize that actions, practices, procedures and controls, as well as human relationships, communication, managerial and organisational policies, incentives, disincentives and reward and penalization system, are all determining factors in the reduction of risk and should consequently all be taken into account in the logical process of continuous improvement which is fundamental to any virtuous system.

**Keywords:** risk management, complex organisations, work accidents, accident's analysis, Swiss Cheese Model, resilient system

## 1. Introduction

Since the mid-nineteenth century, with the advent of so-called 'mass industrialisation' and the subsequent birth of capitalist society, in most industrialised countries, we have witnessed the spread of complex political and cultural phenomena. These juxtaposed on the one hand entrepreneurs, who were seeking to maximise the profits made possible by newly introduced production technologies, and the working class, which became a victim of the increased cost of living associated with meagre wages and working conditions that were often inhumane and in which it was forced to operate. This led to the inescapable necessity for workers to join forces to deal with the 'contractual dictatorship' of entrepreneurs, drawing the attention of the public authorities and stimulating them to intervene in the enactment of laws to protect workers, not only in terms of their wages, but also to safeguard their physical integrity.

This led to the spread in most industrialised countries of the first laws to protect the health and safety of workers. However, for many years, these laws were merely designed to provide compensation to the injured, because they were intended as a tool to 'repair the damage' rather than focusing on implementing preventive measures aimed at reducing the number and severity of accidents [1]. The latter in fact were mostly considered, according to the culture of the time, as the result of fatality, or in other words, as the unavoidable consequence of the advent of mass industrialisation that had led to the spread of increasingly effective but potentially hazardous production technologies.

It would take another few decades (starting from the middle of the last century) for the public authorities to realise the need to implement a preventive system that would aim to reduce the occurrence of injuries rather than to limit their consequences in a mere logic of compensation. The new preventive approach paved the way for the creation of that branch of knowledge related to the study of injury dynamics linked to the spread of increasingly sophisticated production technologies, as well as increasingly complex business management and organisational systems. The aim of these was to find new operational tools of management that could contribute significantly to reducing accidents and thereby ensuring healthier and more efficient working conditions in terms of costs and expected results.

The same preventive approach has undergone a deep conceptual evolution since its birth, shifting from considering injuries as a consequence of the spread of increasingly sophisticated and dangerous production technologies (1950s–1960s) towards a view of injuries as the result of man's failure, unable to operate safely (1970s–1980s), and eventually leading to more modern interpretations of organisational and cultural kind (1990s–2000s). The latter were implemented in more recent years based on analyses that focused on the interactions between man and the environment in which he operates, and between technology and its practical use. This was done by establishing rules and procedures where man is considered the main actor who conceives, designs, implements and manages the whole organisation, interacting with machines and with others through communication and social and interpersonal relationships [2].

The 1950s, therefore, have seen the birth of the first major series of studies on risk and safety management in the workplace, as a result of the rapid industrial and technological developments of the post-war period. As we have seen, these studies focused mainly on the failure of technology and then on the design and construction of technological artefacts [3]. According to this perspective, preventing injury was to be achieved by improving the reliability of industrial machines and their accessibility, and by reducing the 'residual risk', that is, making them ultimately safer. The machine, placed at the centre of the organisational and productive system, guaranteed the success of the corporate mission. In other words, it was seen as having the ability to stand up to the market, to be competitive and to progress. Based on this viewpoint where everything revolved around the technology, man also became less important. The worker was seen as the mere executor of practical actions that were often repetitive and insignificant compared to what the machine was capable of doing. As a result, even with regard to injuries, man was rarely seen as being directly responsible for what happened, whereas the responsibility was rather attributed to the machine in terms of their lack of reliability and safety. Accidents, as a matter of fact, were often a result of malfunctioning equipment and technologies, as well as deficiencies in the working methods associated with them. Fires, explosions and failures were frequent and the 'residual risk' inherent in each machine was very high. Reducing injuries would have been possible by improving the design and development of new technologies, making them more reliable and safe by implementing their reliability and by adopting effective security systems that would make the occurrence of accidents less likely.

Twenty years later, starting with the 1970s, scholars began to look for the root causes of accidents no longer solely in the dangerousness of machinery and work equipment, but rather in the failure (mistakes) of workers [3]. In fact, technological progress had evolved very quickly and the result was that more reliable and safer machinery, equipment and systems had become available. Even the so-called 'residual risk', i.e. the one considered acceptable in relation to the context and the complexity of the machine operated, had dropped considerably, thanks to the introduction of technical standards that required a careful analysis of the potential hazards during the design stage.

Then why were accidents still occurring? The answer according to scholars was to blame man: if an accident happened, it is because man operated the machine improperly. This shift of perspective was favoured by the idea that only by looking at the individual causes, it is possible to come to an understanding of the factors that caused the accident [4]. This approach is based on the *individual blame logic*, namely on a '*accusatory perspective that focuses on the errors and omissions of individuals, with the belief that expert actors wouldn't make mistakes*' [5]. Human nature, therefore, is not seen as inherently fallible, but the idea is that we can think that the action of man can be compared to that of machines, and as such demand the same reliability from it. If an operator has been properly trained and has sufficient experience, he will not make mistakes!

This claim, as will be seen later, is partly true because man by his very nature will never be infallible. However, he will approach infallibility to the extent that he is put in the position to operate at his best, which will be linked not only to his level of training and familiarity in operating with the machine, but also become possible by having considered the production system as a whole, designing machines in line with the operational needs of those who operate them. However, in those years, the efforts to deal with injuries (which were seen mainly as a result of human errors) were directed at people 'in the front line', pinning the 'blame' on someone and then removing the 'bad apples'. The person approach followed a logic of disciplinary type that did not involve any intervention at a systemic level, but that rather triggered a '*blame culture*', which as we will see later is not conducive to the identification of errors, and as such, '*prevents the system to monitor the critical issues, learn from its mistakes and improve as a whole*' [5]. The *person approach*, therefore, focuses on actions, the direct source of the errors, whereas these same actions are seen as stemming from aberrant mental processes such as forgetfulness, carelessness, negligence or imprudence [6]. In this way, the errors and accidents, but also the so-called '*near misses*', are read and investigated as the direct result of the characteristics of human nature.

Later, in the late 1980s, a different perspective emerged, referred to as the socio-technical culture. This is an innovative approach that investigates the causes of errors and accidents by analysing the interaction between human and technical factors [7, 8].

The underlying assumption of the new socio-technical perspective is that accidents are caused by shortcomings and flaws in the controls and measures are put in place by the organisation to curb risky events [5, 9–11]. According to this approach, individual actors are frequently the heirs of the system's flaws. It follows that efforts to remedy those errors should be directed to the organisation as a whole, in order to improve the defences and remove its pitfalls [5]. The socio-technical perspective also helps the spread of the culture of safety, promoting organisational learning and improving the organisation through both reactive and proactive methods [12]. Organisational accidents result from the concatenation of several latent factors that contribute to the event and that originate at the different levels of the system [13, 14]. The socio-technical approach, therefore, considers *'the system'*, understood as the manufacturing organisation as a whole, as the real culprit of failure, and within the system man is seen only as one of several elements that have contributed to causing the error. With this perspective, we understand how actions, practices, procedures and controls, but also human relationships, communication, managerial and organisational policies, incentives and disincentives, and rewarding and punitive systems, are all factors that determine the success or failure of the entire organisation. This is true both in terms of production and the safety of workers, and these factors all need to be considered in the logic of continuous improvement that underlies every virtuous system.

This logic is the defining backbone of the latest studies, which examine the cognitive and organisational processes, here seen as favouring the reliability of organisations, reducing errors and improving safety conditions [15, 16] but also, as we will see later, of other studies linked to *Resilient Engineering*, where the organisational systems are seen as dynamic processes that must continually adapt to achieve their goals, react to changes in the environment and manage unexpected events.

In this sense, safety is conceived as an emergent property of the system, as well as derived from the interactions between the social and technical components of an organisation [17, 18].

Ultimately, it is impossible not to note that the study of injury dynamics has led, hand in hand with the evolution of socio-technical organisational models, to the growth of the science known as 'Ergonomics', into which several fields of study have converged, such as psychology, medicine and engineering [19].

More specifically, we have seen the evolution of 'cognitive ergonomics', which aims to study the interaction between individuals and technology through the development of models and tools for predicting human error, reducing mental workload and to provide guidelines for designing machines that take into account the limited possibilities of the human cognitive system [20].

We ultimately have returned to focusing the attention on the machines, not only to improve their inherent reliability, eliminating the hazards and reducing 'residual risks', but also rather trying to design them with the person who will then use them in mind, so that the user cannot make mistakes. In other words, the machine, operating tool and equipment, but also the organisation as a whole, are designed and manufactured to suit the needs of man, starting with what are his weaknesses and helping him prevent mistakes. ,se 'intelligent' designs are, for example, equipment items made with clearly different shapes and colours, depending on how they are to be used; tools that do not allow their operation in potentially hazardous or unsuitable sites; conspicuously diversified bottles designed to contain different substances; machines that do not permit incorrect settings or operation; and robots that are an aid to man in production, helping him to carry out operations with pinpoint precision, well beyond common human potential. In this case the tool, equipment, housing, container and the very machine mould and characterise themselves, seeking to meet the needs of their users as closely as possible, considering such weaknesses as: distraction, tiredness, stress, limited vision or inattention.

## 2. Risk management in complex socio-technical systems

A socio-technical system is nothing more than an organisation made up of people (human resources, which are the company's personal organism) and technologies (instrumental and production means). It is the interaction between man and production technology, as well as between man and man, that allows the system to operate, transforming input elements into outputs of potential interest to the market [21, 22].

We could compare the business to a human body, where there is a brain that plans, manages and sorts, in order to pursue the objectives, ideas and 'mission'. The inputs are given by the brain and through neural connections they reach the various organs that make up the human being, thus allowing the action, namely the achievement of planned goals. This synergy between the decision-making organ and the executive organs allows the human being to act, performing tasks of various kinds, from seemingly simple ones such as eating, walking or talking, to more complex ones such as practising a sport, learning a trade or playing a musical instrument.

In corporate organisations, the brain can be represented by the senior management, or by the person or people who being at the top of the organisation define the production strategies, ultimately defining the 'what', 'how', 'how much' and 'when' to produce a specific output, which is the result of the business objectives and strategies. From senior management, orders branch out and through a variety of communication channels (similar to neuronal connections), the inputs reach the departments responsible for the action, i.e. the production units. By doing so, through a series of actions involving men, tools, technologies and know-how, the idea takes shape and materialises in the finished product, i.e. in that output that reflects, or should reflect, the goals, ideas and design requirements established by senior management.

A system organised in this way can therefore be more or less complex, and its complexity will have repercussions on the organisational, procedural and management capacities that allow it to operate, including those responsible for the safety and well-being of the people who work there [23].

Where does the complexity of a socio-technical system of this kind reside? It resides in the number and typology of the elements that form it and in their interactions, staring from the people who are part of it, as well as the technologies involved and the manufacturing system used.

However, there is a special feature that clearly distinguishes a complex socio-technical system from a human being. Man's complexity is not tied to the probability of making mistakes, as is the case, instead, in the socio-technical system, in which the interdependence between the various component parts is often the cause of the sometimes unexpected failure of his work.

As a matter of fact, it is precisely the intrinsic complexity of the human being, the result of millions of years of evolution, that gives him a better chance to succeed in carrying out complex actions and that ultimately is what sets him apart from animals and makes him a superior being, guaranteeing the conservation and dominance of the species.

Humans make mistakes, of course, but the mistakes they make (which are the fruit as we will see later several factors such as: inexperience, distraction, incompetence, difficulty of the action to be taken, etc.) do not stem from the intrinsic complexity of the human being. Playing the piano, for example, is an obviously complex action that requires coordination, motor skills, rhythmic sense, concentration and knowledge of music and the instrument. However, what enables man to play the piano is precisely the fact that he has evolved into a complex organism, in which the parts of the system participate in unison to the action's execution. Each element moves in harmony with the others, thereby ensuring the same level of concentration, coordination and commitment, and reacting similarly to external stimuli.

The human being is composed of a myriad of different cells and the brain itself is made up of cells grouped into several distinct parts, each of which is responsible for overseeing different functions. However, everything is coordinated properly in order to ensure that the processes responsible for carrying out an action move in unison in the same direction, pursuing the same objectives and eliminating the unpredictability of the consequences of their behaviour, have a clear understanding of the 'what', 'how', 'how much' and 'when' to take a specific action. This complexity ultimately becomes a successful weapon.

The parts of a socio-technical system are also designed to work together to achieve the common goal. This interdependence, however, as mentioned, despite being led by a single vertex consisting of a small group of elements, cannot reach the same levels of synergy, harmony and coordination in sharing a common purpose, intent and goals, as is the case for the elements that form the human body. Therefore, the number of elements making up a socio-technical system increases their complexity and, consequently, the likelihood of failure, including the unpredictability of the consequences of its actions.

In literature, we find three types of interdependencies between the components of a socio-technical system: the 'generic', the 'sequential' and the 'reciprocal'. These refer to exchanges of the inputs and outputs of each unit and in a way also define their complexity and the potential organisational interferences [24]. With the exception of generic interdependence, in which a compartment or a branch of an organisation typically has full spatial, organisational and decisional independence, and its dependence to the system lies objectively only in the fact that its survival in the market is linked to the survival of the parent company or the partner, in the other two, instead, the interdependence is more restrictive. This is because in the sequential kind, a system unit's output becomes an input for another, and in the reciprocal kind, they continuously exchange inputs and outputs in order to add value to the finished product. In

http://dx.doi.org/10.5772/intechopen.70762

this order, the three types of interdependencies have increasing difficulties of coordination, as they contain an increasing degree of complexity, constraints and uncertainties.

In this light, it is clear how often the safety level guaranteed to a process cannot be separated from what is established in a different process and how ultimately many of the errors that cause a failure in terms of safety can arise from objectives that are not shared in common by the different parts that make up the system, or from the difficulty of working in synergy, as is the case, instead, between the parts that form the human body.

The time factor, for example, is one of the elements that most often contributes to the occurrence of an injury. Operating in a hurry on a tight schedule reduces the safety margins, causing an increased likelihood of making mistakes. As such, it is essential to guarantee each component of the system the time needed to safely perform its actions under a single supervisor/manager that manages time as a resource fairly, effectively and efficiently. However, in reality, this often does not happen and some processes are carried out in a time that is not appropriate to guarantee that they are accomplished safely.

Leaving aside the case where a worker or group of workers decide independently and without a valid reason to act in a hurry, the reason why this happens is to be found in any of the following: the former might lie in the fact that the time estimated by management to complete the process has not been properly planned out from the beginning; the second that upsets have altered the schedule for the process, which are in turn due to predictable or totally unexpected events.

Ultimately, therefore, in the first case, the resource was incorrectly quantified, while in the second case, an unexpected or unpredictable event occurred during the course of operations that has changed the surrounding conditions.

Looking at the first case, the question is: why did management miscalculate the timing of the process? The answer is often to be sought precisely in the inherent complexity of socio-technical systems, starting from the analysis of the people and processes that influenced the final decision.

Contrary to what happens to the constituent parts of the human body, which as we have seen, work in perfect synergy with the objective of ensuring the proper functioning of the whole organism, in the case of socio-technical systems the parts involved, at times, pull in opposite or conflicting directions, without being able to find an effective compromise. Reducing the production time, for example, is a goal of the management to cut down on costs. Therefore, the final decision may have been biased towards the goal of maximising profits, rather than to act according to high safety standards, underestimating the consequences that this may involve for the entire system.

Obviously, the greater the number of components in the system acting to pursue different objectives, the greater is the possible errors caused by incorrect mediation between the interests at stake. It is not uncommon that different objectives of the parts of the same system lead to inaccurate choices regarding the resources made available, which, therefore, will be to the benefit of one or more of the parts and to the detriment of others, with obvious repercussions on the management, organisation and safety of the latter [25].

The second reason that may have led to the shortage of time necessary to complete the process safely, instead, stems from the sequential or reciprocal interdependence between processes.

This is the case in which, for example, an unexpected delay in procurement has led to less time being available to complete the processes, i.e. disruptions have occurred that have changed the conditions originally forecast when designing and analysing the work stages.

The complexity of the system as a whole such as the number of procurement sources required for the completion of the process, the number of variables which in turn have influenced the timing of implementation of each of the procurement processes, as well as the procedure for the exchange of output and input elements between processes are just some of the factors that may have determined the problem.

As mentioned, sometimes the problem is predictable and therefore the system should be aware of the consequences it entails and should have already planned the corrective actions. At other times, instead, the problem is not predictable or is considered so remote as not to require a detailed plan of the actions envisioned to tackle it. This is especially true of complex systems, in which not everything can be planned in detail, since many of the elements that make up the system involve a high degree of behavioural unpredictability. In this case, the key to success for the organisation lies in being able to react quickly to the unexpected event by enforcing immediate compensatory measures to avoid the error, even if these measures were not originally planned and it is not possible to act according to a predetermined and well-tested pattern [18].

It is evident, however, that the complexity of the system, influenced not only by the number of elements that form it, but also by the number of variables that it handles, increases the probability in generating unexpected events. For these reasons, it is strategically important that the same system be able to react quickly to the unexpected events, applying those compensatory strategies that enable it to prevent the deviance from becoming a source of failure.

This is what, as will be seen later, is identified as the organisation's resilience or, in other words, the capacity of complex socio-technical systems that makes them capable, besides planning in detail the processes and procedures that can ensure that synergy of purposes and intent necessary to coordinate the various parts that form it, also to respond effectively to the unexpected event, being flexible and able to adapt to changing operating conditions, in order to find the key to success even when external elements intervene to disrupt the plan.

Sometimes the risk in a complex organisation can arise due to the fact that its components are ignorant of what the other production units have done, in spite of having shared resources such as the workspace or technology, and as such spurring harmful interferences.

It is well-known, in fact, that most workplaces operating simultaneously increase system performance as a whole, allowing it to manufacture the product in less time to the benefit the majority of stakeholders. It is therefore normal that one or more system components will push for streamlining the workspaces and production times, opting for actions that are carried out simultaneously in the same space. However, it is also a known fact that the sources of many events of injury are overlapping stages that occur in confined spaces, which consequently increase the risk of interference.

This is the case, for example, in large construction sites, places where multiple businesses often are working at the same time and that belong to the same socio-technical system responsible for completing the work, but each focused on its own activity and often unaware of what the other companies on site are doing. A general supervision that allows for the best coordination of these activities and that is able to mediate between diversified needs, guaranteeing that each has appropriate resources in order to produce its best, is what is necessary to ensure that the parts involved are driven to act in the common interest, avoiding that the actions of one can adversely affect the other [26].

The simultaneous presence of companies with different interests that are competing in the choice of technologies can sometimes generate system failures and increase the risk for the safety of those who work there. A technological resource that is not appropriate to execute a specific process is, indeed, another factor that often contributes to the occurrence of accidents.

Leaving aside the hypothesis of deliberately reckless choices, for which the identification of technological resources is the result of flagrant violations of safety rules, a department could find itself working with equipment that is unsuitable for the work and therefore potentially dangerous, seeing it was chosen based on considerations arising from conflicting interests within the organisation.

There are several examples of this. Sometimes a technology is inadequate simply because it is outdated. Conflicting interests of the system's parts, in fact, could lead to the company not replacing old equipment, which although possessing the minimal security requirements to ensure compliance with the applicable technical laws and regulations, do not guarantee the same levels of reliability, safety and ergonomics of more modern machines. The reasons for not replacing the fleet of machines are often economic in nature and sometimes linked to investment planning, depreciation or the accumulation of capital, or the occasional use of the resource does not justify the upgrade to new machinery, or more simply, because the company does not have the necessary funds.

In other cases, which find their source in the mediation of different interests of the system's parts, a technological resource is pushed to produce to the limit of its operational capacity precisely because of the need to maximise the yield. This brings the process to always work on the borderline of what is an acceptable residual risk and a high risk, a bit like what happens with Formula 1 race cars, that are driven always to the limit.

At other times, instead, the error is generated by sharing a technological resource and the simultaneous need of the different system's components to use it. This is the case for example of machines or equipment that are used for multiple tasks and that consequently require continuous changes to the settings, programming and configuration, because of the need from time to time to adapt their operation to the needs of those who use them. There is no doubt that this type of choices could prove to be strategic in the intent of streamlining the use of resources, meeting the needs of some of the system's stakeholders. But, on the other hand, the need for frequent actions to change the settings or operating modes represents another source of possible error and the characteristic elements of a complex production system that is constantly called to mediate between different needs at the risk of increasing the likelihood of a mistake and ultimately its vulnerability.

The above are only a few examples of how the complexity of the socio-technical system—understood not only as number and variability of the elements that form it, but also as different concurring objectives, as well as the unpredictability of the actions and reactions of the system's elements—might increase the instability of the system as a whole, generating those internal deviations that are the source of the error [27].

Ultimately, we cannot fail to mention the complex nature of the people belonging to the socio-technical system, though this time not as elements of the system responsible for monitoring parts of it or, in other words, as those who act in the interests of the process they manage, pursuing objectives that are sometimes common and sometimes conflicting, but to the people as such, understood as human beings and as workers.

We have mentioned that the complexity of the socio-technical system arises from the complexity of the elements that form it. As is known, there is nothing more complex and unpredictable than human behaviour.

In the comparison between the human body and a corporate system, the people who form a company are likened to the parts of the body, each of them specialized in performing a certain function and in a continuous relationship of exchanging with the other. The parts of the human body, however, always act as planned, not introducing random variables when carrying out the actions planned. A system of this kind, therefore, achieves that harmony between the parts that allows it more easily to operate in full synergy.

Every individual, by contrast, is different from another. Everyone has his/her own personality traits that influence the behaviours and relationships with others and has different needs and reacts differently to external stimuli. It is true that a person's actions in practice follow what has been established by the organisation, with the aim of achieving the common goal. It is true that training, a culture of safety, work practices and procedures, as well as orders and controls, incentives and disincentives, besides practice, simulations and exercises, represent elements that guarantee that every human resource of the system acts in accordance with the established rules, in the time and manner required, and in harmony with the surrounding environment. But it is equally true that man is neither a robot nor his is behaviour similar to that of a cell in the body. Man acts always preserving that intellectual independence, the result of complex biochemical reactions at work within him, which are in turn influenced by past experience, personal convictions, the culture in which he grew up, the society in which he operates and the behaviours of others, and which ultimately make him unique and unpredictable.

It is also true that in the uniqueness and unpredictability of the human being also lies his genius, the ability to tackle new problems, to handle the unexpected and to act outside the box to find effective solutions even in the event of a problem or emergency [28]. Man, therefore, is the strength in the management of complex systems, the key asset to rely on in order to develop systems that are able to ensure success, regardless of their degree of internal complexity and of the number of constraints or instability.

## 3. Seeking the causes of failure: reason's model

The analysis of the dynamics at the origin of an injury shows that a single error is rarely sufficient to generate the failure, i.e. to lead to serious consequences for the system. Rather, a failure is the result of a chain of latent errors, which can be attributed directly to the human resources that form the system or the technologies available to it, but which are often underestimated by those who make them or who monitor them, not paying due attention to find a solution to them [29]. The theory of errors has evolved hand in hand with the development of complex socio-technical systems, trying to explain, at least ideally, how they are not merely the fruit of incomplete or incorrect human activities, but rather the end result of a series of events (latent errors), accepting the existence for each accident that occurs of many near miss events.

It is based on these principles that in 1990, James Reason proposed a model to explain the dynamics of accidents, adopting a logic which, although a bit outdated nowadays, continues to provide a good pattern for understanding the genesis of accidents in complex systems [30]. Reason's model is also called the Swiss Cheese Model, as it is portrayed as a set of slices of cheese, each of which represents a defensive layer put in place by the organisation to prevent the occurrence of an adverse event. These barriers, however, contain holes, typical of Swiss cheese, which represent the errors, both active (slips and lapses) or run-time errors related primarily to man, and latent (mistakes), which originate from bugs in the design or organisation and overall from the systemic management of the organisational reality [31]. The adverse event will occur only if the holes of the various slices of cheese are aligned along a 'trajectory of opportunity' or, in other words, if we encounter a series of active and latent errors that will prevent the barriers to be really effective, thereby leaving room for the occurrence of the 'top event', understood as an accident. This model, although not immune to criticism, is able to represent what happens in most circumstances when an adverse event occurs, linking the responsibility for what happened to a deeper analysis of the entire organisational system. The model groups the barriers (slices of cheese) in four different types that are associated with four different 'mistakes'. The last barriers are bypassed by those mistakes that directly caused the accident or so-called 'Unsafe Acts', final errors typically made by the operator or caused by the technology malfunctioning at the end of the process.

If we run the chain of events from end to top, we find that these errors are often favoured by 'Preconditions for Unsafe Acts' or conditions that favoured the performance of the unsafe act or the malfunction in general, and for which insufficient barriers were placed, i.e. the necessary organisational and technical measures to prevent their occurrence. We then have so-called 'Unsafe Supervisions', that is, errors made by those who were supposed to supervise the actions of others, guaranteeing success, to arrive at 'Organisational Influences' i.e. those systemic deficiencies inherent within the complex organisation.

The application of this model, therefore, allows us to have a clearer view of what happened in the event of an accident, according to the logic of not stopping to consider only recent episodes of the process that led to the failure, but pushing the analyst to conduct a deeper search, leading him backwards in the investigation of the systemic preconditions inherent in the organisation itself.

In light of the above, below I provide an example of the application of Reason's model to a case that actually occurred, in which the analysis of the origins of the error allows broader considerations as to the evolution of the events, and consequently identifying also the remote causes of the failure.

The case: a lady is hospitalised due to a neoplasia with ovarian metastasis at a clinic specialising in the treatment of cancer. Following analyses in the Department of Gynaecology-Oncology, the doctors suggest an ovarian surgery followed by chemotherapy. The lady accepts and after the surgery, which was also perfectly successful, begins the first cycle of chemotherapy. The trainee specialist issues the prescription for the chemotherapy drug based on instructions from the doctor in charge of the unit, who, in addition to the therapeutic treatment, also prescribes a protective drug to reduce the toxic effects of chemotherapy. The head doctor makes a mistake writing the dose of the protective drug, confusing a comma and prescribing a dose equal to one tenth of the one prescribed by the trainee specialist. In the ward, the prescription is not checked and the lady starts the therapy. Four days later, she is hospitalised due to chronic kidney failure, reporting deafness and displaying obvious signs of intoxication from the chemotherapy drug. She is immediately put on dialysis and dies 18 months later due to complications with the dialysis. A superficial analysis of the accident would lead to focus solely on the events directly related to the occurrence of the error, i.e. the fact that the trainee specialist wrote the wrong dose indicated to him by the head doctor and therefore making the mistake that proved fatal.

Reason's model applied to this case allows us to identify, by tracing the events backwards, all those barriers in the system that did not work, either because not working, not present, removed or purposely bypassed by the chain of events.

The model starts by considering the 'Unsafe act', that is, the last mistake, which in this case was made by the trainee specialist. This error can be caused, as we shall see in detail later, by a number of reasons related to the individual, including lack of training, negligence, incompetence or wilful misconduct. In any case, these reasons are tied with the last action, i.e. the one directly involved in the failure, but that does not consider the accident as a whole. The model, as said, looks deeper, considering the 'Preconditions for Unsafe Acts', that is, the preconditions of the health socio-technical system that were supposed to limit the occurrence of 'unsafe acts' but that instead have determined the conditions that led to the error. In the case we have just examined, the trainee specialist evidently is still not sufficiently familiar with the correct doses of the drug to prescribe for him to notice the error that he is committing and therefore cannot correct it, so that his actions should have been monitored by someone who had more experience. Another precondition that encourages the mistake, typical of the academic world (but other fields as well), is that the trainee specialist (a student) probably did not feel like asking the head doctor for an explanation about the dose communicated to him a second time. It is typical of students, in fact, to think that asking for an explanation about something that is likely considered 'common ground' or 'obvious' is a 'dumb question', because they would see it as tarnishing their reputation, as well as proving an embarrassment in the eyes of fellow specialists. This means that to the simple question: 'excuse me, did you say grams or milligrams?' out of shyness, awe or fear to lose face, the student prefers to keep silent, thus causing the error to propagate.

The relationship between a student and professor, considered a 'precondition for unsafe acts' would broaden the analysis to sociological, psychological and communicative considerations, also taking into account human relationships and psychosocial risks that could affect the evolution of events.

These dynamics related to interpersonal relationships and effective communication are well known to air force operators, who since the 1970s have created and adopted a system called Crew Resource Management (CRM) [32] with the aim of facilitating communication at all levels, in view of the fact that the analysis of injury events pointed to the fact that in several cases of accidents, at least one member of the crew had spotted an abnormal event, but without, however, being able to communicate it effectively to the commander [33]. This management system focuses on the prevention of human error and consists of a set of principles and behavioural and attitudinal models that offer everyone the opportunity to examine and improve their behaviour. Specifically, CRM aims to foster a climate or culture where authority may be respectfully questioned. This is a delicate subject for many organisations, especially ones with traditional hierarchies, so appropriate communication techniques must be taught to supervisors and their subordinates, so that supervisors understand that the questioning of authority need not be threatening, and subordinates understand the correct way to question orders. Another possible 'precondition for unsafe acts' is the one connected to the analysis of the environment in which the actions that caused the error were carried out. It is well known that noisy environments where there is a lot of confusion or there are possible distractions tend to be conducive to errors. In light of this fact, the Sentara Hospital in Virginia introduced an area near the drug dispensers called 'no interruption zone', in which it is absolutely forbidden to speak. This is meant to avoid possible distractions from the nurses responsible for choosing the drug and is a solution that significantly reduces errors due to the improper administration of drugs [34]. Tracing our way back in the analysis of events, we find barriers in the system regarding 'Supervisions', which are the tasks involved in monitoring actions. The head doctor should have double-checked the prescription filled out by the trainee specialist before signing it, as he should have been aware of the chances of him making a mistake, precisely because he is a student. Double-checking also happens to be an effective and widespread tool in a complex socio-technical system because of how efficient it is at avoiding mistakes. This is a standard procedure in the most critical processes, where the consequences of an error may have a strong impact in terms of damage. A doctor or other specialist double-checking the prescription would alone have been enough to put in place an effective barrier to the error, thus avoiding its spread. Even in this case, though, probably due to lack of time, incompetence, or excessive confidence in the trainee's skills, this was not done and the prescription was passed on to the ward as if it were correct. At the level of 'Unsafe Supervisions', we need also to underline the lack of controls in the ward. Here too, the 'barrier to error', i.e. the safety procedure, would require a doctor in the ward to check the prescribed therapy a second time when defining the health protocol. But again, unfortunately, no one did, leading the nurse to administer the wrong dose of the drug. At the level of 'Organisational Influences', that is, with reference to those systemic deficiencies inherent in the organisation, as mentioned, we have to seek the remote causes that led to the primary conditions for the occurrence of the accident. In this case, it should be noted that the head doctor is the only person authorised by the organisation to issue prescriptions for medical treatment, as he is the person who signs the protocol. This procedure is aimed to avoid possible errors of misunderstanding on the quantities or the names of prescribed drugs from the get-go. As we have seen, however, this did not happen. One possible reason can likely be the constraint to save time. Independently of considerations that are specific to our example, in which the doctor may have deliberately committed the irresponsible acts he is blamed for solely due to negligence or because he intentionally decided to break the rules, from a systemic point of view, not having enough time to perform one's duties, as we have already discussed, is a symptom of an inefficient organisation that is often understaffed and forces employees to work on taxing shifts and with excessive workloads.

A corporate reorganisation that better factors the activities to be performed together with the resources to be used, encouraging mobility, dynamism, waste reduction and resource optimisation, and that ultimately promotes 'continuous improvement' through the continued implementation of its processes, is often the decisive response in limiting the conditions that are conducive to the errors. Probably, in a more relaxed environment and with more time, the treatment plan would have been filled out correctly or would have been checked by the head doctor, avoiding the error's occurrence. At a systemic level, we again need to consider the 'wrong practice' or approval of inappropriate behaviours as if they were normal, which, as we will see later in the systemic and organisational theories related to failure, is what often takes the name of 'normalisation of deviance'.

Everyone knows it is not allowed to have trainees issue prescriptions, much less (but as we unfortunately see happening in some healthcare facilities) to have them carry out complex tasks such as making diagnoses or prescribing treatments without proper supervision from the head doctor. But, again, we unfortunately see this happen in many healthcare facilities around the world, where interns replace doctors, as a way to cope with the chronic shortage of staff that often plagues many healthcare systems.

A risk inherent in the blind acceptance of the Swiss Cheese Model, however, is to make the assumption that the more slices of cheese are placed in sequence, the better the chance of avoiding errors. This assumption translates into the organisation having as a priority the need to implement as many defences as possible without, though, considering the repercussions that this can entail for the reliability of the whole system, such as the emergence of new problems (failure mode, directed or caused by interactions between the barriers) [35]. For example, theories on the reliability of regularly servicing machines and plants view disassembling the machine or plant and then inspecting and reassembling its components as being certainly effective and therefore a barrier to error (failure/breakage). Performing such maintenance tasks undoubtedly increases the number of barriers (slices of cheese), as it allows staff to thoroughly inspect the machine/plant for possible faults or conditions that could lead to a possible system failure. According to the assumptions inherent in Reason's model, this practice ensures greater reliability to the system. In reality, however, the result is not always an increase in reliability, because when disassembling and inspecting the components, the errors that can occur when having to reassemble them also increase. The new barrier (slice of cheese) must indeed be viewed as a defence, but also as a possible source of holes on other slices of cheese, holes that initially were not present but that were caused by the introduction of the new slice. Another limitation of Reason's model is to consider errors as resident system pathogens, similarly to how we consider viruses in the human body when they have not yet manifested their harmfulness or, in other words, as harmful entities that can actually wake up suddenly and cause a disease in the body [35]. This assumption does not actually always reflect what is happening in the genesis of the error in complex socio-technical systems, because viruses are purely harmful agents, while the actions that the system performs and that can lead to failure are, as stated by Reason himself [36], the same that sometimes lead to success. The consequence of this is that we cannot always consider the error as necessarily harmful.

## 4. Man as an essential resource for solving accidents: the rise of resilient systems

As a matter of fact, machines do not make mistakes as long as no unexpected events happen against which they are often not able to establish a viable alternative compared to the pre-set programme. 'Man, then, is certainly less reliable than computers in some respects, but if the automatic system goes haywire, who but man can intervene to solve an unexpected problem?' [37].

The Apollo 13 mission has passed down in history as one of the most difficult missions of conquering the Moon, in which a number of malfunctions and unexpected events brought engineers and crew members to make corrective actions, often without following a pre-planned procedure and breaching the mission protocol [38]. The genius of man, in his ability to cope with unforeseen problems also by committing actions that technically had been classified as 'mistakes', was the key to the mission's success, allowing him to avert disaster.

Already when the spaceship was coming out from the Earth's orbit, there was a first dangerous accident. One of the five second-stage Saturn V carrier's engines stopped working, forcing mission control to make the decision to use the remaining four engines longer than planned and thus acting contrary to the rules, in order to allow the mission to continue. Even the third-stage rocket engine was operated beyond its capacity in order to correct the deviation from the trajectory that occurred after experiencing the first problem. Shortly after the successful lunar module docking manoeuvre, 321,860 km from Earth, one of the four oxygen tanks of the command and service module (CSM) exploded, causing serious damage. At this point, it was decided to cancel the descent on the Moon and the engineers decided to focus their efforts on the re-entry manoeuvres. The damaged CSM module included a command module, which hosted the three crew members and which was in fact the space shuttle that would have allowed the astronauts to return to Earth. It also contained a service module needed for the first module to function, which in turn housed, besides the thrusters for re-entry manoeuvres, even the technical equipment to allow life on board and the oxygen tanks. Such a massive malfunction in the service module prevented the astronauts from using the command module for return to Earth, condemning them to a slow and unrelenting agony in space. However, because the lunar module (LEM), fortunately had already been hooked on the CSM when the plan was still to perform the Moon landing, it was decided to transfer the entire crew within the LEM and to use it to return to Earth. The lunar module, though, had been designed only to land and take off from the moon. As a matter of fact, once the astronauts had been allowed to return in the CSM, it would have been abandoned in space orbiting the Moon, having in fact completed its mission. The problems encountered clearly were not trivial. Everything had been turned upside down and the crew was now forced to navigate by sight, without relying on proven mission protocols. To begin with, a crew of three people was occupying a module that had only two seats. In addition, the module itself was preparing to face a 4-day long return journey to Earth, whereas it was intended to land and take-off from the Moon in a total time of 2 days at the most. As such, all service equipment designed to allow life on board was undersized compared to the needs. The crew, in fact, was forced to assemble the new carbon dioxide filters to be installed in the LEM while orbiting and with very few tools available, considering that those provided were not sufficient for three people and for twice as much time than originally expected. The same engines of the LEM, which had been designed solely for landing and taking off from the Moon, did not guarantee success if used for different purposes, such as to return to Earth. These factors sparked long and exhausting discussions between the engineers responsible for monitoring the engines at the base in Houston. Some of them, in fact, did not agree to use the engines of the lunar landing module for the shuttle to return to Earth, believing such a manoeuvre posed a serious risk to the crew's safety. The lunar module engines were specifically designed to complete the landing mission, which meant there was no guarantee of the mission succeeding if they had been used for different purposes. Also, when returning to Earth, there would have been the need to turn on these engines several times to correct the trajectory of the LEM, but the engines had been designed to be switched on only once during lift-off from the moon. Other engineers, instead, considering the serious risk of explosion in attempting to reignite the engines of the SEM due to the damages consequent the explosion of the oxygen tank, were opting for taking the risk of using the LEM's engines. Either way, a decision needed to be made quickly and in the end, the decision was to authorise the use of the lunar module engines to return to Earth, although this operation was not contemplated in any of the mission's security protocols.

The re-entry operation fortunately was a success and astronauts Jim Lovell, Jack Swigert and Fred Haise managed to return to Earth safe and sound.

This experience is an example of how man's ability to cope with unexpected situations, even acting against protocol and therefore committing 'errors' compared to the original plan, can sometimes prove decisive for success. The group of engineers who were pending for not using the lunar module engines basically reasoned on a 'rule' level, i.e. advocating the peremptory enforcement of rules. The procedures said that these engines were designed for a specific purpose, that there was no guarantee of success and that using them for different purposes was not provided for in the protocols, which means it would have been a mistake. Those who were in favour of using the engines applied a 'knowledge'-based approach, that is, to use human intelligence to cope with the unexpected event, not following written rules, breaking some of them and ultimately inventing new ones.

This experience shows how, especially for highly complex socio-technical systems, in which a complicated process cannot always be planned in every detail and in full compliance with a series of strict rules and preconceived algorithms, since the variables involved can be endless and as such, can easily give rise to unexpected events, man's ability to decide to breach the rules can sometimes be decisive.

'Complex systems are always in dynamic equilibrium. They seek variety because it is only through variety that they evolve [...]; sometimes it is positive variety, i.e. 'good' information that enables the system to succeed. At other times, it will be negative information ('error') and could make the system collapse or change it' [18]. An adaptive system, therefore, also learns from the negative information, which is 'news' that increases the system's resilience.

We must not, however, think that to improve the reliability of a system, we have to wait for the occurrence of a serious accident. Improvement comes from analysing the errors, true, but considerably important among them, as we have seen, are latent errors, those holes in the system that caused minor malfunctions, weak signals for which the organisation should fine-tune its sensitivity to know how to recognise and correct them, capitalising on the information it gathers from them. Thanks to a safety culture (correct self-analysis and reporting procedures), we can then build on the information gathered from the analysis of the system and consequently correct deviations, sometimes even improving the rules and considering new scenarios. Risk management based on fixed procedures and established practice, in other words based on a "static management" in which everything is pre-planned, is not able to represent continually evolving systems such as highly complex socio-technical ones are [39]. This is where the theory of 'resilient systems' stems from.

The term 'resilience', its original meaning at least, belongs to the physics of matter and means 'the ability of an object to regain its initial shape after undergoing a deformation caused by an impact'. The Help Center at the APA (American Psychologists Association) defines resilience as 'the process of adaptation in the face of adversity, trauma, tragedy and other significant sources of stress'. It is therefore 'an individual's ability to overcome unusual circumstances of difficulty due to his behaviour and mental qualities of adaptation' [40]. It means "'finding a way out' from difficult experiences" [41], in other words, find the smartest and fastest way to solve the problems. According to computer science, it is 'the quality of a system that allows it to continue to function properly in the presence of faults in one or more of its constituent elements. It therefore indicates the system's tolerance to faults, malfunctions and breakage'. Resilience is typical of high-risk organisations and their adaptive capacity is decisive against threats to the organisation's integrity and hence, in avoiding failure [28]. Some of the main features of a resilient system are: (revised by the publication of Steven and Sybil Wollin [42]; drawn from Fichera [28]):

1. Insight: that is, the system's ability to examine itself, i.e. to gather the information that led to the traumatic event, analyse the problem, know how to read the elements relating to the context and seek alternative solutions.

2. Independence: the ability to maintain its own identity, while not isolating itself—autonomy and independence, without feeling totally bound by other systems or resources that they are able to provide.

3. Interaction and communication: the ability to establish satisfactory relations with its surroundings, entertaining more or less close relationships that, however, if necessary allow to develop the system's aim.

4. Proactivity: the system's ability to acquire and process enough information so that it can read its environment consistently and use it for its own purposes.

5. Creativity: the ability to create the 'organisation', i.e. a new functional order starting from the chaotic and depressive situation following the destabilising event.

6. Humour and irony: a disposition of the person who is part of the system to break away from the problem by belittling its importance through a 'language game' that allows an analysis and positive re-evaluation of the event.

7. Ethics: refers to the ability to feel part of a macro socio-political system and share its values and dynamics.

To clarify the meaning of the unique aspects of resilient systems, we will contextualise their underlying principles below. First of all, we ought to point out that it would be enough to effectively implement the principle of insight to achieve substantial improvements in the vast majority of cases in terms of the organisation's reliability at large. The system's ability to examine itself, to find elements that may have caused or led to the errors, lies in this principle. There are a number of methods to practically implement these analyses, but the starting point is always the systematic collection of information. Unfortunately, organisations are often lacking a valid system for the collection of data that are useful to monitoring their behaviour. Collecting data, in fact, is costly, both in terms of time and resources. It is also a boring job, which is often done by people who normally are engaged in other tasks and consider the needs to fill out forms and records as waste of time that hinders their job. Hence, records are often not kept or are filed only partially and sometimes retrospectively, trying to trace the chain of events and approximating most of the information they contain. A resilient system, instead, needs data and must be able to identify, through a careful analysis of the key elements inherent in the dynamics of the system itself, the strategic information deemed necessary for its implementation, and trying then to systematise and possibly standardise and computerise the data collected. If we were to make a comparison with organisations operating in aviation, it would be like having a black box for each event that occurred. In the event of a system malfunction, therefore, it would be enough to draw information from recordings of the events to get a clearer picture of what happened, and accordingly establishing what should be the improvement actions to be implemented to ensure that the event no longer occurs.

This requires that all the system's resources are aware of the importance of participating actively in this collection of information, especially by establishing reporting elements that can be easily filed (preferably electronic) and that mention the key data as to what is taking place within the environment in which the resource operates. Ultimately, it is necessary that the system is able to identify the crucial points that require a record of the events, such as the correct filing of a medical record, but also records about data that is helpful in analysing the timing of the process, the results achieved vs. those budgeted and customer complaints or ones filed by other stakeholders, rather than reports on so-called near misses. Implementing the above enables the system to collect valuable data about its operation, but alone is not enough. It is equally important that the system has the resources and skills necessary to effectively analyse the collected records whenever it is necessary to investigate an adverse event or when we want to tap into statistical data that can help improve it. This is linked to the principle of proactivity, another feature typical of resilient systems.

Proactivity means moving ahead of events, predicting what will happen in the future through a careful analysis of what happened in the past. What has happened in the past is represented by a set of records, but it is then necessary that within individual organisations, operating units are established/implemented that will analyse and process the collected data and that are able to draw information from the records that is useful to achieve improvement. In fact, what would be the point of collecting and analysing such a huge amount of information if we then did not act to solve the problems identified? To do this, the organisation needs to have a unit that works as much as possible in contact with the individual departments and not only at a centralised level, in order to be able to thoroughly understand its needs and the singularities associated with the individual processes. It is then necessary that the unit be able to critically analyse the results of monitoring the process indicators, suggest solutions for improvement that are valid and applicable but above all, communicate effectively with those who then will have to take the necessary decisions and authorise the actions proposed. The system must ultimately be able to operate according to the logic and principles of total quality management, where a 'senior management' periodically reviews all information collected by individual 'process owners' through a set of business process 'indicators' appropriately chosen to monitor the system and take proactive actions aimed at achieving 'continual improvement' of the organisation.

The concept of 'independence' of a resilient system is less agreeable, as it apparently conflicts with what is happening in the world today. In fact, we live in an increasingly globalised world where each subject, whether we are talking about a country, organisation or market, is increasingly linked to the existence of others, because it constantly requires interaction and support. While recognising in the clustering effect of globalisation an indisputable advantage when it comes to streamlining resources, it is important that the 'satellite' organisation, whichever that may be, preserve its own identity within the macro system in which it operates and especially the possibility to manage the resources necessary for its operation as much as possible independently. The implementation of the above makes the organisation accountable and more prone to engage in ensuring the productivity of what it produces, in essence making it directly responsible for the success or failure, avoiding the fear of belittling the recognition of success and conversely discouraging the idea of being able to impute to others the causes of its failure. These aspects tied to autonomy and independence are the same that act at the level of the individual and that greatly influence his behaviour, making him more participatory or otherwise unresponsive towards the organisation. Systems that give incentives and reward performance, when combined with guaranteed freedom of action and thought, stimulate creativity and innovation, which are the basic ingredients of resilient systems. Independence, however, should not prevent the system from 'interacting and communicating' with its surroundings, or with other units of the same system or with other organisations. We find an example of how communication is implemented within a complex organisation to limit the occurrence of errors in healthcare organisations. A hospital, in fact, must be able to accommodate its patients and receive their relatives, limiting as much as possible situations of uneasiness and interacting with them with professionalism and competence, but also frankly and politely. In this regard, there are, for example, several hospitals that use different colours in the hallways and wards to help visitors find the correct path to follow. With reference to communication between the medical/nursing staff and patients, conversely, specialised courses on communication techniques are becoming increasingly popular and are especially effective in cases where there is a need to communicate with clarity and professionalism and even with discretion and tact. As for internal communication between members of the same system, it is now a known fact that it is at the core of the 'best management' of any organisation. Any system consisting of more than one individual needs to ensure the effective delivery of information between the different individuals that are a part of it. In this respect, an example of the procedures that some organisations have adopted to facilitate the exchange of information is to have the shifts of incoming and outgoing personnel slightly overlap. This favours the chances for colleagues to meet physically in short briefing sessions to exchange information on what happened during the previous shift. Some abnormal events, in fact, might have changed the system's conditions compared to ordinary ones, and as such involve the need to communicate the changes to a colleague. The architectural design of the working environment should then promote the logic of briefing, providing an enabling environment where workers can organise such meetings.

Creativity is perhaps the true essence of resilient systems. Whenever an adverse event tends to destabilise a system, the latter's resilience opposes failure by generating the necessary self-defence mechanisms necessary for its survival. This is the case for example of emergencies or situations that diverge from those imagined, for which, as we have seen already, the resource does not have the time or ability to apply the protocol as planned. In these cases, creativity, both personal and of the work team, has a role at the level of 'knowledge', leveraging their prior knowledge and its own intelligence to resolve situations not attributable to familiar contexts. It is important, however, that when these circumstances occur, we are able to capitalise on the information resulting from those circumstances, treasuring them and exploiting them if necessary to change the rules for the organisation to function. In the air force, for example, as well as in Formula 1 or motorcycle racing, at the end of each qualifying lap or race, drivers are used to meeting up to review together with the engineers, team managers or personal coach how the events unfolded, analysing them, exchanging opinions and information and ultimately learning new scenarios that may occur again in the future. A critical analysis of these events allows them to make the right 'corrections' to the system, capitalising the information received and encouraging its growth.

Humour and irony are two characteristics of resilient systems that refer to the individual more than the system as a whole. As previously mentioned, they are actions meant to enable a breakaway from the problem by belittling it through an analysis that can lead to a positive re-evaluation of the accident. Being able to apply humour and irony to a problem is something that pertains to the psychological framework of each of us and refers to our ability to address the problem with a positive attitude, knowing the pitfalls, but without being overly intimidated by them. In borderline cases, breaking away from the problem results in what we call 'cold blood', that is, the person's ability to exorcise the problem preventing it from overpowering him and thus throw him into a state of panic and as such, hindering his ability to react effectively. The right attitude to deal with situations that involve high emotional stress sometimes stems from innate personal aptitudes, but more often than not these skills can be learned, following the advice of more experienced staff and participating in training courses and simulations. Everybody knows about the widespread use of flight simulators in aviation, aids that are useful to teach the trainee pilot how to cope with unexpected situations, sometimes pushing him to the limits of what is solvable and getting him used to situations where it is necessary to maintain the proper concentration and cold blood. Even in the medical field, emergencies in the emergency room require staff on site possessing the same skills, both technical and in terms of aptitude. For this reason, even among medical staffs, in recent years, realistic simulation techniques have become more and more popular in the training of doctors and nurses. For example, the spread of sophisticated mannequins, able to emulate what could happen in real emergency situations, get used them to developing the ability to cope, with the required preparation and with all due 'detachment', to emergencies they will likely find themselves handling in their job.

It would also be desirable, although hardly feasible, that coordinators know how to identify within each organisation, the resources that are most predisposed to operate in situations involving high emotional stress. This is the case, for example, in the choice of emergency response personnel. Doing so would enhance the possibility of establishing, even within a same shift, a team of people who can work together, each supporting the other's idiosyncrasies, as happens in operating units of the armed forces, where everyone has their own role determined by referring not only to the individual's aspirations, but also to what are his skills and abilities to take action.

Ultimately, ethics represents that additional property of resilient systems of defining its own values and policies shared by all its members. Ethics ultimately is part of the concept of 'corporate policy', which has always been among the cornerstones of quality management systems that have nowadays become the standard. It recalls the need for the organisation to define, inform and especially share with all of the system's members the principles that inspire and guide the business, both internally and within the social context in which it operates. The pivotal point of this principle is contained in the following question: how can we expect rules and procedures to be followed if those who are called to do so do not agree with the strategies and policies that have generated them?

The principle should be familiar to those who deal with security and prevention and its essence can be clearly explained by the following example. If we were called upon to ensure that the staffs of our operating unit wear a helmet, rather than fasten their seat belts or proceed according to our instructions, we could act by enforcing corporate security procedures requiring staffs to comply with these requirements. We also could define the penalties applicable for those who fail to perform as required, and ultimately, we could ensure efficient surveillance, checking that everyone behaves as prescribed.

All this would not be as effective in assuring us the expected results as being certain that our unit really agrees with the need to resort to the safeguards we have devised to avoid damaging repercussions for their own safety. This is what we refer to as a 'safety culture', i.e. endorsing certain convictions, certain principles, certain values of our own and, therefore, behaving accordingly, in accordance with the procedures given, not because they have been imposed on us, but because we are aware of their actual effectiveness.

## 5. Human error and the ability of the system to limit it: correlation between man-machine-environment

Man, as it turns out, is the main resource we need to focus on to improve the reliability of the organisation; through him, it evolves and becomes capable of coping with unexpected events, thus reducing the likelihood of making mistakes. It is, therefore, essential that man be put in the position to be able to operate at his best, taking advantage of his strengths and limiting the consequences of his weaknesses. Human error is commonly referred to as a 'performance that deviates from a prescribed and specified sequence of actions' [43]. And properly handling human error is, after all, a difficult and delicate operation, because the whole operation of fully securing a system depends on it.

To understand how the system can limit the occurrence of human error, we must go back to its genesis, i.e. retrace the process that starts with man's thought and leads to the execution of an action.

Jens Rasmussen in 1983 proposed a classification of human behaviour [44], in which he illustrates a model that is still used today as a basis for understanding the genesis of human errors and for accordingly implementing appropriate actions in trying to limit them.

According to Rasmussen, human behaviour can be broken down into three different types:

- skill-based behaviour,
- rule-based behaviour, and
- knowledge-based behaviour.

'Skill-based behaviours' are automatic behaviours that the subject has already acquired and that he therefore has fully mastered. They include actions that man carries out without having to think or reason about them, nor pay particular attention to during their execution. For example, brushing our teeth, tying our shoes or pressing the brake of a car are all behaviours acquired and implemented automatically, which do not require special attention or concentration. In this case, an error in their execution, as indicated by J. Reason in his book Human Error, 1990 [30] happens simply because of carelessness (slips) or forgetfulness (lapses). The person who performs the action automatically makes a mistake not because he does not know how to respond properly to the situation or cannot remember how to perform the task, but simply because he is momentarily distracted. No grown man and in full mental and physical capacity, in fact, needs to make it a point of remembering to fasten his shoes, let alone follow a particular procedure to remember how to do that. However, we may sometimes leave the house with our shoes untied.

'Rule-based behaviours', instead, are behaviours for which the person is acting according to a rule, whether it be a procedure, practice or regulation. He, therefore, focuses on carrying out actions in the way he has been instructed. The error in this case can be determined by lack of knowledge of the rule ('mistakes'), which thus leads to a wrong action. For example, if a fault indicator of a plant is switched on but the operator does not remember or recognise its significance, he may not implement the proper corrective action. This error, attributable to man as its mere executor, however, is not to be confused with the error that might take place if we were to follow a procedure correctly and nonetheless end up with nefarious results. In this case, in fact, although the behaviour of man is still based on rules, it turned out to be wrong as it is the very procedure that is wrong. Therefore, the mistake will not affect the person who made it, but rather the person(s) who invented the rule.

Finally, 'knowledge-based behaviours' concern actions in which the subject is faced with situations completely or partially unknown to him and must implement predetermined choices to address them, relying on his knowledge of this kind of situation or, in other words, drawing from his cultural and personal experiences. This is the case of innovation or of unexpected events, situations in which, as already seen, only man's ability to making choices that break the mode can be decisive and lead to success.

The consequences of these distinctions are numerous and it is therefore important that the organisation knows how to recognise them, in order to help its members execute actions in harmony with the system that surrounds them. We have repeatedly hinted that knowledge of the 'rule' by the worker is an effective weapon in preventing mistakes. The system designs procedures and work practices, planning their goals, implementing rules for their execution, the resources to be used, the timing of implementation and identifying and correcting possible deficiencies over time. Therefore, assuming that the system has generated a correct procedure and assuming that no unforeseen changes have occurred to change its validity or jeopardise its results, a worker properly trained on how to execute it should not make mistakes. However, we know that in reality, this is not always true. There are multiple causes to errors, but one of the most common lies in the worker shifting in attitude from a 'rule-based' to a 'skill-based' behaviour. As time progresses, especially for those routine actions that the worker carries out automatically and without having to think about them that much, his threshold of attention in tackling them lowers, leaving room for possible distractions that could induce him to make the mistake. When it comes to driving a car, for example, we carry out a series of routine actions in sequence, like closing the doors, putting the transmission into neutral or engaging the clutch, turning the key, or disengaging the parking brake. No one, except for novice drivers, runs over what he has to do or how to do it before carrying out the task, but carries it out automatically. Sometimes, though, we will start the car without disengaging the brake or without closing the door.

How can the system limit these distractions?

The answers are varied and are found in how the human brain works. It is well known, in fact, that the brain cannot process all the information surrounding it with the same attention span, and to avoid being overwhelmed, it tends to ignore the information it believes is unnecessary or of secondary importance. Routine behaviours, precisely because they demand less concentration of the brain, run the risk of receiving less attention, especially if an external stimulus, intended as a distracting event, is introduced and 'seizes' the brain's attention. After all, if we think about it, the human brain needs to feel stimulated. Contrary to what happens, for example, in the CPU of a computer, which can work even for hours at very low utilisation rates, the human brain tends to always work within an optimum range of 'engagement'. If we overload it with information, it will get tired and will tend to lose track of some of the information, but on the contrary, if we do not engage it enough, it will look for other stimuli, focusing the attention elsewhere and trying to process other information from the surrounding environment. If the other information it finds does not stimulate it enough, the brain will draw from its own memory and emotions, ultimately generating those mental processes that we call thoughts, or more correctly refer to as overthinking, understood as mental processes that stray from the context and that are distracting.

The strategy, therefore, is to help man maintain a high level of attention on what he is doing, trying to not overload him with information that might wear him out, but at the same time preventing that the information become ordinary, dull, or uninspiring, with the risk that the brain lowers the level of concentration, ultimately leaving room for distractions. Workplaces, machines or infrastructure should therefore be designed to require that the resources that use them keep a level of attention above the minimal threshold that leads to their distraction. An example is the design of highways, in which the tendency has long been, whenever possible, to avoid long stretches of road on a straight path. In fact, such straight and long stretches make driving monotonous and thus are likely to distract drivers, including gentle curves, although not necessary for purely technical reasons, engages the brain more actively, forcing it to concentrate on the road.

Very popular are also cases where the worker is forced to act in accordance with 'rule-based' behaviours, thus avoiding the risk of the tasks becoming routine and instinctive at the root. Such is the case, for instance, of the growing use of technical checklists like those used in aviation before take-off. The peremptory inspection of all piloting and emergency devices executed in accordance with a list of points addressing the tasks that need to be performed undoubtedly lowers the likelihood of errors, which in this case are due not only to possible distractions but especially to forgetfulness. In this regard, however, there is a need to be cautious and to resort to the use of the checklist only when they are actually necessary, limiting the list of steps to follow. A disproportionate use of checklists beyond what are the real needs would lead man to consider them overly long, boring and probably of little use, and therefore not to apply them with due care and thus reducing their positive effects.

In other cases, the working environment or the machine is designed to 'physically impose' to the person who works there (or who operates the machine) that he act in compliance with the safety standards designed by the system, preventing him from making mistakes. One example is the safety lever on hydraulic excavators that disables movement of the excavation arm. It is located on the side of the driver's seat, forcing the worker to raise it physically in order to climb out of the machine. Or the drive control of some industrial presses that is twofold and installed at a distance, which forces the operator to use both hands to operate the press, forcing him to keep them away from the danger zone.

Sometimes the mistakes man makes due to distractions involve involuntary movements, such as bumping into an object or inadvertently operating a command. In the case of the use of machinery and work equipment, a standard practice to avoid such mistakes is to design controls with 'double consent'. These controls require that the worker carry out at least two actions in sequence in order to operate the device, thus avoiding involuntary actions.

The above examples help us understand just how crucial it is that the whole production system, starting with the design of the working environment, as well as the machinery and equipment installed within it, takes into due account the needs of man, in order to help him carry out the actions without leading him into error. The working environment accordingly is moulded around the needs of the individual, promoting (as we have mentioned) his strengths and minimising his weaknesses. Knowing the causes that lead to human error is also crucial when planning the best corrective action. As we have seen, one of the key prerequisites for improving the corporate safety is to educate and inform employees on performing the right action, in other words, they should always know 'what to do' and 'how to do'. But in cases where, for example, an expert nurse needs to perform a blood test without having previously disinfected the patient's skin, it would be entirely useless to impose that she attends another training course. Her behaviour in performing the blood test, in fact, is 'skill-based', so that only her inattention or forgetfulness may have caused the failure, and certainly not a lack of knowledge of the rule.

Errors due to 'violations', instead, i.e. where man purposely carries out a wrong action, breaking the rules, are a whole different subject. Violations essentially can be broken down into: 'routine', 'necessary', or 'due to wilful misconduct' [30]. Aside from these, which are intentional, that is, made by a person who is fully aware of their outcomes, which is sometimes disastrous, and that involve deliberately acting against the rules, 'routine' violations are known both to workers and supervisors/managers and are normally 'accepted' by the system, as they are sometimes deemed useful to optimise/accelerate the execution of an action. The violation is also not viewed as possibly impairing the system's operation or as really increasing the risk to the worker. A routine violation, in fact, is always linked to its perpetrator (more or less) underestimating its consequences, both in terms of the probability of its occurrence and of the damage. A routine violation is a 'short cut', a way to simplify a rule which is seen as unnecessary, because it is considered too complex to be respected/redundant or overly cautionary to be complied with in full. Removing a crankcase from a machine just for speeding up work operations, not using personal protective equipment against falls from a height considering the occurrence unlikely, or not wearing a seatbelt only because uncomfortable, are some of the many examples that lead man to expose himself to the risk, breaking the safety rules.

One of the best strategies applied to avoid these violations is that of 'setting the example', especially by employees who have a role of responsibility in the company. Man, in fact, tends to emulate others and modifies his behaviour based on the example set by others, be it positive or negative. In other words, he acts in tune with what is considered the standard practice within the system, often regardless of the behaviour expected of a rule. If, for instance, we were to transfer some workers operating in a virtuous system and who diligently follow the rules to a similar system that is not as virtuous and where transgression, instead, is the norm, many of them after a certain time would begin to conform to the new system, breaking the rules. The system should then set the example, encouraging the appropriate behavior of its resources, also through the use of rewards for proper conduct and not merely punishing transgressors. In this way, each of the system's components will receive fair recognition for his work and will be driven to apply its virtuous behavior, advancing the entire organisation.

The urge to break a rule not only stems from considering it overly cautious or not fully agreeing with its purpose, but also often stems from benefits (sometimes even financial) for the company or, as we have already seen, for some stakeholders within the system. However, we need to consider that the 'normalisation of deviance'—the process that generates a steady erosion of standard procedures, where minor violations and irregularities are accepted and tolerated, in the long run could prove destructive to the system's integrity. In the absence of accidents, in fact, these deviations will become 'normalised' and part of the routine. This will lead to people within the system slipping into danger without being fully aware of it.

Such is the case, for instance, of 'reverential bows', which have become famous after the tragic events that involved the cruise ship Costa Concordia, which on 13 January 2012, helmed by Captain Francesco Schettino, while preparing to perform a flyby along the coast, hit a reef off the island of Giglio and flipped over on its side, causing the deaths of 32 people [45]. I do not want to make direct references to the case in question, in which the trial ended with the conviction of Captain Schettino, whom the Court judged to be the sole person responsible for the heinous choice to perform a dangerous manoeuvre for selfish reasons alone. However, multiple sources report that this ill-considered action was not the only one of its kind in the history of navigation, but on the contrary, as many witnesses confirm, would seem to be a widespread practice. The reasons that may lead a ship captain to make a 'bow' are not exclusive to a particular type of person. For some, the motivation is strictly personal, i.e. the desire to sail near the places where he was born or grew up or where his loved ones live. For others, it is simply the taste of doing something daring, while some people suggest that economic interests spur some ship captains, for whom the majestic view on a cruise ship just a few metres from the shore offers considerable publicity. The fact is that one or more parts of the system are driven to make a potentially risky manoeuvre, reassured in their choices by what we already called 'normalisation of deviance'. The management thought that transgressing the rule was possible and that the outcome of their actions was well worth, determining only a little increase of the risk, considered wrongly to be minimal and controllable. The story of the Costa Concordia, instead, sadly reminds us that the often positive outcomes of risky behaviours should not lead us to consider them as legitimate. The fact that we often avoided disaster must not lead to an acceptance of the risk. Approving risky behaviours to pursue personal agendas, in fact, produces countless negative effects: 'it damages the culture of safety; it stretches the boundaries of the risk; it increases the tolerance of errors that do not cause damage; it increases the level of acceptance of risks in favour of interests tied with efficiency and productivity; it ultimately leads the safety system down a slippery slope in which accidents are increasingly possible' [46].

Ultimately, 'necessary' violations, as we have seen, involve a deliberate and motivated steering away from a well-devised procedure because the actual conditions make it impossible to

http://dx.doi.org/10.5772/intechopen.70762

follow or lead us, after meticulous thought, to prefer different solutions. Apart from the case of the Apollo mission that we discussed earlier, in which the head of the mission authorised the use of propulsion systems to return to Earth against protocol because in the given state of emergency, this solution was the only means of salvation; if we again consider the case of the Costa Concordia, it is important to note how the manoeuvre made by Captain Schettino immediately after the disaster, which was to dock the ship on a rocky reef just North of the port of Giglio, although such a manoeuvre was not contemplated in the emergency procedures, ended up facilitating rescue operations, avoiding to have to perform them in open sea, with all the consequences that would have involved. The impressive gash on the left side of the ship, in fact, would have caused it to sink, considering that as many as five compartments had filled up with water (while protocol established three at the most). 'Necessary' violations, in short, occur whenever an unexpected event occurs, i.e. in those conditions where time, the context's conditions and compelling necessity impose a breach of protocol, which, therefore, in that moment is seen as inadequate, leading us to prefer unplanned actions that are the fruit of the 'knowledge' of the person who carries them out. These violations represent a 'treasure' that must be capitalised by the organisation, whether or not they lead to the mission's success. Every time that 'necessary' violations occur, the organisation must ask itself how it got to that point, what did not work, what it failed to forecast, and what did not go according to plan. Consequently, one or more briefings after the event will allow it to evaluate the pros and cons of the actions that broke the rules and will ensure that the organisation can learn from the situation by improving and implementing its own rules and procedures, and as such, displaying that proactive behaviour that is typical of resilient systems.

## 6. Error reporting incentives: the importance of communication and information-sharing at all levels

The currently most credited method for improving the level of safety in an organisation is to push human resources to encourage error reporting as the only way to reduce those latent factors that, if not corrected early on, can lead to serious consequences for the system, causing potential accidents.

This perspective, as it turns out, is based on the idea that accidents are seldom caused by a single event, be it technological or generated by human error, but more often by a chain of events related to past failures inherent in the system, designed and built by man, and therefore in itself just as fallible.

In aviation, it is estimated that for every serious accident or catastrophically damaging event, no less than 30 inconveniences of medium importance occur, and these correspond to no less than 600 minor inconveniences [47]. Therefore, since the catastrophic accident always originates from a broad base of malfunctions and minor inconveniences, if we could provide the system with a tool capable of having a sizeable impact on basic malfunctions, we could drastically reduce the overall number of catastrophic events. It is vital therefore that anyone at any level and whatever his role within the organisation is committed to pursuing an active participation on flagging any malfunctions or errors discovered in their job, not underestimating the consequences that they can generate in the dynamics of events leading up to the failure.

Latent errors, understood as malfunctioning technology, are often overlooked, especially when the error has not produced harmful consequences for the system. There are several reasons why this happens, but they all involve underestimating the possible consequences of the malfunction, both in terms of the probability that it can generate a loss and the magnitude of that loss. There is also often a generalised 'laziness' on the part of the system's members in reporting what they know, especially if they were not directly involved with what happened, thus preventing the organisation to solve the problem.

Therefore, sometimes minor malfunctions due to wear and tear, flawed materials, or lack of experience in controls or repairs, such as a broken mirror, a worn rope, a burnt-out LED or a protective cover that was not reassembled on the machine, are not given the due attention, leaving minor latent and potentially harmful bugs within the system. Man, on top of it all, has the extraordinary, but in this case, has counterproductive ability to adapt quickly to mutations in the surrounding environment. After a while, malfunctions, bugs, or errors are no longer being perceived as potentially dangerous: we get used to them, we no longer notice them. However, it remains inherent in the system, ready to take part in the chain of events that could lead to a failure.

Experience also shows, especially in the case of human errors, that they are often concealed by the person who committed them, or by people who are aware of another person's errors, in a logic of avoiding negative repercussions or sanctions.

Unfortunately, faulty actions, understood as potentially dangerous mistakes made by man and that have harmful repercussions for the system, are often dealt with by applying only a logic of punishing the action to solve the problem. The organisation seeks to single-out the culprit, as he is considered the only person responsible for what happened, as the last actor in the process or action that led to the error. The search for the culprit also triggers a natural defensive stance by the person who 'made the mistake', who will tend to hide the actions for fear of being punished. In this culture of blame, someone who says, 'I was wrong', is as if he were admitting that 'I am wrong' [2], essentially claiming to be allegedly unfit to be part of a system in which he sees himself as a possible cause of failure.

The person guilty of making the mistake consequently has a poorly constructive attitude or one that is not constructive at all in resolving the issue, because he feels responsible and vulnerable, fearing the loss of esteem by his colleagues. He thinks he has betrayed the organization's trust, so he will do everything he can for hiding his responsibility, hampering the analysis of what happened and consequently the possibility for the system to solve the problem. The culture of isolation and punishment of culprits, therefore, does not generate those organisational defences that help solve the failure, but instead encourages the adoption of defensive behaviours that are often inappropriate.

http://dx.doi.org/10.5772/intechopen.70762

In this regard, the official report by Edgar Cortright on the causes of the explosion of the oxygen tank during the Apollo 13 mission, we discussed earlier, made it clear that it was generated by an impact that the tank had suffered 2 years before when performing certain maintenance tasks. As a matter of fact, the report specifies that the inconvenience had been duly reported in the maintenance records, but that it had been judged to be of minor importance. The records, in fact, referred to a 'slight collision', playing down the consequences that it might have on the proper functioning of the tank.

It is unclear whether the collision was indeed 'slight' nor as to what was the basis for judging it as such, but the fact is that it is certain that it was the origin of the malfunction, having caused damage inside the tank. It is therefore legitimate to ask whether a more alarming report than one downplaying the seriousness of the inconvenience by maintenance staff as to what had really happened would have alerted more the engineers in charge of the operating unit, inducing them to further inspect the possible damage suffered by the tank. The dynamics of the events suggest that the staff assigned to maintenance or to transport the tank, feeling somehow responsible for the mistake, acted in the logic of downplaying what happened for fear of negative repercussions from the organisation.

Even in the previously mentioned case of the Costa Concordia ship, the dynamics of the events could have had less serious consequences if communication errors had not been made when handling the emergency. Captain Schettino repeatedly pointed out during the trial that, following the collision with the rock and the order that followed to shut the watertight compartments, nobody had been able to tell him exactly how many of them were flooded. The investigation later clarified that the flooded compartments were five, while recordings of the conversations between the captain and the crew members responsible for their control stated only three. The difference, as we have already seen, is not trivial, because with only three compartments flooded, the ship would not have sunk.

During the investigations carried out by the magistracy, it was later revealed that one of the crew members, during a wiretap following the accident, told a friend that he realised that water was leaking from one of the hatches he was supposed to monitor. This important piece of information, for reasons that have not been clarified yet, did not reach the bridge, causing a delay in the implementation of the evacuation procedures.

The dissemination of knowledge at all levels, though, is not important only in an emergency or when a malfunction or unusual circumstance is uncovered; it is an effective weapon also in reducing injuries during the system's normal operation. On the other hand, this is something that is well-known in the air force, where the systematic exchange of information and flight experiences by organising regular meetings at the end of a drill involving not only the pilots directly participating in the mission, but also all the other squad members or at times, even members of other units, significantly contributes to the dissemination of knowledge. This allows the system to learn by itself based on the mistakes made, consequently fostering the development of that new wealth of knowledge that enables it to self-implement and become more efficient and safer.

## 7. Conclusions

Safety management in highly complex socio-technical systems cannot be narrowed down to merely comply with corporate procedures and regulations, assuming they are well-structured, or even regular inspections and maintenance on plants, machinery and equipment. It is rather the result of a synergy of all the system's components, both human and technological, whose interaction, based on shared process objectives and strategies, is an essential prerequisite to ensure its success. Every element that can play a part at any level in creating the risk must be taken into account, starting from the human resources.

It is therefore necessary that the working environment is designed to meet the needs of man, bearing in mind his weaknesses or carelessness, fatigue, or stress, factors that can lead him to make a mistake. But at the same time, he must be put in the conditions of exploiting his strengths, stimulating his creative and decision-making qualities in implementing the system and managing accidents. The latter, moreover, are always present in highly complex systems, which evolve and change their conditions because they driven to innovation and progress.

Therefore, it is necessary that the organisation can recognise them and correct them promptly, especially when it comes to small anomalies that have not yet demonstrated their potential for damage, capitalising on the information it has received and at times, managing to improve the rules and considering new scenarios.

## Author details

Andrea Antonucci

Address all correspondence to: andrea.antonucci@unich.it

Department of Medical, Oral and Biotechnological Sciences, University G.d'Annunzio, Chieti-Pescara, Italy

## References

[1] Navarro AM. The Art of Preventive Health and Safety in Europe. ETUI, European Trade Union Institute: 2015. 57 p

[2] Catino M. Da Chernobyl a Linate. Incidenti tecnologici o errori organizzativi? Bruno Mondadori: Milano; 2006

[3] Wiegmann DA, Shappell SA. Human error analysis of commercial aviation accidents: Application of the human factors analysis and classification system (HFACS). Aviation Space and Environmental Medicine. 2011;**11**(72):1006-1016

[4] Winch GM, Maytorena E. Making good sense: Assessing the quality of risky decision-making. Organization Studies. 2009;**30**(2-3):181-203. DOI: 10.1177/0170840608101476

[5] Orabona G. La sicurezza come pratica e network eterogeneo.La sala operatoria e l'allineamento della sociomaterialità [thesis]. Università degli studi di Trento; 2011. 261 p. Available from: eprints-phd.biblio.unitn.it/696/1/Giusi_Orabona_TesiPhD.pdf

[6] Pierre MS, Hofinger G, Buerschaper C, Simon R. Crisis Management in Acute Care Settings. Human Factors, Team Psychology, and Patient Safety in a High Stakes Environment. 2nd ed. Berlin Heidelberg: Springer-Verlag; 2011. DOI: 10.1007/978-88-470-2799-2

[7] Hendrick HW. Ergonomics in organizational design and management. Ergonomics. 1991;**34**(6):743-756. DOI: 10.1080/00140139108967348

[8] Rasmussen J. Information Processing and Human-Machine Interaction: An Approach to Cognitive Engineering. Amsterdam: North-Holland; 1986. ISBN-13:978-0444009876

[9] Turner B, Pidgeon N. Man-Made Disaster. 2nd ed. Boston: Butterworth-Heinemann; 1997. 200 p. ISBN-13:978-0750620871

[10] Vaughan D. The Dark Syde of Organizations: Mistake, Misconduct, and Disaster. Annual Review of Sociology. 1999;**25**:271-305. DOI: 10.1146/annurev.soc.25.1.271

[11] Vaughan D. The Challenger Launch Decision. Risky Technology, Culture and Deviance at Nasa. 2016th ed. Chicago: The University of Chicago Press Books; 1996. 620 p. ISBN-13:978-0226851761

[12] Gephart RP, Van Maanen J, Oberlechner T. Organizations and risk in late modernity. Organization Studies. 2009;**30**(2-3):141-155. DOI: 10.1177/0170840608101474

[13] Carthey J, de Leval MR, Reason J. Intitutional resilience in healthcare systems. Quality in Health Care. 2001;**10**(1):29-32. DOI: 10.1136/qhc.10.1.29

[14] Reason J. The contribution of latent human failures to the breakdown of complex systems. The Royal Society. 1990;**327**(1241):475-484. DOI: 10.1098/rstb.1990.0090

[15] Weick K, Sutcliffe KM, Obstfeld D. Organizing and the process of sensemaking. Organization Science. 2005;**16**(4):409-421. DOI: 10.1287/orsc.1050.0133

[16] Weick KE. Sensemaking in Organizations. Michigan, USA: SAGE; 1995. 248 p. ISBN-13:978-0803971776

[17] Hollnagel E. Barriers and Accident Prevention. Aldershot, UK. Routledge; 2004. 242 p

[18] Hollnagel E, Woods DD, Leveson NC. Resilience Engineering. Concepts and Precepts. Aldershot, UK: Ashgate; 2006

[19] Cacha CA. Ergonomics and Safety in Hand Tool Design. Lewis; 1999. 128 p. ISBN: 1-56670-308-5

[20] Hollnagel E. Cognitive ergonomics: It's all in the mind. Ergonomics. 1997;**40**(10):1170-1182. DOI: 10.1080/001401397187685

[21] Cherns A. The principles of sociotechinical design. Human Relations. 1976;**29**(8):783-792

[22] Trist E. The Evolution of Socio-Technical Systems: A Conceptual Framework and an Action Research Program. Toronto: Ontario Quality of Working Life Centre; 1981. 67 p. ISBN: 0-7743-6286-3

[23] Bonazzi G. Storia Ragionata di Sociologia dell'Organizzazione. 1st ed. Milano: Franco Angeli; 1987. 416 p

[24] Linstead S, Linstead A. Thinking Organization. London, New York: Routledge; 2005. 235 p. ISBN: 0-203-68492-3

[25] Hasenfeld Y. Human Services as Complex Organizations. 2nd ed. SAGE: Los Angeles; 2010. 597 p. ISBN: 978-1-4129-5693-2

[26] Smith NJ, Merna T, Jobling P. Managing Risk in Construction Projects. 3rd ed. Wiley Blackwell: United Kingdom; 2014. 252 p. ISBN: 978-1-118-34723-2

[27] Ridley J, Channing J. Safety at Work. 6th ed. Oxford (UK): Butterworth-Heinemann; 2003. 1064 p. ISBN: 0750654937

[28] Fichera G. Individui, sistemi e gruppi resilienti: quando il trauma è motivo d'orgoglio. Ticonzero. 2006;**96**:1-9

[29] Turner BA. The development of disasters–A sequence model for the analysis of the origins of disasters. The Sociological Review. 1976;**24**(4):753-774. DOI: 10.1111/j.1467-954X.1976.tb00583.x

[30] Reason J. Human Error. New York: CambridgeUniversity Press; 1990. 305 p. ISBN: 0521314194

[31] Reason J. Human error: Models and management. BMJ. 2000;**320**:768-770. DOI: 10.1136/bmj.320.7237.768

[32] Helmreich RL. On error management: Lessons from aviation. Education and Debate. 2000;**320**(7237):781-785. DOI: 10.1136/bmj.320.7237.781

[33] Helmreich RL, Merritt AC, Wilhelm JA. The evaluation of crew resource management training in commercial aviation. International Journal of Aerospace Psychology. 1999;**9**(1):19-32. DOI: 10.1207/s15327108ijap0901_2

[34] Rhonda L. Quiet Zones. PSQH; 2010 Available from: https://www.psqh.com/analysis/distractions-and-interruptions-impact-on-nursing/

[35] Bracco F. In cerca di nuovi modelli per la sicurezza organizzativa: La resilience engineering. Ticonzero. 2007;**71**:0-9, **72**:0-9

[36] Reason J. Managing the Risks of Organizational Accidents. 1st ed. Aldershot (UK): Ashgate; 1997. 272 p. ISBN: 978-1840141054

[37] Bracco F. Approccio sistemico alla sicurezza nelle organizzazioni complesse: i sistemi ad alta affidabilità. Ticonzero. 2005;**61**:1-12

[38] Zelon H. The Apollo 13 Mission. Surviving an Explosion in Space. 1st ed. New York: PowerKids Press; 2002. 24 p. ISBN: 1282212702

[39] Rasmussen J, Svedung I. Proactive Risk Management in a Dynamic Society. 1st ed. Raddningsverket; 2000. 160 p. ISBN: 91-7253-084-7

[40] Kreisler L. Resilience mise en spirale. Spirale. 1996;**1**:162-167

[41] Castelletti P. La metafora della resilienza: dalla psicologia clinica alla psicologia dell'assistenza umanitaria e della cooperazione. Nuove Tendenze della Psicologia. 2006;**4**(2):211-233

[42] Wolin SJ, Wolin S. The Resilient Self: How Survivors of Troubled Families Rise Above Adversity. Paperback; 1993

[43] Leveson LG. Safety in integrated systems health engineering and management. In: NASA Ames Integrated System Health Engineering and Management Forum (ISHEM); November; Napa, CA. 2005

[44] Rasmussen J. Skills, rules, and knowledge; signals, signs, and symbols, and other distinctions in human performance models. IEEE Transactions on Systems, Man, and Cybernetics. 1983;**SMC13**(3):257-266. DOI: 10.1109/TSMC.1983.6313160

[45] Catino M. Analisi preliminare sui fattori umani e organizzativi del disastro della nave Costa Concordia [Internet]. April 2012. Available from: download.repubblica.it/pdf/20127rapporto-progresss-concordia.pdf

[46] Catino M. Il caso concordia: alcune riflessioni su un incidente. Emergency Care Journal. 2012;**8**(1):3-6

[47] Quarto M. La qualità nel campo della manutenzione. Sicurezza del Volo. 2007;**262**:21-25